Chromatin Structure and Function

Molecular and Cellular Biophysical Methods

Part A

NATO ADVANCED STUDY INSTITUTES SERIES

A series of edited volumes comprising multifaceted studies of contemporary scientific issues by some of the best scientific minds in the world, assembled in cooperation with NATO Scientific Affairs Division.

Series A: Life Sciences

Recent Volumes in this Series

The series is published by an international board of publishers in conjunction with NATO Scientific Affairs Division

A	Life Sciences	Plenum Publishing Corporation
B	Physics	New York and London
C	Mathematical and Physical Sciences	D. Reidel Publishing Company Dordrecht and Boston
D	Behavioral and Social Sciences	Sijthoff International Publishing Company Leiden
E	Applied Sciences	Noordhoff International Publishing Leiden

Chromatin Structure and Function

Molecular and Cellular Biophysical Methods

Part A

Edited by

Claudio A. Nicolini

Temple University
Philadelphia, Pennsylvania

PLENUM PRESS • NEW YORK AND LONDON
Published in cooperation with NATO Scientific Affairs Division

Library of Congress Cataloging in Publication Data

Nato Advanced Study Institute, Erice, Italy, 1978.
  Chromatin structure and function.

  (NATO advanced study institutes series: Series A, Life sciences; v. 21)
  Includes bibliographical references and indexes.
  CONTENTS: pt. A. Molecular and cellular biophysical methods. – pt. B. Levels
of organization and cell function.
    1. Chromatin – Congresses. 2. Carcinogenesis – Congresses. I. Nicolini, Claudio A.
II. Title. III. Series.
QH599.N37 1978   574.8'732   78-24268
  ISBN-13: 978-1-4684-0975-8   e-ISBN-13: 978-1-4684-0973-4
  DOI: 10.1007/978-1-4684-0973-4

First half of the Proceedings of the NATO Advanced Study
Institute, held at Erice, Italy, April 12–26, 1978

© 1979 Plenum Press, New York
Softcover reprint of the hardcover 1st edition 1979
A Division of Plenum Publishing Corporation
227 West 17th Street, New York, N.Y. 10011

All rights reserved

No part of this book may be reproduced, stored in a retrieval system, or transmitted,
in any form or by any means, electronic, mechanical, photocopying, microfilming,
recording, or otherwise, without written permission from the Publisher

To
My Wife Julia
and
My Sons Davide and Christian

PREFACE

This book, divided in two volumes, is the result of a NATO-Advanced Study Institute, held at Erice during 1978 and aims to approach in a widely interdisciplinary fashion the problem of chromatin structure, both at molecular and cellular levels.

It has been edited in an organic and tutorial format, with the contributions of several leading scientists; to cite a few, authors range from physical chemists (as K. F. Van Holde and P. O. Ts'o) biologists (as J. Bonner and J. Gilmour) molecular biophysicists (as M. E. Bradbury and D. Olins) to cytologists (as T. Casperson and M. Mendelsohn) and geneticists (as L. Sachs).

This first volume, following an *introduction* to the properties of isolated *chromatin* (section I), contains *basic chapters* which presents the *theory and instrumentation of all techniques* (section II) applicable to the study of nucleosome and chromatin (as electron microscopy, circular dichroism, hydrodynamics, Raman Spectroscopy, nuclear magnetic resonance, neutron and x-ray diffraction, equilibrium binding studies, flow birefringence, transcriptional assays) and of nuclei and chromosome (cytochemistry, automated image analysis, microfluorimetry, scanning and flow cytometry).

In order to make this volume comprehensive and accessible to a wider scientific community (particularly graduate and post-graduate *students* in *physical biosciences*) equal emphasis is placed in the presentation both for the breadth, going from *molecular biophysics* to *biophysical cytology,* and the depth, where the numerous techniques are treated in detail.

Even if particular reference is made to the genetic apparatus and its constituents, an attempt is made to present the relevant advantages and disadvantages of each biophysical method in general terms, as applicable also to the study of *any* large biomolecule; towards this end, brief summaries of the relevant and fundamental physical principles are frequently given.

Claudio Nicolini

CONTENTS OF PART A

CONTENTS OF PART B

INTRODUCTION

During April 12-26, 1978, the eighth course of the International School of Biophysics, a NATO - Advanced Study Institute, was held at the "Ettore Majorana Center for Scientific Culture" in Erice, Sicily, co-sponsored by the North Atlantic Treaty Organization, National Science Foundation (USA), The Italian Government and the European Molecular Biology Organization.

The subject of the course was *"Chromatin Structure and Function"* with 91 participants (from 15 different countries) selected world-wide.

The current high level of interest in the structure and function of chromatin·is adequately testified by the thousands of manuscripts which have appeared in the literature during the past five years which have pertained to areas directly related to these subjects. The scope and depth of knowledge and range of disciplines which have been brought to bear in the study of chromatin structure and its relation to cell function are indicated in several recent review articles.

One of the objectives that the Erice course has successfully accomplished has been to promote the close communication and colla-boration among scientists active in this field of "chromatin" with different backgrounds and expertise, such as: biologists, physicists, biophysicists, biochemists, engineers, and physicians toward an advancement of knowledge in this basic and interdiscipli-nary field of life sciences.

The implications of a definite characterization of chromatin structure and function are now obvious since they bear directly on the mechanisms of cancer, aging, medical genetics, chemical carcinogenesis, and cell proliferation.

During the Advanced Study Institute and consequent proceedings, now published by PLENUM, we adopted a *structured,* organic and comprehensive *approach to the problem of chromatin structure and function (both at the molecular and cellular level) with focus on*

*the methodologies, techniques and on the various levels of chromatin
organization, stressing their implications for cell function.*

Today new knowledge, not only in *biophysics* which is at the
crossing of several "hard" and "soft" sciences, is frequently
produced by deeply interdisciplinary interactions among scientists
of different backgrounds. In this respect, chromatin constitutes a
unique example since we may identify at least three dimensions where
research is conducted: one (X-axis), along the level of chromatin
organization studied from the Angstrom (histone protein octamers
and the nucleosome) through the multimeters and solenoid, up to the
micron level, i.e. intact interphase nuclei and metaphase chromosome; the second (Y-axis), along the methodology and technology
utilized, from biology through chemistry up to physics and engineering; the third one (Z-axis), along the specific biological system
or mechanism, approached from the concept of the cell cycle, through
aging and carcinogenesis, up to differentiation. Each investigator,
has his own X-Y-Z coordinates in such a *"three dimensional configuration"* and frequently conducts his search in an isolated environment with occasional and superficial contacts with the remaining
"scientific space". As occurs   also in all other human endeavors,
this frequently leads to an acritical intellectual inertia or at
best to self-perpetuating inner circles, whose primary functions
are to produce an avalanche of "papers", some of which do fulfill
a need for exchange of new findings, but some of which are generated
to satisfy personal, academic or economic imperatives. Looking at
the rate at which the scientific "literature" is growing, one
wonders whether knowledge is growing at the same rate, or whether
intellectual energy and economical resources (of finite amount in
any society) are wasted because proper *"value criteria and
channels of communications"* are not open among scientists active
in parallel approaches toward the solution of the same problems.
Need exists, therefore, for the adoption of an absolute reference
system where findings and efforts are to be judged and/or comprehensive approaches developed. This should also help to decrease
the so frequently encountered intellectual arrogance (due to
cultural "isolation" or lack of sophistication) and increase the
sense of self-criticism and humility (in terms of a more open
attitude toward new technology or ideas) in studying the complex
mechanisms determining the structure and function of living systems.
In the twentieth century any significant conquest of the human race
(as splitting the atom or reaching the moon) has been the *rigorous
(step-by-step, without miraculous shortcuts,* as attempted unsuccess-
fully over the past 20 years in cancer research) *and analytical
work of teams of scientists with different "hard science" backgrounds
and expertise.* Even if knowledge is transmitted to younger
generations (in the University) through traditionally separated
disciplines such as engineering, physics, chemistry, biology or
medicine, this surely does <u>not</u> correspond to the way new knowledge
is acquired in all fields of sciences, and particularly in life
science.

To contribute toward the filling of such gaps, participants and lecturers of the Erice Advanced Study Institute and contributions to this book on chromatin have been chosen in such a manner as to warrant spherically isotropic distribution in the three-dimensional space outlined above.

The simultaneous contribution of several outstanding scientists, each one a world-wide leader in his own specialization, has permitted me to edit this comprehensive book, which hopefully respects such interdisciplinary aims. Several books are available in the area, but they usually cover specific topics, focusing mostly either on a given technique, biological problem, chromatin constituents, or level of organization, but few are covering the extremely broad field in an organic and tutorial format (i.e. comprehensive and accessible with profit to a wider scientific community) from histone proteins to intact nuclei, from molecular to cytological approaches.

Within the inherent limitations of any conference proceeding (such as this) I have attempted to structure the entire book in an organic and tutorial format, such as to have not a scattered collection of research papers, incoherent and with frequent unnecessary overlap, but a sequential series of chapters dealt in depth, from the basic properties of chromatin throughout all the numerous techniques employed (occasionally treated in details, including a brief summary of their basic physical principles), through the various levels of chromatin organization, up to their implications for cell function.

The Institute's content did *not* reflect the volume of literature pertaining to a particular technique or chromatin component, but how they are uniquely useful in providing additional and complementary information on chromatin structure and its relation to cell function.

Specifically the book consists of four parts, each one followed by a chapter on the pertinent discussion which occurred at the time of oral presentation.

I) an introduction to the physical, chemical and biological properties of isolated chromatin and their relationship to chromatin of living cells (Janes Bonner, USA).

II) basic chapters which present the theory and instrumentation of all the numerous physical, chemical, functional, morphological techniques and methodology applicable to the study of chromatin, both IN SITU and isolated from living cells (Stuart Gilmour, UK; G. Harrington, USA; Gerald Fasman, USA; Ada Olins, USA; Donald Olins, USA; Ian Walker, UK; Frank Kendall, USA; John Pardon, UK; Edwin M. Bradbury, UK; Tobjorn Caspersson, Sweden; Claudio Nicolini, USA; Mortimer Mendelsohn, USA: B. Shaw, USA; Paul Ts'o, USA).

III)   various levels of chromatin organization as determined
        by the above techniques, i.e. nucleosome, multimers,
        chromatin, chromosomal proteins and their enzymatic
        modifications, such as acetylation, methylation, and
        phosphorylations in determining gene expression and
        chromatin organization (Kensel Van Holde, USA; I. O. Walker,
        UK; A. Prunell, USA; John Ploem, The Netherlands; Joel
        Gottesfield, UK; S. Bram, France; G. Dixon, Canada;
        Donald Olins, USA; B. Shaw, USA; S. Gilmour, UK).

IV)   structure and function of the genetic apparatus in the
        mammalian cell, stressing their relationship to neoplastic
        transformation, aging, cell cycle, medical genetics
        differentiation, and chemical carcinogenesis (Edwin
        Bradbury, UK; Louis Smets, The Netherlands; Silvio Parodi,
        Italy; W. Sawicki, Poland; Walfried Linden, West Germany;
        Leo Sachs, Israel; Paul Ts'o, USA; Claudio Nicolini, USA;
        D. S. Sarma, Canada; Ian Walker, UK; Ferruccio Ritossa,
        Italy; F. X. Wilhelm, France; G. Verly, France).

At the end of the book, (part V) I have included a final review
and synthesis of the genetic apparatus dealing with clarifications
of specific topics, or focusing on controversial issues as models
for chromatin structure and in new avenues as biophysical cytology
or neutron diffraction. The course was of such interdisciplinary
nature that the scientists specialized in one field have been
teaching scientists highly qualified in a different area. The role
of lecturer and student was frequently interchanged during the
meeting as the theme of common interest (chromatin study) was
developed from the viewpoint of different sciences, in a beautiful
small town on top of a mountain overlooking the Mediterranean (that,
according to a legend, was founded by Erice, son of Venus, more
than three thousand years ago). In synthetic analytical terms we
could say, with L. Sachs, that *SC + AA = LE, that is Science in
Chromatin plus Art in Archaeology equal Life in Erice*. It is not
paradox then to state that the Chromatin Institute was held in the
same geographical region where a *few thousand years before the Greek
Leucippus and Democritus and later on the Roman Lucretius* (in his
poem "De Rerum Natura") *gave the foundation of biophysics, describing
how the atoms, after various interactions, acquire stable configurations, corresponding to the living and inanimate worlds*. This simple
and unitary theory, which brings *life science into the realm of
physical science,* remarkably maintains its validity even after
several centuries of alternative vicissitudes.

To follow the evolution of such fundamental ideas in successive
steps, is quite impossible in such context: I like however only to
recall that the content of this Erice Institute (and therefore of
this book) which relates chromatin structure to cell function,
represents one of the most recent developments of that old idea.

Following the earlier discovery of the *direct relationship between spatial structures of such molecules such as methane and benzene and chemical activity, the discovery in 1953 of the structure of the double helix of DNA represents the turning point for a similar relationship between three dimensional structure and biology.* It is indeed this relationship that emerged as one of the most intriguing *"take home messages"* from the institute: *the relationship between cell function and tertiary (nucleosome) and quaternary (solenoid or other form of superpacking) structures of chromatin DNA, as modulated by interaction with histone and non-histone proteins* (and their enzymatic modifications) *during the cell cycle, cell transformation, aging, and differentiation.* In addition to affirm a more *dynamic view* of DNA organization in isolated chromatin, the Erice Institute raises the question as to whether *tertiary - quaternary structures are specifically linked to a higher order (quinternary) organization* which can now be detected *"In Situ"* by means of recent technological advancements in the area of *biophysical cytology,* to an extent up to now impossible to any human observer or biochemical assay.

In conclusion, I hope that this book will constitute a useful and stimulating guideline to doctoral and post-doctoral students as well as to senior scientists, interested in the most recent developments in the wide interdisciplinary approach to structure and function of the genetic apparatus and its constituents and their relationships to cell function.

Finally, I would like to express my graditude to Professor Antonio Borsellino for giving me the opportunity to direct the eighth course of the International School of Biophysics (which have seen in previous years the active participation also of several Noble-Prize winners, such as Wald, Eccles, Katz) and to Ms. Pinola and Dr. Grabriele of the Majorana Centre for coupling high efficiency and courtesy in a unique cultural setting. My last, but not least, acknowledgement is to my wife Julia and my Uncle Luigi for their constant advice and dedication, considering that to realize and operate within a "three-dimensional scientific space" was a quite difficult and absorbing experience, even if challenging, not only in purely scientific terms, but also for its profound social implications.

Claudio Nicolini

SECTION I:
WHAT IS THE CHROMATIN?

PROPERTIES AND COMPOSITION OF ISOLATED CHROMATIN

James Bonner

California Institute of Technology
Division of Biology
Pasadena, California 91125

Chromatin isolation and properties

We now know that the development of organisms depends on the selective turning off and on of individual genes at particular times during the development of the organism. The understanding of development of higher creatures is therefore the understanding of the control of gene expression. One way to approach the control of gene expression is to study the control of gene expression with isolated interphase chromosomes, the state in which chromosomes express themselves by transcription into messenger RNAs or premessenger RNAs. During the last twenty years the study of isolated interphase chromosomes, called chromatin, has become a major subject of modern biology. We start herewith with the isolation of interphase chromatin.

Chromatin isolation

Chromatin was first prepared in semipure form by Zubay and Doty (1959) who ground calf thymus tissue in low ionic strength buffer and purified it by repeated centrifugation and resuspension of the high molecular weight aggregate followed by solubilization by shearing. Since thymus cells are almost completely composed of nuclei this purification turned out to be sufficient for the purposes of Zubay and Doty and for many purposes since then.

Later methods of preparation of chromatin are based mainly upon the methods suggested by Huang et al. (1960), Huang and Bonner (1962) and Marushige and Bonner (1966). The crude chromatin prepared by the method of Zubay and Doty is layered over 1.7 M sucrose in 0.01 M Tris buffer, pH 8. The Chromatin is then pelleted for 2 hours at 22 k rpm in a Spinco #25 rotor.

3

Membranes and ribosomes bound to membranes as well as adventitious
proteins remain in the gradient while the chromatin pellets. Most
chromatin preparations prepared as outlined above are of very high
molecular weight and form turbid solutions. To overcome this
problem it has been suggested by Marushige and Bonner (1966) that
the chromatin be sheared in the Virtis 45 homogenizer, 30 volts,
90 seconds. This reduces the DNA of the chromatin to about 10 kb
(1 kb = 1000 base pairs). The sheared chromatin is then pelleted
at 10 k rpm in the SS34 rotor of the Sorvall RC2B for 15 minutes.
The soluble supernatant is sheared chromatin.

Almost every investigator or group of investigators of chromatin have produced variants on the above procedure. Two variants
are of particular importance. The first is the preparation of
chromatin from purified nuclei. It is not difficult to prepare
nuclei from liver, frozen liver, or from any other of a host of
animal cells and tissues by the method suggested by Wallace et al.
(1977). The nuclei are then lysed with detergent, the chromatin
pelleted by centrifugation at 20 k rpm, and resuspended. The
chromatin thus prepared from isolated nuclei is virtually as pure
by criteria, to be outlined below, as that isolated by the method
of Marushige and Bonner, even without centrifugation through
sucrose. This method has the virtue that it eliminates the need
for the sucrose sedimentation step and the ultracentrifuge and
uses only lower speed preparative centrifuges such as the Sorvall
RC2B.

A second modification of procedures invented by protein
investigators is that introduced by Chong et al. (1974), Chae
(1975) and Douvas et al. (1975). It has become apparent that
interphase chromatin contains a powerful serine protease and that
this protease degrades not only histones of chromatin but also nonhistone chromosomal proteins. It is well-known from earlier
chromatin studies that calf thymus chromatin stored in a refrig-
erator at 4°C degrades itself so rapidly that after one week as
chromatin it can no longer be considered to be of even semi native
composition. This is due to proteolytic degradation of chromosomal proteins. In order to avoid such proteolytic degradation
it is necessary to use serine protease inhibitors such as phenyl-
methanesulfonyl fluoride (PMSF) or di-isopropylfluor phosphate
(DFP). Use of these serine protease inhibitors from the very
beginning of the grinding process inhibits proteolytic degradation
of chromosomal proteins and results in chromatin from which proteins can be extracted (myosin, for example) which are not
isolatable from chromatin prepared in the absence of protease
inhibitors.

One important problem and question in the isolation of chro-
tin has been the extent to which chromatin DNA and its associated
proteins is contaminated by nucleoplasmic or cytoplasmic proteins

which are only adventitiously associated with the isolated chromatin. This is particularly an important question since chromatin
isolated by the methods outlined above is isolated at very low
ionic strength, much lower than that of the nucleus which is
about 0.25 M. A considerable number of investigators have isolated chromatin in the presence of separately labeled nucleoplasmic or cytoplasmic proteins. It has turned out that the
adventitiously added labeled proteins in such experiments constitute about 5% of total chromosomal proteins and consist of course
of nonhistone chromosomal proteins only (Garrard et al., 1974).

    Furthermore some adventitiously bound proteins isolated with
chromatin, pelleted in low ionic strength buffer are proteins
which are very disadvantageous for the further study of chromatin,
for example, ribonuclease. We have shown that ribonuclease may be
largely removed from chromatin by the precipitation of the latter
(purified chromatin) from low ionic strength buffer, namely, 0.15
or even 0.3 M NaCl. This pellets the chromatin, leaves adventitious proteins in solution and the pelleted chromatin when resuspended in low ionic strength buffer, proves to be greatly
depleted in the adventitiously bound ribonuclease. Three such
serial precipitations will substantially remove ribonuclease from
many kinds of chromatin.

Composition of chromatin

    Chromatin as isolated from rat liver, which is a model for
chromatins in general, possesses a histone to DNA ratio of about
1:1. It also possesses a nonhistone chromosomal protein to DNA
ratio of 0.6:1. It possesses an RNA:DNA ratio of .1:1.

    Chromatin which shows a composition substantially different
from the proportions outlined above may be suspected of contamination. For example, in earlier days it was often found that the
RNA:DNA ratio of isolated chromatin would be of the order of 2 or
3:1. This almost surely indicated the contamination of the chromatin by ribosomes and therefore ribosomal RNA. Similarly a very
high protein to DNA ratio, higher than 2:1, should be suspected as
caused by contamination of nonchromosomal proteins adventitiously
adhered to chromatin. Criteria for purity of isolated chromatin
have been outlined in Methods in Enzymology, volume 12B, page 3
and further, 1968 (Bonner et al. 1968).

Histones and histone chemistry

    In the early days of the study of isolated chromatin, let us
say, in the years 1950 to 1965, it was not known how many kinds of
histones there are even though histones had been discovered by
Miescher in 1871, the same year and the same Miescher who discovered DNA. Early views concerning the composition of chromatin

and the composition and number of histones are to be found in the work The Nucleohistones by Bonner and Ts'o (1964). Serious work on the separation of histones from one another and the study of their chemistry was done by Rasmussen et al. in the laboratories of J. Murray Luck in 1958 and continued by Rasmussen and Murray in 1962. These workers found out how to separate three classes of histones from one another. There were the lysine-rich, slightly lysine-rich and the arginine-rich histones. (The three classes of histones all possess approximately 24 mole % basic amino acids). These studies were continued by Fambrough et al. (1966, 1968, 1969). Fambrough's studies used the same kind of BioRex P70 column chromatography with development by guanidine chloride as used by Murray et al. Fambrough added preparative polyacrylamide gel electrophoresis as a tool for the analysis of purity of individual fractions. He showed that there are five histone classes: H1, H2a, H2b, H3 and H4. These classes are found in all of the higher eukaryotes and, with some slight alterations, in all lower eukaryotes as well. Histones are absent from prokaryotes.

DeLange, Fambrough and colleagues (1969a,b) found that the primary structure of histone 4 is almost totally conserved as between peas and cows, an evolutionary history of 600 million years or more. This strict conservation of primary structure of histone 4 awakened a great deal of chemical interest in the histones and all of the five histones have now been sequenced from one or more individual species of creatures. Histone 3 is also quite conserved (DeLange et al., 1973; Brand and Van Holte, 1972) Histones 2a and 2b are somewhat less conserved (Iwai et al., 1970; Panyim and Chalkley, 1971; Sautiere et al., 1972). Histone H1, which is the most different in its properties from the other histones is the least conserved of the five and has the most variants as has been shown most recently by Cole (1977). As we will see below histones H4, H3, H2a and H2b play a cooperative role in the production of a chromosomal unit. Histone H1, on the contrary, plays a role as a companion of DNA on the spacer sequence between adjacent chromosomal units.

It will be here noted in passing that the histone genes are repetitive ones, that is, the gene for each histone is repeated several times in the genome. This amounts to 20 or so times per histone species in mammals to several hundred times per histone species in other organisms (Kedes and Birnstiel, 1971). The reason why the histone genes are repeated to this considerable extent may well have to do with the fact that histones must be rapidly synthesized during cell division time to be able to bind with DNA produced during the replicative phase of the cell cycle.

The nonhistone chromosomal proteins

The histones associated with DNA and chromatin have been

studied extensively, particularly since their recognition by Huang
and Bonner (1962) as repressors of DNA transcription. The nonhistone chromosomal proteins have been less thoroughly studied
until recent years. One method of study of nonhistone chromosomal
proteins is to first remove histones from chromatin with 0.2 or
0.4 M H_2SO_4 in which histones are soluble and in which DNA and
nonhistone chromosomal proteins are insoluble. Following this
treatment the DNA containing nonhistone chromosomal proteins can
be treated with SDS, sodium dodecyl sulfate, and the nonhistone
chromosomal protein then studied by SDS chromatography. This was
first done by Marushige et al. in 1968 and studied more extensively
by Elgin and Bonner (1970) and by Elgin et al. in later years. A
second method of attack on the nonhistone chromosomal proteins is
to remove most of the histones and nonhistone proteins from DNA by
dissociation of DNA from proteins with three of four molar sodium
chloride. The dissociated chromosomal proteins are then separated
from DNA either by exclusion chromatography on A 50 or by ultracentrifugation which pellets the DNA and separates it from proteins
(Douvas et al., 1975; Van den Broek et al., 1975). The proteins
thus separated from DNA (which do not include all of the chromosomal proteins since H4 and some actin sticks particularly tightly
to DNA even in 4 M NaCl) are then subjected to dialysis to lower
the salt concentration to 0.4 M NaCl. The histones are then
separated from the nonhistones by ion exchange chromatography on
BioRex P70. At this salt concentration and on this weak anion
exchanger, histones are retained while most nonhistones are
allowed to pass through the column.

The SDS complexes of the nonhistone chromosomal proteins,
electrophoresed on polyacrylamide gels, exhibit a wide variety of
molecular weights, from about 225,000 down to the lower limit of
resolution of such polyacrylamide gels, namely, of molecular weight
of about 10 to 15,000. The major nonhistone chromosomal protein
components are similar in a wide variety of chromatins as those
of HeLa, rat liver, Drosophila, etc. (Elgin and Bonner, 1970).
The reason why the major molecular weight components, a dozen or
so, of the nonhistone chromosomal proteins of the chromatins of
different creatures are similar has been demonstrated now by
Martin et al. (1973) and Douvas et al. (1975). Martin et al.
(1973) showed that two major proteins of the nonhistone category,
those of 38,000 and 40,000 molecular weight respectively, are the
hnRNA packaging proteins. Douvas et al. showed that a major
chromosomal protein of 45,000 molecular weight is actin, while the
the 50,000 and 55,000 molecular weight proteins associated with
all chromatins thus far studied are alpha and beta tubulin, the
components of microtubules. The 225,000 molecular weight protein
of chromatin is myosin. Other minor components of the nonhistone
chromosomal proteins are the serine protease alluded to above as
well as other components of the actomyosin system: actinin,
tropinin, tropomyosin, etc. Approximately half of the mass of all

nonhistone chromosomal proteins are structural proteins as outlined
above.

There has been some discussion of whether or not the actomyosin, tubulin, etc., found in chromatin are artifacts; that is,
artifactually isolated with chromatin, or whether they are de
facto components of interphase chromatin. The bulk of evidence
now suggests that actomyosin and the other structural proteins are
in fact components of chromatin. Thus actomyosin, etc., are not
found in nucleoplasma as free entities. The removal of the nuclear
membrane (and the actomyosin known to be associated with this
membrane) by detergents before the lysis of the nucleus does not
result in any lowered concentration of actomyosin in the subsequently isolated chromatin. It is quite probable therefore that
actin and myosin are in fact components of isolated interphase
chromatin. Speculations are principally concentrated on the
possibility that these components may have to do with chromosome
condensation and the subsequent act of mitosis or meiosis. It is
of interest that it is becoming increasingly clear that wherever
chemomechanical systems are found in living beings it turns out
that the chemomechanical system is an actomyosin powered one.

Fifty percent of the total nonhistone chromosomal protein
consists of structural components. What is the remaining 50%?
About 2 to 4% of the total mass of nonhistone chromosomal proteins
bind sequence specifically to homologous DNA (Sevall et al., 1975).
Further minor components are, in addition to the serine protease
alluded to above, acetyl transferase, RNA polymerase and probably
many other enzymes. We do not yet know how to account for all of
the nonhistone chromosomal proteins except that they are very
considerably in number. A 20 cm gel of the whole nonhistone
chromosomal proteins of rat liver chromatin reveals 115 components
which can be resolved (Garrard et al., 1974). A 12 cm gel of the
nonhistone chromosomal proteins which bind sequence specifically
to homologous DNA reveals at least 100 components (Savage et al.,
unpublished results). There are apparently a vast number of
chromosomal nonhistone proteins varying in abundance and in properties. It will still be a vast task to unscramble what they all
do.

Template activity of chromatin

It was found in 1962 by Huang and Bonner that when the acid
soluble proteins are removed from isolated chromatin the thus deproteinized chromatin becomes a 10- to 20-fold better template for
transcription by purified RNA polymerase than does chromatin itself. When the acid soluble proteins are reconstituted onto the
DNA, the DNA again becomes a poor template for RNA transcription
by purified RNA polymerase. This was in fact the first conclusive

evidence that the acid soluble proteins of chromatin act as regulators of transcribability. On the basis of these findings Marushige and Bonner (1966) have developed something called the template activity assay of isolated chromatin. The way this assay works is as follows: A series of reaction mixtures containing a constant amount of purified RNA polymerase and constant amounts of all the substances required for transcription of DNA by RNA polymerase is supplied with increasing amounts of either a) chromatin, as rat liver chromatin (or any other kind of chromatin), or b) DNA prepared from the same chromatin preparation. In the case of rat liver chromatin a given amount of DNA as pure DNA is approximately five to ten times more efficient as a template for RNA synthesis than is chromatin of the same species. We say, therefore, that the template activity of the chromatin is 10-20% that of pure DNA of the same kind. This is true for a wide variety of chromatins. The template activity of chromatin assay as described above and as described in detail by Marushige and Bonner (1966) has been subsequently found to be independent of whether the RNA polymerase used is homologous, that is, polymerase II of rat liver chromatin (unpublished work) or heterologous polymerase, that of purified E. coli polymerase.

The template activity of chromatin may also be determined by transcription of chromatin with RNA polymerase and subsequent hybridization of the transcripts to denatured deproteinized chromosomal DNA by the methods of RNA-DNA hybridization. Alternatively, whole nuclear RNA may be hybridized to the DNA of isolated chromatin and the template activity of chromatin in vivo thus determined. This has been done by a variety of workers, for example, by Holmes and Bonner. It has turned out that not only the repetitive and single copy sequences of DNA are transcribed in vivo in approximately the same proportion as they are represented in chromosomal DNA, but also, the extent of hybridization of in vitro or in vivo transcribed RNA to DNA approximates the template activity of the chromatin as measured by the template activity assay.

Structure of chromatin

The structure of interphase chromatin has proved to be totally resistant to study by optical microscopy and until recently, resistant also to study by electron microscopy. Only during the years 1974 to present has electron microscopy revealed that the major component of interphase chromosomes, chromatin, is a beaded strand consisting of subunits about 100 Å across and separated from one another by about 40 base pairs of DNA (Olins and Olins, 1974). The Olins and Olins electron micrographs provided convincing evidence that interphase chromatin is, at low ionic strength at least, composed of a "beads on a string" structure; namely, beads about 100 Å in diameter separated from one another by short

spacers. The whole structure was studied in more detail by
Griffith in 1975 who found that the viral SV40 minichromosomes,
that is, the SV40 DNA of HeLa cells, becomes complexed with histones in their eukaryotic host cell and also assumes the beads on
the string structure. With this minichromosome Griffith was able
to show that at physiological ionic strength the beads are pressed
against one another so that linkers between beads are not apparent,
just as he has found previously in interphase chromosomes of
higher eukaryotes (Griffith, 1970). At very low ionic strength
such as 0.01 M buffer, the beads on a string separate and reveal
the DNA linker. The beads have been given the name "nubody" by
Olins and Olins (1974) and "nucleosome" by Oudet et al. (1975).
"Nucleosome" seems to be winning the struggle for survival.

An important discovery concerning the structure of the nucleosome is that of Garrard et al. (1974) that the molecular stoichiometry of the histones, H2A, h2B, H3, and H4 is approximately
1:1:1:1 with a mass ratio of .9 histone to 1.0 DNA. The mass
ratio requires two molecules of each of these four species of
histones to be present for each one mass of DNA. Our knowledge
of the stoichiometry of the histones in chromatin suggests that
the histones must interact with one another. This is suggested
also from our knowledge of the primary structure of the histone.
All of the four histones enumerated above, those present in the
nucleosomal structure, consist of a highly basic N-terminal
peptide and a highly hydrophobic C-terminal peptide. This structure suggests that the N-terminal peptide is for interaction with
nucleic acids, while the C-terminal peptides are for interaction
with other proteins. That this is so has been shown convincingly
by D'Anna and Isenberg beginning in 1974. These workers found
that histones do interact with one another in DNA-free solution.
The interactions are through the hydrophobic C-terminal tails and
the affinities are very great. The work of D'Anna and Isenberg
and also of Kornberg and Thomas and others (1974) has shown that
an octamer of eight histones, two of each species, appear to
constitute a core of the nucleosome structure and that the DNA is
bound around this core.

A particularly imaginative approach to the structure of chromatin was introduced by Hewish and Burgoyne (1973) and followed up
by Kornberg and Thomas in 1974 and by many others since that time.
It was revealed by Hewish and Burgoyne that interphase chromatin
when it is digested by micrococcal nuclease is cleaved into DNA
fragments of a standard length approximately 200 base pairs long.
Further digestion reduces the length of the unit fragment to about
160 base pairs, the so-called "core" fragments. The core fragment
is complexed to the histone octamer. The 40 base pairs which are
not in the core fragment is the linker and this linker is complexed with histone 1. Histone 1 was earlier shown in 1970 by
Griffith and Bonner (unpublished results) to be stationed at

intervals about 160 base pairs apart along the DNA strands. This
was shown by making use of the method of Brutlag et al. (1970) by
which histone 1 is covalently attached to DNA by treatment for 1
hour with 1% formaldehyde at 0°C. The other histones are not
covalently attached to DNA. By this treatment they can be removed
by banding the H1-DNA complex in cesium chloride of the appropriate
density. The resulting DNA-H1 complex was subjected to high resolution electron microscopy and the H1 molecules found to be
regularly distributed, as pointed out above, approximately 160 base
pairs apart. Many other pieces of evidence had previously suggested that H1 does different things to chromatin than do the other
four histones. For example, the removal of H1 has little effect
upon the biophysical properties of chromatin. It does not change
the melting properties of DNA or its hyperchromicity (Tuan and
Bonner, 1969) or the characteristic X-ray diffraction patterns of
chromatin fibers (Richards and Pardon, 1970).

The structure of interphase chromatin is therefore now
relatively clear. The bulk of the chromatin is composed of nucleosomes which are connected by linkers of DNA complexed with H1.
Each nucleosome is composed of an octamer of histone and the DNA
is wound around the outside of the octamer. A vast variety of
enzymological studies on the nucleosome have been carried out.
The DNA is arranged in such a fashion that it can be degraded by
micrococcal nuclease into 10 base pair long fragments by sufficiently long degradative attack.

Nucleosomal structure appears to be a characteristic of
eukaryotic chromosomes. It is found not only in higher plants and
in animals but also in Dictyostelium, a lower protozoan (called a
slime mold), as well as in true fungi as in yeast and Neurospora.
In the lower organisms such as Dictyostelium and Tetrahymena,
histones are present and nucleosomes are found, but the histones
are not electrophoretically identical to the histones of higher
creatures. They differ slightly (or considerably in the case of
H1) in molecular weight and amino acid composition. Nonetheless,
nucleosomal structure and histones and their properties appear to
have been considerably conserved since the beginning of the
eukaryote kingdom.

Chromatin in the nucleus and in the test tube

Chromatin as isolated by the procedures outlined above,
is soluble in low ionic strength buffer and may be worked with as
a soluble solution of a nucleohistone. In 0.15 M to 0.3 M NaCl
or KCl solution, however, chromatin becomes almost completely insoluble. This is the range of ionic strength found in the nucleus
of higher organisms. Therefore, in the nucleus of higher organisms the chromatin must be insoluble. We shall see below that the

template expressed portion of chromatin is soluble in solutions of
such ionic strength and we must therefore envisage the chromatin
of interphase nuclei as consisting of an aggregated, precipitated
mass of nonexpressed nucleosomal chromatin from which protrude a
few loops of soluble chromatin which is in the expressable form,
that is, in a form transcribable by RNA polymerase. The difference between transcribable and nontranscribable chromatin will be
returned to below. It is important for our purposes, however to
note and even to stress that chromatin as we know it in the test
tube is not similar to chromatin as it is found in the nucleus in
life. In all probability chromatin as it is found in the nucleus
in life assumes higher orders of structure, the nucleosomal chain
coiling itself into structures of higher order.

REFERENCES

Zubay, G. & Doty, P. (1959) J. Mol. Biol. $\underline{1}$:1-20.
Huang, R. C. C., Maheshwari, N. & Bonner, J. (1960) Biochem.
    Biophys. Res. Comm. $\underline{3}$:689-694.
Huang, R. C. C. & Bonner, J. (1962) Proc. Nat. Acad. Sci. $\underline{48}$:
    1216-1222.
Marushige, K. & Bonner, J. (1966) J. Mol. Biol. $\underline{15}$:160-174.
Wallace, R. B., Sargent, T., Murphy, R. & Bonner, J. (1977) Proc.
    Nat. Acad. Sci. $\underline{74}$:3244-3248.
Chong, M.T., Garrard, W. T. & Bonner, J. (1974) Biochemistry $\underline{13}$:
    5128-5134.
Chae, C. B. (1975) Biochemistry $\underline{14}$:900-906.
Douvas, A. S., Harrington, C. & Bonner, J. (1975) Proc. Nat. Acad.
    Sci. $\underline{72}$:3902-3906.
Garrard, W. T. & Bonner, J. (1974) J. Biol. Chem. $\underline{249}$:5570-5579..
Bonner, J., Chalkley, G.R., Dahmus, M., Fambrough, D., Fujimura,
    F., Huang, R. C. C., Huberman, J., Jensen, R., Marushige, K.,
    Ohlenbusch, H., Olivera, B., & Widholm, J. (1968) Methods in
    Enzymology, $\underline{12B}$:3-65.
Bonner, J. & Ts'o, P. (Eds.) The Nucleohistones, Holden-Day, Inc.,
    San Francisco, California.
Rasmussen, P., Murray, K. & Luck, M. M. (1962) Biochemistry $\underline{1}$:
    79-89.
Luck, J. M., Rasmussen, P., Sutake, K. & Tsuetikova, A. (1958).
    J. Biol. Chem. $\underline{233}$:1407.
Fambrough, D. & Bonner, J. (1966) Biochemistry $\underline{5}$:2563-2570.
Fambrough, D., Fujimura, F. & Bonner, J. (1968) Biochemistry $\underline{7}$:
    575-584.
Fambrough, D. & Bonner, J. (1968) J. Biol. Chem. $\underline{243}$:4434-4439.
Fambrough, D. & Bonner, J. (1969) Biochim. Biophys. Acta $\underline{175}$:
    113-122.
DeLange, R., Fambrough, D., Smith, E. & Bonner, J. (1969a) J. Biol.
    Chem. $\underline{244}$:319-334.
DeLange, R., Fambrough, D., Smith, E. & Bonner, J. (1969b) J. Biol.
    Chem. $\underline{244}$:5669-5679.

DeLange, R., Hooper, J. & Smith, E. (1973) J. Biol. Chem. 248:
3261-3274.
Brand, T. & VanHolte, C. (1972) FEBS Letters 23:357-360.
Iwai, K., Ishikawa, K. & Hayashi, H. (1970) Nature 226:1056-1058.
Panyim, S. & Chalkley, R. (1971) J. Biol. Chem. 246:7557-7560.
Sautiere, P., Tyrou, D., Laine, B., Mizon, J., Lambelin-Breynaert,
M., Rufin, P. & Biserte,G. (1972) C. R. Acad. Sci., Paris
274:1422-1425.
Cole, R. D. (1977) In: "The Molecular Biology of the Mammalian
Genetic Apparatus", P.O.P Ts'o (Ed.) Vol. I: pp.91-104,
North Holland, Amsterdam.
Kedes, L. & Birnstiel, M. (1971) Nature New Biol. 230:165-169.
Marushige, K., Brutlag, D. & Bonner, J. (1968) Biochemistry 7:
3149-3155.
Elgin, S. C. R. & Bonner, J. (1970) Biochemistry 9:4440-4447.
Douvas, A. S., Harrington, C. A. & Bonner, J. (1975) Proc. Nat.
Acad. Sci. 72:3902-3906.
Van den Broek, H., Nooden, L., Sevall, J. S. & Bonner, J. (1973)
Biochemistry 12:229-236.
Martin, T., Billings, P., Levey, A., Ozarslan, S., Quinlan, T.,
Swift, H. & Urber, L. (1973) Cold Spring Harbor Symp. Quant.
Biol. 39:921.
Sevall, J. S., Cockburn, A., Savage, M. & Bonner, J. (1975)
Biochemistry 14:782-789.
Olins, A. & Olins, D. (1974) Science 183:320-332.
Griffith, J. (1975) Science 187:1202-1203.
Griffith, J. (1970) Ph.D. Thesis California Institute of
Technology.
Oudet, P., Gross-Bellard, M. & Chambon, P. (1975) Cell 4:281-300.
Garrard, W. T., Pearson, W., Wake, S. & Bonner, J. (1974) Biochem.
Biophys. Res. Comm. 58:50-57.
D'Anna, J. & Isenberg, I. (1974) Biochemistry 13:4992-4997.
Kornberg, R. & Thomas, J. (1974) Science 184:865-868.
Hewish, D. & Burgoyne, L. (1973) Biochem. Biophys. Res. Comm.
52:504-510.
Tuan, D. & Bonner, J. (1969) J. Mol. Biol. 45:59-76.
Richards, B. & Pardon, J. (1970) Exp. Cell Res. 62:184-196.

EXPRESSED AND NONEXPRESSED PORTIONS OF THE GENOME:

THEIR SEPARATION AND THEIR CHARACTERIZATION

James Bonner

California Institute of Technology
Division of Biology
Pasadena, California 91125

INTRODUCTION

We have considered the nature of the bulk of the eukaryotic genome. It consists of DNA complexed with histones in nucleosomal configuration. This bulk is not transcribed into RNA. What is the nature and structure of that portion of the genome which is transcribed? Several methods for separation of transcribed from non-transcribed portions of the genome have been described. Some of these are commented upon below.

METHODS

Sucrose density gradient centrifugation

It has been known for many years that when isolated chromatin prepared as described in Lecture 1 is sheared by Virtis shearing (30 volts/90 sec) and subjected to sucrose density centrifugation, the resulting fragments exhibit a wide variety of sedimentation coefficients. This range is from about 30 Svedberg units (S units) to over 100 S units (Chalkley and Jensen, 1969). In more recent work it has often been found that sucrose density gradient centrifugation of chromatin sheared as described by Marushige and Bonner can be separated into two discrete components: A light component of sedimentation coefficient about 30 S, and a heavy component of sedimentation coefficient over 100 S, with little intervening material. Chalkley and Jensen have shown that the heavy component can be transformed to the light component by treatment with 5 M urea and that this transition is not reversible. In addition, it has been suggested from time to time that the light fraction

represents the template active portion of chromatin and that the
heavy fraction represents the template inactive portion of chromatin. To what extent is this in fact true? We have found (Savage
and Bonner, 1978) that the light fraction when it is studied by
the methods of renaturation kinetics contains the complexity of
the whole genome, that is, the light fraction which constitutes
about 10% of the whole genome for rat liver chromatin, is a
random set of the sequences of the whole genome and, therefore,
no sequence fractionation has been accomplished by separation of
light and heavy chromatin by sucrose density gradient centrifugation.

     Other methods for separation of active from inactive chromatin
are reviewed by Savage and Bonner (1978). We have, in addition,
developed a method which appears to successfully separate expressed portions of the genome and to separate them in a quantitative
fashion from the nonexpressed portions of the genome. This method
relies upon attack of chromatin by the enzyme DNase II which makes
double stranded clips of the DNA of chromatin as well as of
purified DNA (Marushige and Bonner, 1971; Billing and Bonner, 1972;
Gottesfeld et al. 1974,1975). Because this nuclease makes double
stranded clips on chromatin DNA, it may be thought of as a shearing agent, even though shearing is in principal a hydrodynamic
concept. Marushige and Bonner incubated unsheared chromatin with
DNase II in pH 6.6, which is rather far from the pH optimum of
the enzyme which is approximately 4.6. At the end of the incubation period the reaction is stopped by raising the pH to 7.5, even
further removed from the pH optimum of the enzyme. Unsheared
chromatin is separated by centrifugation at 15 kg. The rapidly
attacked material remains in the supernatant and is referred to as
supernatant 1, while the pellet is referred to as pellet 1. The
supernatant may be further fractionated by the addition of 0.15 M
NaCl or 2 mM $MgCl_2$. Nucleosomal DNA, particularly that containing
histone 1, is extremely insoluble in NaCl or magnesium chloride at
these concentrations, and therefore precipitates. The resulting
precipitate, removed by centrifugation, is known as pellet 2 and
the remaining supernatant as supernatant 2. The kinetics of
digestion are clear-cut. With sufficient time as much as 80% of
the chromatin is solubilized into supernatant 1. By this time,
pellet 2 grows to about 50% of the total chromosomal DNA and the
supernatant S2, to about 20% of the total chromosomal DNA.

     The final limiting amount of supernatant 2, S2, DNA obtainable
from a given chromatin varies according to the template activity of
the chromatin in question. Thus chromatin of Novikov cells has a
template activity 10% of that of its DNA. It yields an S2 equal
to 10% of total ascites chromosomal DNA. Rat liver chromatin with
a template activity of 20% with respect to its DNA, yields 20% of
its DNA as supernatant 2.

After 5 minutes of incubation with DNase II about 10% of rat liver chromosomal DNA is found in the S2 fraction. After this time of incubation the DNA has a double stranded length of about 700 base pairs and a single strand length of about 200-500 bases. Characterization of the solubilized material has been done with this 5 minute supernatant 2 fraction. That it is indeed the transcriptionally active component in chromatin is shown by the facts 1) that it bears nascent RNA, that is, RNA pulse labeled in vivo with a radioactive tracer; 2) that it is hybridized to a high level by whole cell RNA; and 3) that it is a subset of the single copy sequences of whole rat DNA as well as a subset of the repetitive sequences. The sequences of rat liver DNA released by DNase II are therefore a subset of all of the DNA sequences of the whole rat genome, and these are the sequences which are hybridized by the RNA sequences which are transcribed in rat liver.

Billing and Bonner (1972) found that over 70% of pulse labeled nascent RNA is released from ascites cell chromatin when less than 10% of the DNA has been solubilized by DNase I, that is, the DNA of chromatin which bears nascent RNA is also readily attacked by DNase I which makes single stranded nicks in DNA. The same result has been obtained with DNase II.

The fact that DNase I selectively degrades DNA sequences which are transcribed into RNA has been used extensively by Garel and Axel (1976) and Weintraub and Groudine (1976). Garel and Axel found that oviduct chromatin isolated from chicks induced to produce ovalbumin lost its ability to react with ovalbumin messenger cDNA following partial digestion with DNase I. Weintraub and Groudine obtained similar results with the globin gene system. In our laboratory we have found that if chromatin is first attacked by DNase I the DNA isolated from that chromatin has lost its ability to hybridize to S2 DNA released from chromatin by DNase II digestion (Savage and Bonner, unpublished results). To understand these results one must remember that DNase I rapidly degrades DNA to acid soluble, very small oligonucleotides, while DNase II releases large oligonucleotides and under the conditions used as described above liberates single stranded fragments of an average length of about 300 bases. One might say that DNase I destroys active genes while DNase II releases active genes.

A third enzyme which has been used a great deal in the study of chromosomal structure is Micrococcal nuclease, sometimes known as Staphylococcal nuclease. This enzyme which produces single stranded nicks in double stranded DNA is almost nonselective as between expressed and nonexpressed portions of the genome. When it is used to attack whole chromatin, it reduces approximately 50% of the total exposed DNA to acid soluble fragments. Thus we have an array of three enzymes: DNase II makes double stranded clips in DNA and selectively clips out the expressed portion of the genome,

DNase I by single stranded nicks selectively degrades to acid
soluble material the expressed portion of the genome, while
Micrococcal nuclease indiscriminately attacks all DNA of the
genome with the reservation that that portion most protected by
histones and nucleosomal structure is attacked much more slowly
than is the DNA of linkers, etc. Micrococcal nuclease does not
selectively release or degrade the expressed portions of the
genome.

FINDINGS

Chemical composition of template active chromatin.

     If rat liver chromatin is attacked by DNase II and the
template active segments released as described above, it becomes
possible to compare the chemical composition of the expressed
portion of the genome with that of the nonexpressed portion of the
genome. The histone-DNA ratio of the expressed portion is approximately the same, perhaps slightly lower, than that of whole chromatin. The difference, if any, may be artifactual, due to protease
degradation of histones. The point to be emphasized is that
template active chromatin does contain histones and in nearly the
same proportion to DNA and in nearly the same proportion to one
another as the template active inactive portion of the genome or
as the whole chromatin.

     When chromatin is attacked with DNase II, it is converted into
200 base pair long pieces of DNA. These fragments contain 8
molecules of the core histones and 1 molecule of H1. The template
active portion of the genome is also converted into 200 base pair
long units of DNA containing histones in the same proportion as
described above. Thus, histones do not disappear from DNA during
transcription. Therefore, the DNA of template active chromatin is
organized into subunits which are similar in size to those of
template inactive chromatin. In the limit digest of template
active chromatin about 50% of the DNA is converted to acid soluble
material, either by DNase II or by Micrococcal nuclease. This is
also true of whole chromatin. The physical structure of the
repeating units of template active chromatin is not yet clear.
What is known concerning this matter will be returned to below.

Properties of template active and template inactive chromatin

     The first of these properties has to do with the melting
spectrum of the DNA of the several fractions of chromatin. When
rat liver chromatin is dialyzed to very low ionic strength (0.25
mM EDTA), under which conditions pure rat liver DNA melts at $43°$,
chromatin melts with three transitions. The first at about $56°$;
the second at about $75°$; and the third at about $85°$. That the
latter two transitions are due to the complexing of histones to

DNA is shown by the fact that when histones are selectively removed by dissociation of chromatin with increasing concentrations
of salt, these latter two transitions disappear while the transition at 43° progressively increases. The transition at 55° is not
affected, or at least minimally affected, by the removal of
histones.

The template active S2 portion of chromatin melts quite
differently. The template active portion shows a transition at
43°; that is, some of the DNA of the template active portion of
chromatin melts essentially as does pure DNA. A major portion
melts at 57° and little or none melts at the transitions associated
with histone complexed DNA. The melting properties of the template
inactive portion of chromatin are those of whole chromatin with
major transitions in the histone-DNA complex regions. These
facts imply that in the template active portion of chromatin,
histones do not interact with DNA as they do in template inactive,
or bulk, of chromatin.

Similarly, other types of biophysical investigations have
shown that the DNA of template active chromatin is more like pure
protein-free DNA than is the DNA of whole chromatin. Thus the
cDNA spectrum of template active chromosomal DNA is more like free
DNA than is that of template inactive chromosomal DNA (Gottesfeld
et al., 1974). It may be remembered also that electron microscopy
has shown that there are genes which are clearly expressed and
clearly recognizable as expressed by electron microscopy; namely,
the ribosomal genes of eukaryotes. It has been shown by Hamkalo
et al. (1973) that the contour length of the DNA undergoing transcription by RNA polymerase to produce the ribosomal RNA precursors
is equal to the contour length of the RNA being produced, that is,
the DNA under transcription to form ribosomal RNA is completely
and fully extended. Its packing ratio is 1 as contrasted to the
packing ratio of DNA in whole (beads on a string) chromatin which
is 7 or greater. This finding also implies that the DNA of
expressed chromatin is relaxed, extended, and behaves with respect
to other physical properties as deproteinized DNA.

Transformation of inactive chromatin to active chromatin

We have seen that transcribable chromatin is characterized
by physical properties different from those of nontranscribed
chromatin. 1) The DNA of transcribed chromatin exhibits a
relatively low T_m. 2) The packing ratio of transcribed chromatin
is close to 1. 3) the circular dichroism spectrum of transcribed
chromatin is that of B-form DNA. 4) Transcribed chromatin is
readily attacked by nucleases such as DNase II and DNase I. In
all of these respects the physical properties of isolated template
active chromatin approach the properties of free DNA. What kind
of modification of chromosomal proteins could bring about this

alteration in protein-DNA interaction, an alteration which would
cause histones to cease to stabilize DNA against melting and to
cease to confine DNA to a packing ratio of 7 and to instead allow
the DNA to relax and extend? There has been a great deal of
speculation over the years that histone modification of some kind
might be responsible for the "activation of genes." Thus Allfrey
et al., 1964 have suggested that transcriptional control might be
achieved by histone modification as, for example, by acetylation of
the ε-amino groups of lysines.

    Marushige (1976) has in fact shown that chemical acetylation of
calf thymus chromatin increases its transcribability by purified
RNA polymerase and this without any removal of histones from
chromatin. Histones and chromatin remain associated. They are,
however, associated in someway different from that characteristic
of template inactive chromatin. We have studied this whole phenomenon in considerable detail. We have found (Wallace et al.,
1977) that chemical acetylation of chromatin with acetic anhydride
as described by Marushige not only makes chromatin transcribable
by added RNA polymerase, but also causes the acetylated chromatin
to assume the physical properties of template active chromatin.
As a result of acetylation the DNA of chromatin becomes low melting. It also becomes sensitive to DNase I and to DNase II. It
does not become significantly more susceptible to Micrococcal
nuclease digestion than it was before acetylation. Further, it
has been shown by Marushige and by our group (unpublished) that
acetylation by acetic anhydride causes acetylation of the ε-amino
groups of lysine and that the lysines which are acetylated are all
contained in the N-terminal peptides of the core histones, that is,
the acetylation occurs in those portions of the histones which
interact with DNA rather than in those portions of the histones
which interact with one another to form the nucleosomal core. We
interpret these findings to mean in a general way that acetylation
of chromatin causes the histones of the nucleosomal core to relax
their binding of N-terminal peptides to DNA and thus to allow the
DNA to escape, and to an extent that permits the DNA to relax to
a lower packing ratio. This results in turn in the DNA of the
nucleosome becoming transcribable. Thus acetylation is a possible
candidate for the role of transformation of template inactive
nucleosomes to template active chromosomal configuration.

    We have no firm evidence yet whether acetylation is in fact the
primary cause of such transformation. We do know that pulse
labeling of chromatin with acetate causes template active chromatin
(S2) to always contain a 2-4-fold higher specific activity of
acetyl groups per unit histone than does template inactive chromatin.

There are, therefore, still important unknowns to the understanding of the operation of chromatin and the activation of
specific genes. The first is how sequence specificity is introduced into the transformation. I have described above how
acetylation could, for example, cause the observed changes and
alteration between template inactive and template active conformation of genes. What is it that tells which gene is to be thus
transformed, perhaps by the histone acetylase of chromatin (a chromosomal enzyme)? It has been described in Lecture I how a portion
of the rat liver nonhistone chromosomal protein binds sequence
specifically to homologous DNA and, in fact, to families of
repetitive sequences. These proteins might be the agents which
introduce sequence specificity into histone acetylase-chromatin
interaction, or into some analogous chromosomal transformation
interaction.

A second point of great interest and importance is whether or
not acetylation is in fact the agent which transforms inactive into
active genes. Our two principal questions are,therefore: 1) the
nature of transformation and 2) the nature of the agent which
sequence specifically specifies that the transformation shall take
place here.

All of the studies of how to transform nontranscribable genes
into transcribable genes in the chromatin of higher organisms have
been carried out with whole chromatin; that is, genomes containing
many tens of thousands of genes capable of expression, a few
thousands of which are expressed in any particular kind of chromatin
and a few tens of thousands of others which are expressed in the
chromatin of other organs of the same creature. All described
attempts to turn on particular genes (in the chromatin of higher
organisms) have been done using the methods of reconstitution of
dissociated chromatin. Chromatin is dissolved in 2 M NaCl and
5 M urea. All histone and nonhistone proteins are dissociated from
DNA. Next we gradually dialyze the NaCl away; next the urea.
Chromatin which looks like the original is reconstituted. The
reconstituted chromatin is now transcribed by RNA polymerase and
the resultant transcripts compared by hybridization-competition
with those of native chromatin.

Fidelity of reconstitution (sequence fidelity) has been found
by those whose hybridization studies included either (1) only
repetitive sequences (Bekhor, Kung and Bonner 1969; Huang and
Huang 1969), or (2) have used c-DNA probes to individual messages
(Paul et al., 1973; Stein et al., 1975. Both approaches are
fraught with artifacts.

In addition, those who have sought to show that gene expression
is controlled by the complexing of particular nonhistone chromosomal proteins to DNA have even more problems. Their method is

to dissociate, for example, brain chromatin, add erythropoetic
chromosomal nonhistone proteins, reconstitute the mixture and then
show that globin message is transcribed from the altered brain
chromatin. Problems of endogenous message, super sensitivity of
c-DNA probe, etc., abound. No single reconstitution study is
convincing.

   It seems to me that the difficulty of working with a large
complex of genes simultaneously is an overwhelming one and makes
almost impossible pure and clean experiments. There are questions
of the residual messenger RNA contained in chromatin before it was
transformed and of the messenger RNA possibly contained in the nonhistone chromosomal portion allotted to the reconstituted chromatin, as well as the fact that the gene to be probed as globin or
histone is one gene which constitutes perhaps one part in 10^6, or
at most, one part in 10^4 of the genome, or less. The signals
expected are very small. New methods are needed. They have
appeared in the nick of time.

   Our current approach to the subject of the control of gene expression in eukaryote organisms is to isolate chromosomal fragments 10-20 kb in length and containing particular expressed or
nonexpressed coding sequences from, for example, rat liver chromatin. We clone these coding sequences by recombinant DNA technology.
We then reconstitute them with core histones into minichromosomes.
We now try to find what it is, for example, in liver nuclei which
causes the serum albumin gene of rat liver to become expressed,
transcribable, to produce the messenger RNA characteristic of
serum albumin. We have not yet found out what the element is in
rat liver nucleosomal material which when added to the fragment
of rat liver DNA containing this serum albumin coding sequence and
packaged as a minichromosome causes this minichromosome to become
transcribable. I am confident, however, that this technology, a
much simpler and surer one than any tried heretofore, will in the
long run lead us to discover both the nature of the elements which
confer sequence specificity upon the control of gene expression
and the nature of the alterations in histone-DNA interaction which
cause the alteration in physical conformation of nucleosomes so
that they become both extended, low melting and transcribable. To
my way of thinking, in the isolation and multiplication of individual specific coding sequences and their flanking regulatory sequences lies the secret of the unlocking of the mystery of the
control of gene expression.

REFERENCES

Chalkley, R. & Jensen, R. (1969) Biochemistry 7:4380-4833.
Savage, M. & Bonner, J. (1978) In: "Methods in Cell Biology",
    G. Stein and J. Stein (Eds.) Academic Press, New York, in press.
Marushige, K. & Bonner, J. (1971) Proc. Nat. Acad. Sci. 68:
    2941-2944.
Billing, R. & Bonner, J. (1972) Biochim. Biophys. Acta 281:453-
    462.
Gottesfeld, J., Garrard, W., Bagi, G., Wilson, R. & Bonner, J.
    (1974) Proc. Nat. Acad. Sci. 71:2193-2197.
Gottesfeld, J., Bagi, G., Berg, B. & Bonner, J. (1976) Biochemistry
    15:2742-2482.
Garel, A. & Axel, R. (1976) Proc. Nat. Acad. Sci. 73:3966-3970.
Weintraub, H. & Groudine, H. (1976) Science 193:848-856.
Gottesfeld, J., Bonner, J., Radda, G. & Walker, I. O. (1974)
    Biochemistry 13:2937-2945.   .
Hamkalo, B., Miller, O. & Bakken, A. (1973) Cold Spring Harbor
    Symp. on Quant. Biol. 38:915-919.
Allfrey, V., Faulkner, R. & Mirsky, A. E. (1964) Proc. Nat. Acad.
    Sci. 73:3937-3941.
Marushige, K. (1976) Proc. Nat. Acad. Sci. 73:3937-3941.
Wallace, R. B., Sargent, T., Murphy, R. & Bonner, J. (1977) Proc.
    Nat. Acad. Sci. 74:3244-3248.
Bekhor, I., Kung, G. & Bonner, J. (1969) J. Mol. Biol. 39:351-364.
Huang, R. C. C. & Huang, P. C. (1969) J. Mol. Biol. 39:365-378.
Stein, G., Maus, R., Gabby,E., Stein, J., Davis, J. & Adamadkan, P.
    (1975) Biochemistry 14:1859-1866.
Paul, J., Gilmour, R., Affara, N., Birnie, G., Harrison, P., Hell,
    A., Humphries, S., Windass, J. & Young, B. (1973) Cold Spring
    Harbor Symp. Quant. Biol. 38:885-890.

DISCUSSION (PART I)

DR. WILHELM: Did you try to digest the nuclei first with DNase I
and then with DNase II to see whether the active fraction S_2
disappeared?

DR. BONNER: *We did digest chromatin with DNase I (15-20% made
acid soluble). The remainder yielded no S_2 on DNase II digestion.*

DR. WILHELM: Could you explain why you obtain only an enrichment
of 7 for the globin genes in the active fraction and why the
physical properties - thermal denaturation for example - of
the S_2 fraction are so different from the bulk of chromatin
although there is a full complement of histone bound to the
DNA?

DR. BONNER: *The S_2 of cell chromatin is about 15% of total DNA.
More or less all S_2 is removed from chromatin and all globin
gene is in S_2, so the enrichment in S_2 over whole chromatin
would be seven-fold. This is what is found. The physical
properties of S_2, including its thermal denaturation profile,
all indicate that the N terminal peptides of the histones have
relaxed their bindings to DNA.*

DR. SARMA: If you treat S_2 DNA with DNase I to an extent of
1 to 10% do you digest all the transcribing genes?

DR. BONNER: *Yes, because it is very rapidly attacked by DNase I.*

DR. SARMA: Does this mean that S_2 DNA has lots of non-transcriba-
ble genes?

DR. BONNER: *I do not say this. It has both transcribable and
transcribing genes.*

DR. SARMA: What makes a transcribable gene different from
transcribing gene?

25

DR. BONNER: I do not know. Both ready to be transcribed and transcribing genes seem to be in extended form.

DR. SARMA: What is the biological and functional advantage of the nucleosomal structure?

DR. BONNER: Probably it represents the first step in compacting 2 meters of DNA into the nucleus.

DR. SARMA: How are the two classes of transcribable genes (the multicopy rRNA genes and tRNA genes synthesizing rRNA and tRNA and unique copy gene synthesizing mRNA) distributed in chromatin in terms of nuclease accessibility and conformation?

DR. BONNER: rRNA, tRNA and mRNA genes are all in the S_2 fraction. However, electron microscopy evidence has been accumulated which indicates that rRNA genes appear to lack nucleosomal particles, while mRNA genes possess nucleosomes.

DR. SARMA: DMSO treatment is generally coincident with the release of viral sequences. Could this affect the transcription of hemoglobin genes?

DR. BONNER: It is quite true that DMSO induces expression of the virus. However, we have not studied whether this could affect the expression of hemoglobin genes.

DR. DIXON: You mentioned that in S_2 there is almost three times as much NHCP as in bulk chromatin. Is this NHCP enriched in any particular NHCP species?

DR. BONNER: The non-histone chromosomal proteins of the S_2 fraction contains all of the Hn-RNA packaging protein. Actin, tubulin and actomyosin are also present mainly in S_2 fraction.

DR. SMETS: Does the S_2 chromatin increase during DNA replication? Are there cell cycle dependent changes in the percentage of S_2 DNA?

DR. BONNER: It has not been determined.

DR. VAN HOLDE: How much does the sedimentation coefficient drop when histones are acetylated?

DR. BONNER: The sedimentation coefficient drops to about 7S.

DR. LANGLOIS: Do metaphase chromosomes have a different structure at the nucleosome level than interphase chromatin?

DR. BONNER: I do not know; perhaps people working at the E. M. level will be able to tell us.

DR. NICOLINI: I am disturbed by the lack of interest and precision in defining the object of your analysis and how much it could be different from the original object "in situ". Without such an effort it is difficult to communicate. In studying higher levels of organization such standardization is extremely important.

DR. BONNER: *I think we are communicating here. My view on physiocochemistry of chromatin is that these types of studies are useful only as they explain something. I am aware of the effects of shearing. Chromatin prepared following method one, that I described this morning, is approximately 10 Kbases long.*

DR. NICOLINI: You seem to feel that the main effect of shearing is only the falling off of some terminal protein in the fragment. It was however observed at the E.M. level that unsheared chromatin displays fibers 300Å in diameter, while sheared chromatin has only 100Å diameter fibers (Klug and co-workers).

DR. BONNER: *I am aware of this difference in fiber diameter, but we had no occasion of discussing this subject.*

DR. NICOLINI: Am I correct in summarizing your feelings, that you think efforts in characterizing and standardizing our object of investigation and their relationship with the original object are of minor importance?

DR. BONNER: *I have already stated that my 10 Kbases chromatin is something that does not exist in the cell; however, it is an object simple enough for starting a study. You can do things that you cannot do with an insoluble blob. I did not say that you do not want to characterize the blob, but we have to study the complex structure after we get to know something about the elementary structure.*

DR. NICOLINI: I think however that a strong word of caution should be given on the possible artifacts, since too frequently and superficially investigators did and still do extrapolate their findings on sheared chromatin, to determine possible mechanisms which control gene expression. Major efforts should go both in physically standardizing the object of investigation, and in improving the technology in order to study chromatin as close as possible to its truly native conditions. Higher order of chromatin organization (as solenoid) could indeed be the key for an understanding of control transcription and gene expression.

SECTION II:
PHYSICAL, CHEMICAL AND BIOLOGICAL TECHNIQUES FOR STUDYING NUCLEOSOME, CHROMATIN, CHROMOSOME AND NUCLEI

ELECTRON MICROSCOPY: A TOOL FOR VISUALIZING CHROMATIN*

ADA L. OLINS

The University of Tennessee-Oak Ridge Graduate School
of Biomedical Sciences and Biology Division, Oak Ridge
National Laboratory, Oak Ridge, TN 37830

The molecular biologist is fortunate that many of the macromolecules of interest can be visualized by the electron microscope,
thus helping in the interpretation of biophysical and biochemical
information. The problem arises in finding objective criteria for
evaluating such images; gleaning the meaningful information and
discarding the chaff. It is the goal of this article to set forth
such guidlines to sharpen the student's awareness of the characteristics of acceptable electron microscopic data. The evaluation of
a micrograph, as other biophysical data, can and should be objective and not based on artistic merit. Progress in our understanding
of chromatin structure has been accelerated by electron microscopy
and will serve as an example for this discussion.

CRITERIA FOR GOOD ELECTRON MICROGRAPHS

It requires a good deal of patience to take interesting, high
quality electron micrographs, yet only in very rare cases can I
recommend acceptance of a poor quality micrograph. Electron microscopy, like all other good data, must be reproducible, thus present-
ing the microscopist with many opportunities to improve the quality
of the micrograph. Some characteristics expected in good electron
micrographs are as follows:

---

*Research supported by the National Science Foundation Grants
PCM76-01490 and PCM77-21498 and by the Division of Biological and
Environmental Research, U. S. Department of Energy, under contract
W-7405-eng-26 with the Union Carbide Corporation.

1. A clean, empty background allows rapid identification of
the sample. Amorphous stain deposits, dirt and bacteria can be
confusing and certainly raise questions about the care exerted in
doing the experiments.

2. The contrast in the micrograph should be sufficient to
permit easy recognition of the structure as described by the microscopist. This does not, of course, apply to Xerox copies of micrographs, transpariencies projected in an illuminated room or slides
viewed at a great distance.

Biological specimens are composed of small atoms which have
little intrinsic electron density, therefore most contrast is obtained by the addition of electron dense molecules (stains).

Positive staining is achieved by incubating the specimen with
the stain followed by washing the specimen in order to remove all
unbound stain. This method requires a high affinity between stain
and specimen.

Negative staining is achieved by allowing an excess of stain
to surround the specimen producing a dark outline of a light macromolecule. The stain is not washed off after it is applied to the
grid. Since it does not (usually) have a high affinity for the
specimen, it generally surrounds the structures on the grid. The
goal is to obtain a very delicate outline of the structure. Often
the stain is too thick and obstructs the edges of the specimen.
Sometimes the stain dries in an uneven pattern leaving gaps which
could be misconstrued to represent the location of macromolecules.
In this case, reproducibility and correlation with other known
characteristics of the specimen must be considered by the microscopist. A simple control, rarely shown in the literature, is to
look at a grid covered with film and stain, but having no specimen
on it. Some stain irregularities can be eliminated by filtering
the stain before application to the specimen. Many different stains
should usually be tried, giving the microscopist a good opportunity
to become familiar with the sample structure and its variation with
different stains.

Shadowing a specimen with metal (e.g., Pt, Pd, Au) vapor in
vacuo produces very highly contrasted specimens, but is usually done
in such massive quantities as to greatly increase the size of the
specimen. This method has been extremely useful for studying DNA,
spread in the presence of cytochrome c (Kleinschmidt, 1959), because
the length of these long molecules is not greatly changed by the
layers of cytochrome or metal. Single- and double-stranded DNA can
be distinguished, the contrast is very good, and biochemical data
is usually available to support the interpretation of micrographs
of shadowed specimens. However, this method cannot be reliably used

for high resolution ultrastructural analysis of macromolecules without special precautions to reduce the amount of metal evaporated and
the granularity of the metal deposits. Shadowing can be done on a
rotating sample, or unidirectionally, the latter being especially
useful in determining the height of the molecular specimens.

   Supporting film thickness and uniformity are extremely important, especially for high resolution studies of stained macromolecules. Since the image is formed by absorption or scattering of
electrons, the larger the difference in electron density between
specimen and supporting film (background), the greater the contrast
in the final micrograph.

   Recently, we have adopted the methods of Ottensmeyer et al.
(1975) for tilted beam, darkfield microscopy to achieve very markedly improved contrast. In this method the image is formed by the
scattered electrons, eliminating the "noise" of the bright undeflec-
ted primary electron beam. For this method, more than any of the
others discussed, a thin uniform supporting film is essential for
improving the contrast.

   3. Fine high resolution electron micrographs should be either
in focus or slightly underfocus. At focus, in bright field, the
grain is not visible and contrast is at a minimum; at an underfocus
setting, the grain is sharp or crisp; at an overfocus setting, the
grain is blunt or rounded. It is important to set the correct focus
at a magnification higher than 40,000X so that the grain is visible
and adjustments in the objective lens current can be made accurately.
Proper focus cannot be achieved unless the astigmatism, manifest by
a directional orientation of the grain, of the objective lens is
corrected. The size of the background grain should be smaller than
the structure of the macromolecular detail of interest; i.e. a pattern seen within the image of the macromolecule, which is also seen
in the background, cannot be interpreted to be macromolecular substructure.

   4. Magnification of an electron micrograph should be sufficiently high so that the detail of interest is easily visible.
This magnification should be achieved by the electron microscope
and not a photographic procedure, so that the focus is correct and
a large grain pattern does not dominate the structural detail.

                    SPECIMEN PREPARATION

   Life cannot proceed in the vacuum of the electron microscope.
At best, we can only hope to visualize an aspect of the structure
which reflects the true organization of molecules in their biological roles. How much distortion can be tolerated? How little
distortion can be achieved? These are difficult questions, whose

answers change as new techniques are devised. At present, however,
we can suggest and justify criteria adopted within our laboratory
for preparing chromatin specimens:

    1. Chromatin tends to form gelatinous fibrillar aggregates,
which should be avoided because the underlying substructure of the
fibers is obscured or altered. It is especially helpful to use
freshly isolated nuclei or chromatin. Frozen cells, frozen nuclei
or frozen chromatin all tend to form gelatinous aggregates. This
is not seen, however, in frozen nuclease digest fragments, especially
monomers (ν_1), dimers (ν_2) and trimers (ν_3). We usually apply the
samples to the grids the same day that live cells are harvested.
The forces of air-water or air-buffer interphase spreading seem to
cause similar gelatinous fibrillar networks and this method is,
therefore, avoided in our laboratory.

    2. Many strange structures can be found on an electron micro-
scope grid. Which are artifactual and which reflect the structure
of chromatin? The obvious control of looking at stained grids without sample has already been mentioned. Other correlations with the
known biochemical and biophysical properties of the sample must be
made. For example, if hydrodynamic data (such as analytical ultracentrifugation) and electrophoresis in DNA gels indicate that the
sample has the properties of ν_1 but the sample on the grids consists
of large fibers or aggregates, it can be assumed that the preparation of the chromatin solution for microscopy should be improved.

    If reconstructed samples are to be studied, close comparison
should be made to the original starting material; e.g., higher order
organization of nucleosomes in 20-30 nm fibers should be studied
before and after dissociation with EDTA, if a "native" reorganizing
effect of Mg^{++} is to be demonstrated.

    Another example of a useful comparison is the resemblance of
the ultrastructure of ν_1 with nucleosomes on long chromatin fibers
spilling out of a freshly isolated nucleus, thus establishing their
identity with more certainty.

    3. Is the structure of interest reproducible? This question
should have an affirmative answer for: a regular distribution of
the structure within a single sample; and for the reappearance of
the structure in samples prepared at different times from separate
sources. The rapid acceptance of the universality of the nucleosome
as the fundamental unit of chromatin structure depended, in large
measure, on its ubiquitous distribution in electron micrographs and
its correlation with biochemical data.

    4. Organic solvents such as ethanol destroy the 100 Å low-
angle x-ray reflections (Pooley et al., 1974) and the appearance of

nucleosomes in the electron microscope (Woodcock et al., 1976). For
this reason, critical-point drying, and dehydration with organic
solvents, has been avoided in our laboratory. Salt and pH conditions
must also be carefully controlled and their effects understood. For
example, 0.6 M NaCl is known to remove the lysine-rich histones in
most eucaryotes; therefore, chromatin treated with this level of salt
cannot reflect the structural contribution of the H1 histone class.

CURRENT OPTIMAL TECHNIQUES USED IN OUR LABORATORY

The methods used in our laboratory have been strongly influenced by the important technical breakthrough made by Miller and
Beatty (1969), which permitted spreading of chromatin for visualization of "genes in action". This procedure, which involved isolation of nuclei, centrifugation onto carbon-coated grids and drying
from dilute Kodak Photo-flo, displayed the close proximity of RNA
polymerase molecules, the increasing lengths of RNA transcripts
within a gene and the tandem arrays of ribosomal genes. Although
chromatin within the nucleus must be much more compact than seen in
such preparations, it demonstrates the primary level of association
of RNA, protein and DNA. Higher-order interactions, and internal
structure of DNA-protein and RNA-protein molecules were not devulged
in these early studies.

Our initial observations of nucleosomes (ν bodies) were made on
fibers streaming out of isolated chicken erythrocyte nuclei. The
biochemical data showed that more than 95% of the proteins in these
nuclei are histones. The negative staining indicated that thicker
regions of the chromatin fibers were formed by close association of
several nucleosomes. The particles were visible everywhere. Further
rapid support for the particulate structure of chromatin came from
biochemical data reported in many laboratories (Felsenfeld, 1978).

Several samples of current chromatin micrographs are presented
in this section to illustrate the points raised earlier in this
manuscript. Figure 1 is an electron micrograph of a portion of a
chicken erythrocyte nucleus, with extended chromatin fibers streaming out of the nuclear periphery.

This darkfield electron micrograph is in focus; the grain is
symmetric and the contrast is maximal. Although the stain was not
washed off the grid it seems to bind specifically to certain regions
of the chromatin. This is probably due to the high affinity of the
positively charged uranyl ions for the negatively charged phosphate
and carboxyl groups in the chromatin fibril. For thicker specimens
(see Figure 3) or at higher concentrations of uranyl acetate, the
stain would surround the specimen as a true "negative stain". A
bright field electron micrograph of a similar preparation would have
much less contrast (Olins et al., 1975).

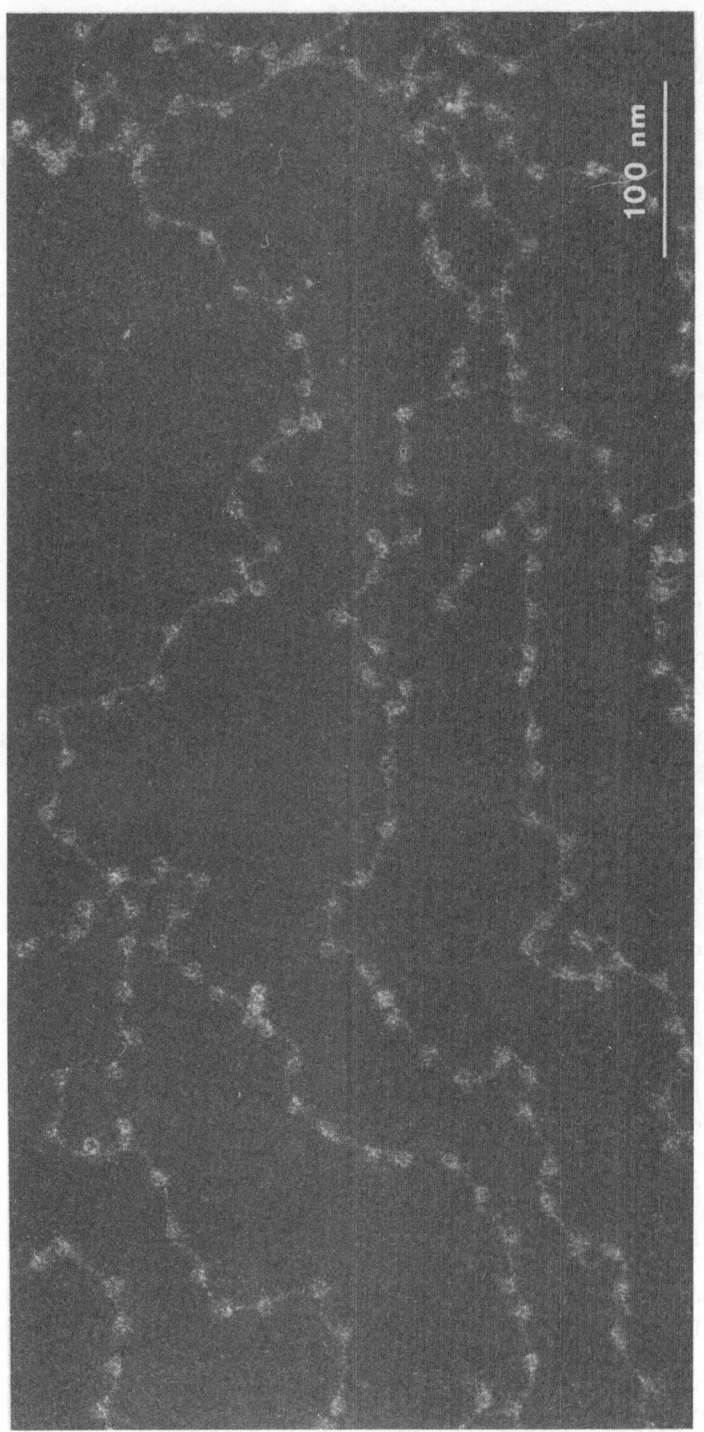

Figure 1. A darkfield electron micrograph of the edge of a chicken erythrocyte nucleus. The freshly isolated nuclei were gently suspended in 0.2 M KCl, then diluted 1:100 with 0.2 mM EDTA. After 10 minutes of swelling the solution was made 0.9% HCHO, and centrifuged through 10% HCHO pH 7 onto a glowed carbon-coated grid. The grid was then washed in dilute Kodak Photo-flo, drained on the edge of Bibulous paper, dried, stained for 30 seconds with 0.1% uranyl acetate, drained and dried again. For details see Olins and Olins (1974).

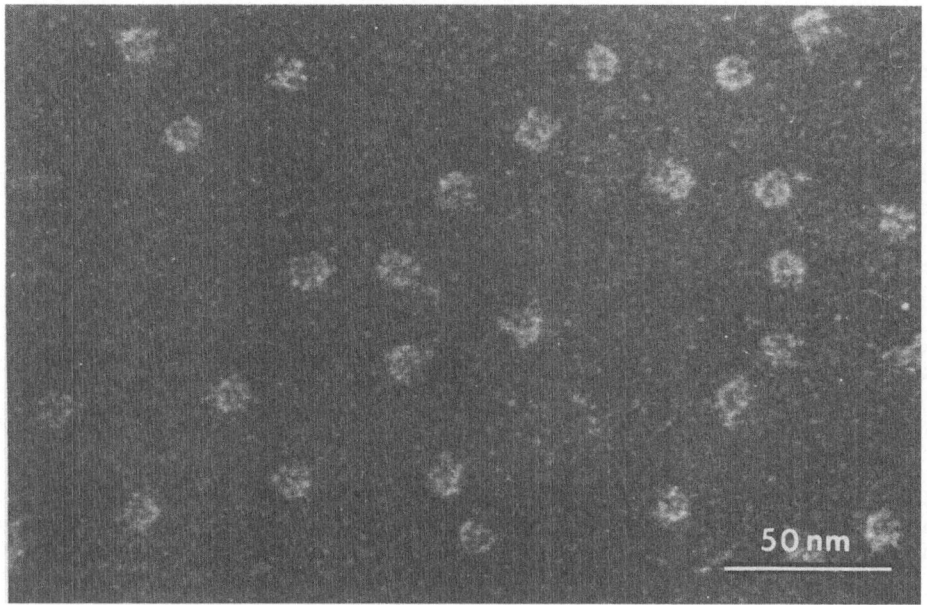

Figure 2. A darkfield electron micrograph of isolated nucleosomes. Chicken erythrocyte nuclei were digested with micrococcal nuclease and fractionated by sucrose gradient centrifugation. A drop of ν_1 A_{260} = 1.0 was placed on a glowed carbon-coated grid for 30 seconds, rinsed with dilute Kodak Photo-flo, drained, dried and stained as in Figure 1.

In order to visualize nucleosomal ultrastructure, a micrograph of isolated monomer nucleosomes (Figure 2) is presented. Again, darkfield is used to achieve contrast. The clarity of detail is due, in part, to the high magnification (200,000X) at which the micrograph was made and, in part, to the low concentration of stain used. The similarity of the particles in the chromosomal fibers of Figure 1 and the nucleosomes (ν_1) isolated after micrococcal nuclease digestion in Figure 2 is demonstrated.

The true negative staining results demonstrated in Figure 3 show the close packing of nucleosomes in the 20-30 nm fiber (Olins, 1977). Stain is clearly surrounding the particles rather than binding to them. These fibers are so thick that the internal structure, i.e. the nucleosomes, cannot be clearly visualized in darkfield microscopy. This is due, in part, to the extra electron density of the negative stain surrounding the fiber and, in part, to the superimposition of nucleosomes on opposite sides of the thick chromatin fiber. This micrograph is slightly underfocus, as can be seen by the increased contrast and the crisp symmetric background grain.

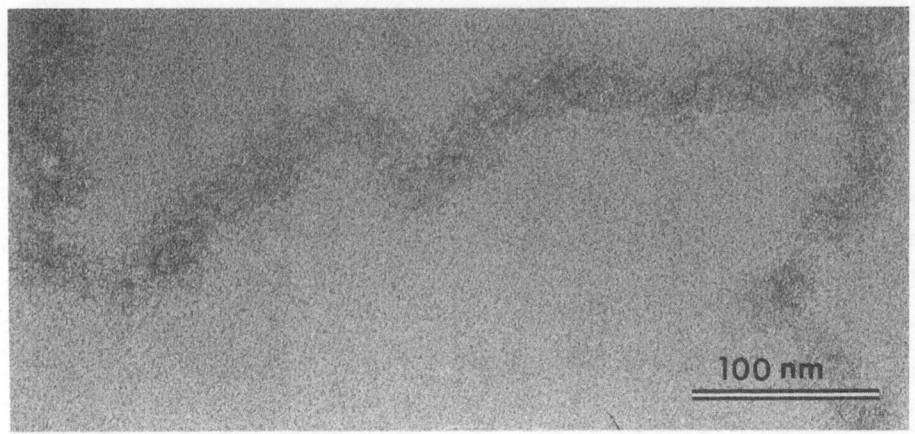

Figure 3. A bright field electron micrograph of a 20-30 nm chromatin fiber at the periphery of a chicken erythrocyte nucleus. Freshly
isolated nuclei are suspended in 0.25 M KCl 0.2 mM $MgCl_2$ then diluted
1:100 in 0.2 M KCl. After 10 minutes the nuclei are fixed and processed as described in the legend for Figure 1.

The final figure in this demonstration is a darkfield micrograph of part of an extrachromosomal nucleolar ribosomal gene as it
is being transcribed. Although this method of preparing the sample
preserves the nucleosomal structure of the inactive chromatin fibril,
it is not yet possible to describe the structure of either the RNA
polymerase or the RNP fibrils. It is evident that new preparative
methods must be devised which preserve the structure of these important chromatin components. The carbon film used in this preparation
is exceptionally uniform and thin, producing fine contrast of a delicate specimen.

The challenge of preparing high-quality micrographs was elo-
quently presented 300 years ago.

"Tis not unlikely, but that there may be yet invented
several other helps for the eye, as much exceeding
those already found, as those do the bare eye, such
as by which we may perhaps be able to discover living
Creatures in the Moon, or other Planets, the figures
of the compounding Particles of matter, and the particular Schematisms and Textures of Bodies." (R. Hooke,
Micrographia, 1664.)

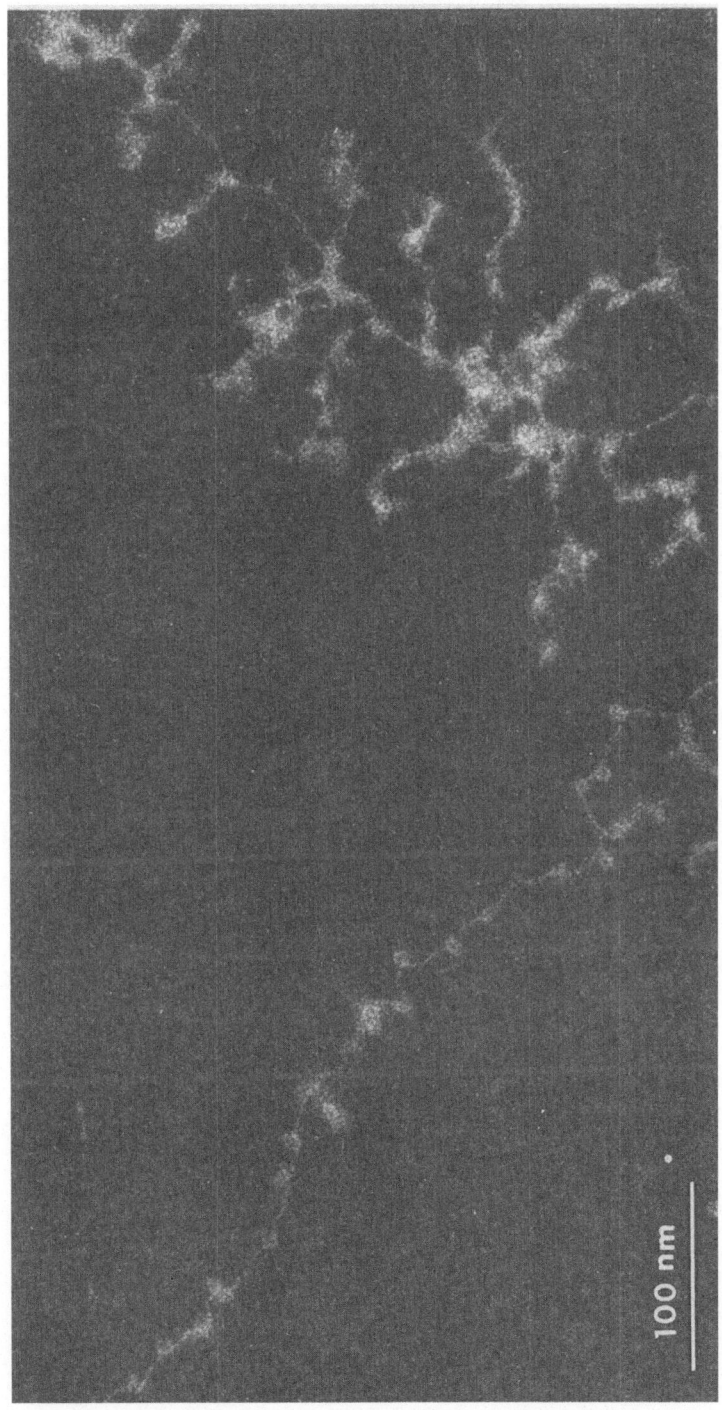

Figure 4. A darkfield electron micrograph displaying part of a ribosomal gene in transcription and an inactive chromatin fiber. Freshly dissected Triturus viridescens oocytes were opened in 0.15 M KCl to isolate the germinal vesicle (g.v.); in less than 2 minutes the g.v. was placed in a drop of water pH 9, opened with forceps and allowed to disperse. The contents of a single g.v. were centrifuged through 3.7% HCHO, 0.1 M sucrose pH 8.8 onto a glowed carbon-coated grid and processed as in Figure 1.

ACKNOWLEDGEMENT

I thank Mayphoon H. Hsie and Elizabeth A. Wilkinson for excellent assistance, G. J. Bunick for v_1 and D. E. Olins for helpful discussions.

REFERENCES

Felsenfeld, G. (1978) Nature 271, 115-121.

Hooke, R. (1664) "Micrographia", Printers to the Royal Society, London.

Kleinschmidt, A., and Zahn, R. K. (1959) Z. Naturforschg. 14b, 770-779.

Miller, O. L., and Beatty, B. R. (1969) Science 164, 955-957.

Olins, A. L. (1977) Cold Spring Harbor Symposium on Quantitative Biology 42, in press.

Olins, A. L., Carlson, R. D., and Olins, E. E. (1975) J. Cell Biol. 64, 528-537.

Olins, A. L., and Olins, D. E. (1974) Science 183, 330-332.

Ottensmeyer, F. P., Whiting, R. F., Schmidt, E. E., and Clemens, R. S. (1975) J. Ultrastruct. Res. 52, 193-201.

Pooley, A. S., Pardon, J. F., and Richards, B. M. (1974) J. Mol. Biol. 85, 533-549.

Woodcock, C. L. F., Safer, J. P., and Stanchfield, J. E. (1976) Exp. Cell Res. 97, 101-110.

TRANSCRIPTIONAL CONTROL OF NATIVE CHROMATIN

R. Stewart Gilmour

Beatson Institute for Cancer Research
Wolfson Laboratory for Molecular Pathology
Garscube Estate, Switchback Road, Bearsden
Glasgow, G61 1BD, Scotland

WHAT IS CHROMATIN?

Most work which has been done with chromatin refers to the dispersed chromosomes of interphase cells. Isolated chromatin consists of DNA, histones, non-histone proteins (NHP) and small quantities of RNA. Numerous comparisons have been made of the relative concentrations of these components in chromatins from a wide variety of sources (see review by MacGillivray and Rickwood, 1974). Although in higher eukaryotes histones and DNA maintain an approximate one to one ratio, in lower eukaryotes this is not the case. The levels of NHP vary considerably throughout the species. There is some circumstantial evidence to suggest that the amount of NHP is directly related to the amount of transcriptionally active DNA in chromatin. The main definitional problem is to decide which of the many types of molecule within the nucleus should be considered as a genuine component of chromatin. The original preparative procedures simply involved the extraction of whole cells with isotonic saline; however, with the recognition of cytoplasmic proteins, membrane and ribosomes, procedures were altered to exclude these contaminants. Obviously a molecule which never associates at any time with the DNA might not be considered as a part of chromatin, for example, RNA processing enzymes; but what about proteins which are transiently associated with the DNA and furthermore may shuttle between nucleus and cytoplasm? Clearly chromatin is not a static entity but can undergo a variety of chemical and structural changes in response to the metabolic status of the cell. The purpose of this chapter is to outline the in vitro experimental approach that has led to the idea that chromatin has built in transcriptional controls which regulate the availabity of genetic

41

information.   In a subsequent chapter the structural features
responsible for this phenomenon will be considered.

EVIDENCE FOR DIFFERENTIAL GENE TRANSCRIPTION

One of the strongest arguments for differential gene activity
has 'come from the study of puffing in the polytene chromosomes of
larval Diptera.   It has been firmly established that puffs are
chromosome sites undergoing active transcription and detailed
analysis of puffing patterns have revealed the presence of tissue
specific puffs as well as characteristic changes with development
(see reviews by Ashburner 1972 and Panitz 1972).   These results
which have been obtained from studies of several Chironomus and
Drosophila species suggest differential gene activity during
development and in functionally different cell types.   Although
puffs have not been detected in the chromosomes of higher animals,
evidence suggests that chromatin can exist in different states of
condensation.   Electron microscopy studies on interphase nuclei
of calf thymus lymphocytes described heterochromatin (condensed)
and euchromatin (diffuse) forms.   Incorporation of radioactive
precursors in vitro showed that the bulk of the rapidly labelled
RNA was associated mainly with euchromatin (Allfrey and Mirsky
1962; Allfrey et al., 1963; Frenster, 1963).

The application and development of DNA-RNA hybridization
methods to eukaryotic systems has contributed more than any
other single technique to our understanding of the relationships
between genomic DNA and its transcription product.

A demonstration of differential gene expression requires
evidence for a selection of DNA sequences transcribed into RNA.
Early investigations of the sequence homologies of total cellular
RNA isolated from a variety of species demonstrated that
qualitatively different RNA molecules are present in different
cell types of the same organism; for example in mouse tissues
(McCarthy and Hoyer, 1964), developmental stages of sea urchin
embryos (Glisen et al., 1966) and embryonic mouse liver (Church
and McCarthy 1967), Xenopus oocytes and blastulae (Davidson
et al., 1968) and mouse liver uterus before and after oestrogen
stimulation (Church and McCarthy, 1970).

The finding that eukaryotic chromatin can support the
transcription of RNA in vitro in the presence of added RNA
polymerase, usually bacterial, suggested a cell-free approach to
the study of this phenomenon.   Using the same hybridization
techniques to compare the homologies of in vitro transcribed RNA
and total genomic DNA it was concluded that the ability of chromatin
to support RNA synthesis is restricted compared with that of DNA.
This appeared to be due to a limitation in the availability of
sites in DNA which can be transcribed so that a restricted set of

RNA sequences is produced (Paul and Gilmour, 1966, 1968; Bekhor et al., 1969; Huang and Huang, 1969; Smith et al., 1969, Spelsberg and Hnilica, 1970; Sawada et al., 1972). In some of these experiments it was shown that in vitro RNA transcripts were indistinguishable from the in vivo RNA of the organ from which the chromatin was isolated, as judged by competitive hybridization. (Paul and Gilmour, 1968; Smith et al., 1969; Tan and Miyagi, 1970). This led to comparative analyses of chromatin transcripts from different organs of the same animal and it was generally concluded that while there were many common RNA sequences transcribed tissue specific RNA sequences were also synthesized.

However, with the discovery that eukaryotic DNA is composed of both repetitive and unique nucleotide sequences it became clear that in all these hybridization experiments the concentrations of hybridizing DNA and RNA sequences were such that only the homologies between RNA and the reiterated sequences in the DNA were detected. No information about unique sequence transcripts can be obtained using these hybridization conditions. There is now a growing body of evidence which indicates that structural gene sequences in eukaryotes probably occur once per haploid genome. It is likely that globin (Bishop and Freeman 1973; Bishop and Rosbash, 1973; Harrison et al., 1974), ovalbumin (Harris et al., 1973) and fibroin (Suzuki et al., 1972) are coded by single genes although in the case of sea urchin histones multiple gene copies are present (Weinberg et al., 1972). That this conclusion can be extended to total cellular mRNA has been determined by following the hybridization of labelled mRNA or the cDNA derived from it by reverse transcription in the presence of excess cellular DNA. Studies of this nature which have been carried out on Friend cells ([B]irnie et al, 1974), HeLa cells (Spradling et al., 1974; Klein et al., 1974; Bishop et al., 1974), rat myloblast (Campo and Bishop 1974), mouse L cells (Greenberg and Perry, 1971) sea urchin embryos (Goldberg et al., 1973; Davidson et al., 1975) and Dictyostelium (Lodish et al., 1973) indicate that almost all mRNA sequences are transcribed from non-repetitive regions of the genome. For further discussion the reader is referred to the review by Lewin (1975).

Clearly this generalisation challenges the inference of the original hybridisation experiments with in vivo made RNAs and chromatin primed RNAs. Where tissue specific hybridisation was equated with the presence of specific mRNA species. While it is obvious that these would not have been detected, the fact remains that these experiments show that broad tissue specificities exist within the repetitive DNA transcripts and as yet remain unexplained. The whole question of whether selective gene expression can be seen in in vivo made RNA has been re-examined from the point of view of messenger sequences. This has been

done by hybridizing RNA to the unique DNA sequences isolated from
the total genomic DNA or to cDNA made to specific mRNA.   Under
these hybridization conditions the concentrations of the unique
sequences is effectively concentrated and hence can undergo
hybridization.   For example, transcripts derived from unique DNA
sequences were analysed by hybridising isolated unique DNA to a
large excess of total cellular RNA or in some cases nuclear RNA.
There again distinctly different populations of unique sequences
are present in the RNA's of different tissues as demonstrated in
adult mouse tissues (Brown and Church, 1972; Grouse et al.,
1972),mouse embryonic stages (Church and Brown, 1972),
Dictyostelium developmental stages (Firtel, 1972) and chick
oviduct before and after oestrogen stimulation (Liarakos et al.,
1973).

      The recent use of globin cDNA hybridization probes by a
number of workers has extended this conclusion for the particular
case of globin mRNA in erythroid versus non-erythroid cellular
RNA (Axel et al., 1973; Leder et al., 1973; Steggles et al.,
1974; Humphries et al., 1974; Gilmour et al., 1974).

      These results, from a wide variety of sources, attest to the
generality and significance of differential gene transcription.
Selective sequences of both repetitive and unique DNA are
transcribed in vivo in a tissue specific fashion.

      The problem of analysing in vitro synthesised chromatin
transcripts for specific mRNAs has been greatly facilitated by the
use of specific cDNA hybridization probes.   In particular a
number of groups have reported the in vitro transcription of
globin mRNA from erythroid chromatin in a variety of tissues and
have contrasted this with a corresponding lack of these sequences
in transcripts from non-erythroid chromatin of the same species.
(Axel et al., 1973; Gilmour and Paul, 1973; Steggles et al.,
1974; Barrett et al., 1974).   Harris et al., (1975) has found a
substantial concentration of ovalbumin mRNA sequences in the
transcripts from oestrogen stimulated chick oviduct chromatin
as compared with unstimulated oviduct chromatin.   In HeLa cells
Stein et al. (1975) have described the in vitro transcription of
histone genes in chromatin from S-phase cells but not in G_1-phase
chromatin.   Since then the validity of these and subsequent
results has been challenged by a number of criticisms applicable
to in vitro chromatin transcription in general.   It seems
appropriate at this stage to consider the justification of these
criticisms and the attempts that have been made to answer them.

VALIDITY OF IN VITRO TRANSCRIPTION - (a) Endogenous RNA

      The problems associated with endogenous RNA are best
illustrated by examining the author's results with mouse foetal

liver chromatin. This tissue attains maximum haemopoietic activity after 14 days *in vitro* at which time about 70% of the cells are erythroid. Chromatin was prepared from 14 day foetal livers and adult mouse brain and incubated with E. coli RNA polymerase. Transcripts were isolated and hybridized to globin cDNA to estimate the concentration of globin mRNA sequences. The details of these steps are given below.

Chromatin preparation: 1 g foetal livers are hand homogenized (Dounce homogenizer, 0.001 inch pestel clearance) in 36 mls 2 mM Tris; HCl pH 7.5; 5 mM $MgCl_2$; 1 mM Dithiothreitol (DTT); 3 mM $CaCl_2$ and left on ice for 5 mins. After adding 4 mls of 2 M sucrose another 30 secs. additional homogenization is carried out. 20 ml aliquots of homogenate are layered on 15 mls of 2.0M sucrose; 2 mM Tris pH 7.5; 1 mM DTT; 5 mM $MgCl_2$; 0.28 m NaCl and centrifuged in a swing-out rotor at 30,000 g for 1 hr. at 4°C. The pellet is suspended in 20 mls 1 mM Tris; 0.1 mM EDTA; 0.28 m NaCl; 1% Triton-X-100 and centrifuged at 12,000 g for 10 mins. The pellet is washed three times with 1 mM Tris; HCl pH 7.9; 0.1 mM EDTA. The viscous gel constitutes chromatin.

Transcription: Reaction mixture (2 mls) contain 0.04 M Tris pH 7.9; 2.5 mM $MnCl_2$; 0.1 mM EDTA; 0.1 mM DTT; 0.8 mM ATP, CTP, GTP and UTP; 0.5 - 1 mg DNA as chromatin and 100 Burgess units of E. coli RNA polymerase.

RNA isolation: After incubation for 1 hr at 37°C the reaction is cooled to 4°C and 20 µg RNAse free DNAse added. After a further hr. at 4°C sodium sarcosinate and EDTA are added to final concentrations of 0.2% and 30 mM resp. and the mixture incubated with 200 µg Proteinase K (International Enzymes) for 30 mins at 37°C. A single phenol:chloroform (50/50:v/v) extraction is carried out and the aqueous phase passed over Sephadex G-50. Where unmercurated UTP is used columns are run in 0.05 m NaCl; 10 mM Tris pH 7.6. The excluded material is precipitated with 2.5 vols. ethanol re-dissolved in 1 ml HEPES pH 7, 5 mM $MgCl_2$; 0.25 m NaCl; 3 mM $CaCl_2$ and incubated with 20 µg RNAse free for 15 mins. at 37°C. After re-extraction with phenol: chloroform and chromatography on Sephadex G50 the RNA is precipitated with ethanol; dried and hybridised.

Hybridization and estimation of globin sequences: In this study we used a titration technique in which a fixed amount of $[^3H]$ cDNA (1 ng) is annealed at 43°C with increasing amounts of RNA in 10 µl hybridization buffer (0.5 M NaCl; 25 mM HEPES; 1 mM EDTA; 50% formamide, pH 6.7). The theoretical considerations of this approach and the treatment of results are described elsewhere by by Young et al. (1974). Hybrid was measured by the sensitivity of cDNA to single-stranded nuclease (S_1) prepared from Takadiastase

by the method of Sutton (1971).   The method is demonstrated in
Figure 1 which shows the titration of pure 9S globin mRNA and
reticulocyte polysomal RNA to 1 ng globin cDNA.   Complete
hybridization of the pure globin mRNA is achieved at input ratios
of RNA:cDNA = 1.6:1, consistent with the finding that cDNA
represents a 60% copy of the mRNA.   The input concentration of
polysomal RNA required to achieve saturation is 50 times that
needed for pure mRNA, thus giving an estimate of 2% for globin
mRNA content.   A small percentage of cDNA (5%) is normally
resistant to S_1 nuclease in the absence of RNA while an
additional 10–15% does not appear to be homologous to 9S RNA and
is degraded.

    An alternative method of estimating globin mRNA content is to
carry out a kinetic or Rot analysis.   This method will be
demonstrated in later data.

    Figure 2 shows the hybridization analysis of RNA transcribed
from foetal liver and adult mouse brain chromatins.   There is an
apparent absolute difference between them;   about 0.002% of the
foetal liver transcript is globin specific.   In a control
incubation RNA polymerase was omitted and an appropriate amount
of t-RNA added.   Re-isolation of the t-RNA was carried out in
parallel with the transcripts;   significant hybridisation was seen
in this case.   The globin RNA sequences detected in the control
have come from the endogenous RNA that co-isolates with the
chromatin and contributes an estimated 50% of the observed
hybridisation in actively transcribing incubations.   The most
serious criticism for the present argument is that any _in vivo_
synthesized endogenous RNA isolated with the chromatin template
will be indistinguishable from _in vitro_ RNA synthesized by the
exogenously added polymerase unless careful controls are
included.

    There seems to be some variance in the literature as to
the amount of endogenous RNA found in the chromatin of different
systems, as judged by control incubations containing chromatin
but no polymerase.   Nearly all the data referred to are derived
from hybridizations with excess RNA, where even minute levels of
endogenous RNA contamination of the _in vitro_ transcripts will
produce misleading results.   It is not certain that actively
transcribing incubations and control incubations are exactly
equivalent in all other respects except for the presence of
RNA polymerase.   In view of the discrepancies it is difficult
to assess the reliability of 'enzyme minus' controls _per se_
unless they are supplemented by more rigorous data.

    A more reliable control has been employed for measuring
both endogenous RNA and newly synthesised RNA simultaneously in

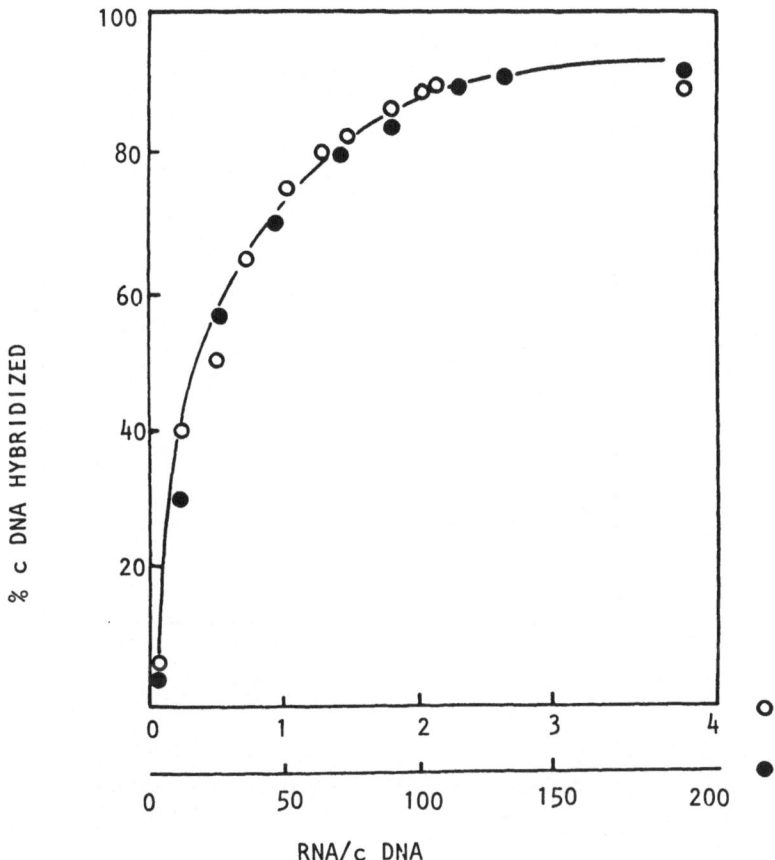

Legend to Figure 1: Titration to 1 ng [³H] globin cDNA of pure 9S globin mRNA (o —— o) amd reticulocyte polysomal RNA (• —— •).

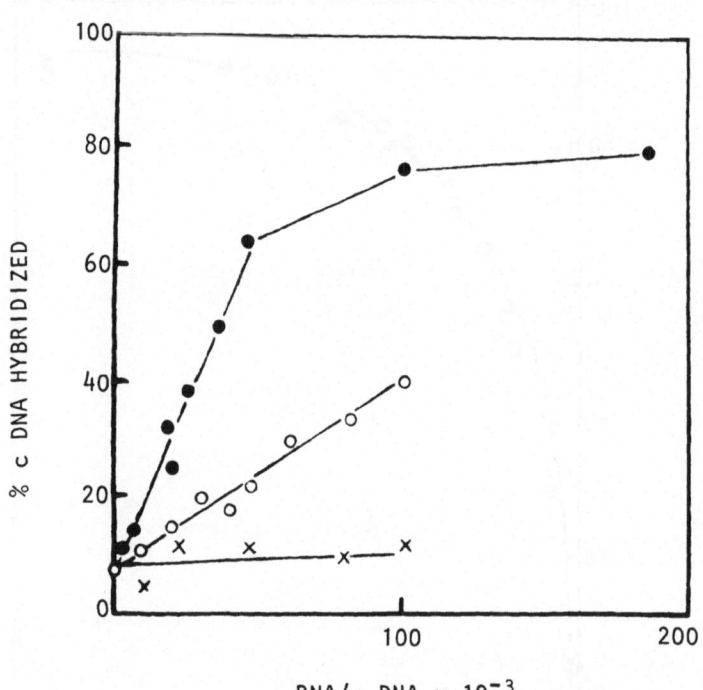

Legend to Figure 2:   Titration of RNA transcribed from chromatin
with E. Coli polymerase, against globin cDNA.   Mouse foetal
liver chromatin, (● ── ●); mouse foetal liver chromatin
incubated without RNA polymerase, (o ── o); mouse brain
chromatin (x ── x).   In incubations where no polymerase was
present E. Coli tRNA was added and aliquots hybridized in the
same proportions as for complete incubations.

the same incubation.   RNA was transcribed from foetal liver
chromatin incorporating highly labelled [^{32}P] ATP (1 - 2 Ci/mM).
After hybridization as described to [^{3}H] cDNA, the reaction was
diluted with 0.5 ml 0.2M NaCl;   0.05 M Tris-HCl pH 7.5 and
treated with 20 μg/ml pancreatic and T$_1$ ribonucleases at 30°C
for 2 hrs.   The digest was passed through Sephadex G50 in 0.2 M
NaCl.   The excluded material which contained less than 0.1% of
the original ^{32}P counts in RNA was centrifuged to equilibrium in
CsCl gradients according to the method of Szybalski (1968) to
separate hybridized cDNA from unhybridized cDNA and [^{32}P] RNA.
Figure 3 shows [^{3}H] counts of cDNA appearing at CsCl densities
of 1.78 corresponding to RNA/DNA hybrid and 1.71 corresponding
to unhybridized cDNA.   From the [^{32}P] counts associated with
the hybridized cDNA it can be calculated that [^{32}P] accounted for
60% of the hybridized RNA.   Similar experiments have been published
by Wilson et al.(1975a);   it is also concluded that about 50% of the
hybridizing RNA is derived from de novo transcription of the
chromatin by the bacterial polymerase and 50% of the hybrid arises
from contaminating endogenous RNA sequences.

It should be pointed out that these experiments suffer
from the difficulty of recovering a relatively small amount of
^{32}P-label in hybrid from a huge background of unhybridized
^{32}P-labelled RNA;   however, when it can be demonstrated that
this background can be reduced to an acceptable level the data
provide the strongest single argument for the in vitro
transcription of a specific gene.

(b)   Transcription with mercurated nucleotides

The recent introduction of mercury-substituted ribonucleotide
triphosphates as substrates for RNA polymerase offers in theory
a method for isolating newly synthesized in vitro RNA from
endogenous RNA.   In practice, chemically mercurated UTP (Hg-UTP)
is prepared by incubating UTP with mercuric acetate as described
by Dale et al., 1975.   The acetate salt of Hg-UTP however cannot
be polymerased in this form by E. Coli RNA polymerase, an enzyme
sensitive to thiol modification.   The mercaptoethanol derivative
on the other hand is an excellent analogy for UTP and exhibits a
K$_m$ value slightly higher than the natural substrate (Dale and Ward
1975).   The inclusion of 15-20 mM mercaptoethanol in transcription
incubations containing Hg-UTP (acetate) ensures optimal
incorporation.   After synthesis RNA purification is performed as
usual (details are given later) and the product passed over thiol
agarose prepared according to Cuatrecasas (1970).   Only those RNA
molecules containing Hg-UTP are retained;   only a few percent of
the UMP residues in the RNA need to be substituted for binding to
occur.   After washing the column free of contaminating nucleic
acids, the purified Hg-RNA is eluted with a buffer containing
mercaptoethanol.   This approach has been applied to isolate

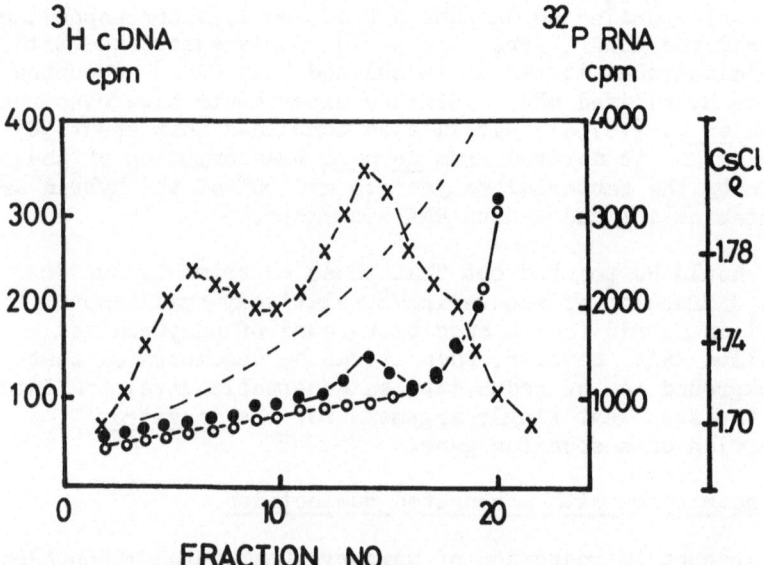

Legend to Figure 3:   Isopycnic banding in CsCl of the hybrid
formed between [3H] globin cDNA and [32P] RNA transcribed from
mouse foetal liver chromatin.   Counts in [3H], (x —— x);
counts in [32P], (●—— ●); density of CsCl (- - - -).
A sample of [32P] RNA treated in an identical fashion but
without hybridization to cDNA was run in a parallel gradient,
(o —— o).

in vitro transcripts of chromatin from contaminating endogenous RNA, with subsequent hybridization analysis of the transcripts. (Smith and Huang, 1976; Crouse et al., 1976; Beissmann et al., 1976; Beebee and Butterworth, 1976; Towle et al., 1977).

Recently a number of serious criticisms of this method have been raised. In a careful analysis of the transcription of duck reticulocyte chromatin by E. coli RNA polymerase Zasloff and Felsenfeld (1977a,b) arrived at the conclusion summarised in the following paragraphs.

a)   Using Hg-UTP as the sole source of UTP, Hg-transcripts from duck reticulocyte chromatin purified on thiol agarose was found to contain 0.02% of globin specific sequences on analysis with ^{3}H-globin cDNA.

b)   In a control experiment ØX 174 DNA was transcribed with coli polymerase in the presence of varying amounts of Hg-UTP and unsubstituted UTP. Transcripts containing up to 25% of their UMP, residues mercurated were found to hybridise back to the template DNA with identical kinetics. Fully substituted RNA however failed to hybridise to any great extent while 50% substituted RNA hybridised only partially. This result raises doubts as to whether the globin sequences detected in fully mercurated chromatin transcripts are due to newly synthesised RNA. It also confirms the observation of Beebee and Butterworth (1976) who used fully substituted Hg-UTP for the in vitro transcription of ribosomal RNA in rat liver nucleoli. The isolated RNA failed to hybridise to ribosomal DNA unless it was first demercurated.

c)   Zasloff and Felsenfeld further demonstrated that the hybridizing globin sequences arose not from newly synthesized mercurated RNA but from a carry over of endogenous globin sequences present in duck reticulocyte chromatin. They propose that the bacterial polymerase transcribes endogenous globin RNA to form a hybrid which by virtue of its mercurated complementary strand is isolated by thiol agarose chromatography. Several lines of evidence support this conclusion.

i.   Artificial chromatin templates reconstituted from E. coli RNA and erythrocyte histones were supplemented with globin mRNA to the same proportions found in duck reticulocyte chromatin. A substantial fraction of the mRNA could be bound to thiol agarose after transcription with E. coli RNA polymerase in the presence of Hg-UTP. It has been known for some time that RNA primed synthesis of RNA by bacterial RNA polymerases can occur (Krakow and Ochoa, 1963; Fox et al., 1964; Maitra et al., 1967; Melli and Pemberton, 1972; Wilson et al., 1975b) and in a separate experiment Zasloff and Felsenfeld showed that under the conditions used for chromatin transcription, E. Coli RNA polymerase will copy globin mRNA directly to form a hybrid.

ii. The recovery of globin sequences in transcripts from duck reticulocyte chromatin is relatively insensitive to high levels of actinomycin D. Even at concentrations sufficient to inhibit 95% of the total RNA synthesis, less than a 30% reduction in the levels of globin sequences is observed. It is known that the priming of RNA with single stranded RNA primers is relatively insensitive to actinomycin D inhibition (Fox et al., 1964).

iii. When Hg-transcripts from duck reticulocyte chromatin are heated to dissociate hybrids prior to thiol agarose chromatography, over 90% of the Hg-RNA normally bound in the absence of heating is still retained on the column. However the level of globin sequences recovered in the bound fraction dropped to 5% of that found with unheated RNA. This suggests that Hg-transcripts are contaminated with unmercurated globin sequences in hybrid form. This conclusion was further confirmed using anti-strand globin cDNA as a probe. While little hybridization was observed between heated Hg-transcripts and cDNA efficient annealing to anti-strand cDNA occurred. The hybrids could also be bound to thiol agarose in vitro. Recently similar artefacts due to mRNA copying by polymerase have also been reported by Giesecke et al. (1977) during the transcription of the ovalbumin gene in hen oviduct chromatin.

These results raise a number of questions about chromatin transcription data that have already been published. In the studies of Biessmann et al. (1976) with Drosophila chromatin and Smith and Huang (1976) with myeloma cell chromatin where fully substituted Hg-UTP was incorporated into transcripts it is difficult to escape the conclusion that the bulk of the observed hybridization is due to co-purified endogenous sequences. In the case of the data of Crouse et al. (1976) with chicken reticulocyte chromatin and Towle et al. (1977) with oestrogen stimulated chicken oviduct chromatin where partial mercury substitution is employed it is not clear whether the results are due to de novo transcripts or contaminating endogenous RNA sequences. Clearly the findings of Zasloff and Felsenfeld (1977a,b) demand that more stringent controls should be included when using mercurated nucleotides. These authors suggest that hybridising sequences should be shown (a) to be actinomycin sensitive during synthesis; (b) to be retained on thiol agarose after heat treatment and (c) to be retained on thiol agarose after hybridization to cDNA.

In addition to the carry over of endogenous RNA sequences by hybrid formation, Crouse et al. (1976) and Konkel and Ingram

(1977) have shown that mercurated RNA can aggregate non-specifically with endogenous RNA during certain isolation procedures and thereby carry contaminating sequences through the thiol agarose purification step.   Several procedures are described which minimise this effect.   It is also pointed out that some studies check for non-specific absorption of unmercurated RNA to thiol agarose and others check for non-specific interactions between mercurated and unmercurated RNA by mixing the purified components just before thiol agarose purification.   However, controls in which mixing occurs prior to RNA isolation should be carried out if the major source of non-specific contamination is to be estimated.

Recently we have made a thorough examination of the in vitro transcription of the globin gene in foetal liver chromatin using mercurated UTP.   The procedures used are as described earlier with the following modifications: transcription reactions contain 15 mM mercaptoethanol and 0.6 mM UTP/0.2 mM Hg-UTP.   After deproteinisation the reaction mixture is passed through Sephadex G-50 in CB (1% SDS; 0.2 m NaCl; 10 mM Tris: HCl pH 7.5) and the excluded material precipitated with 2.5 volumes ethanol, 16 hrs. at -20°C.   Pelleted material is taken up in 1 ml 10 mM Tris pH 7.5 and heated at 100°C for 7 mins.   After rapid cooling an equal volume of 2 x CB and the sample run into a 3 ml column of thiol sepharose 6B (8 μmole thiol gp. per ml) previously washed with CB + 0.3 M mercaptoethanol and re-equilibrated with CB.   The material is held in the column for 30 mins at room temperature and then the column washed with 100 mls CB.   The Hg-transcript is eluted with CB + 0.3 M mercaptoethanol, precipitated with ethanol and finally desalted by passing through Sephadex G-50 in distilled water.

Figure 4 shows the hybridization to globin cDNA of foetal liver transcripts containing Hg-UTP which were purified with and without prior heat denaturation.   In these and subsequent experiments, hybridizations were carried out by the kinetic or Rot analysis.   Unlike the titration analysis described earlier where all the available globin sequences in increasing aliquots of RNA are hybridized to completion, in this method all points have identical inputs of cDNA and RNA and the effect of time on the extent of the reaction is measured.   This value called $R_{o}t$ is expressed by the product of Ro (initial concentration of RNA (moles nucleotide/l) x time , t (secs). The value of $R_{o}t$ at 50% hybridisation of cDNA, $R_{o}t\frac{1}{2}$, depends on the concentration of available globin mRNA sequences.   For pure globin mRNA hybridised to globin cDNA under the conditions described the $R_{o}t\frac{1}{2}$ is 4×10^{-3}; the concentration of globin sequences in an unknown RNA is proportional to the experimentally determined $R_{o}t\frac{1}{2}$ for the sample.

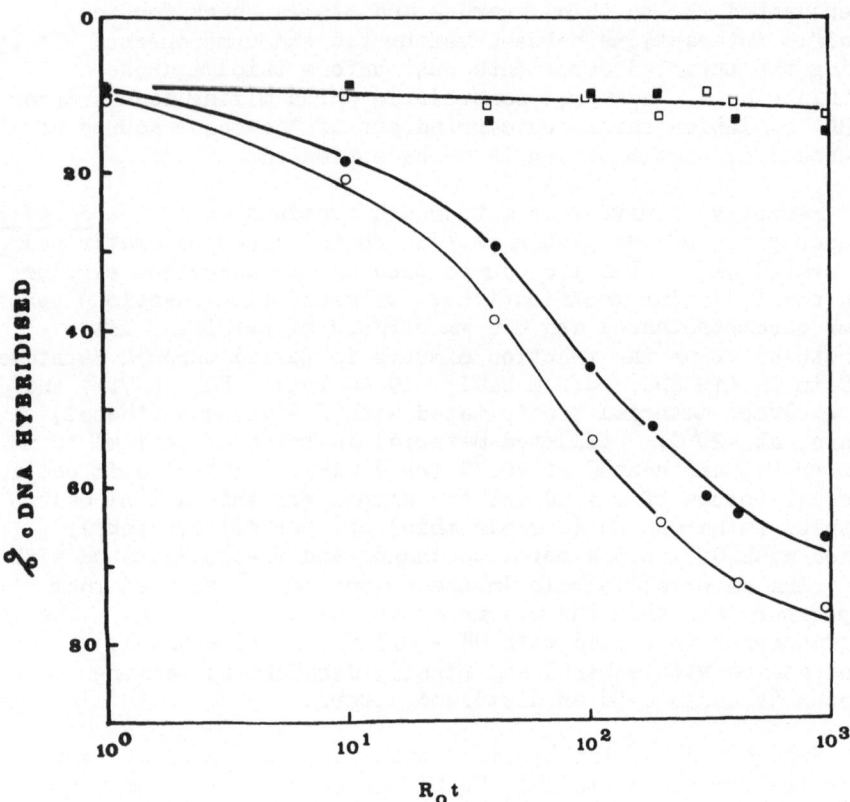

Legend to Figure 4: Hybridization of $[^{3}H]$ globin cDNA to Hg-
transcripts from mouse foetal liver either with (● — ●) or
without (o — o) heat treatment prior to thiol agarose
purification. Also included are Hg-transcripts from adult liver
chromatin, (□ — □) and the hybridization obtained when mouse
foetal liver chromatin is incubated without polymerase and an
equivalent amount of "transcript" as carrier Hg-t-RNA added
(■ — ■).

It was found that both heated and unheated Hg-RNA transcripts
hybridized to globin cDNA; the effect of heating was to reduce the
level of hybridization to some extent.   However in this case
0.004% of the RNA was still found to represent globin mRNA
sequences.   This result is not in agreement with the findings
of Zasloff and Felsenfeld (1977a, b) who found that with mercurated
transcripts of duck reticulocyte chromatin prior heating abolished
the ability of purified transcripts to hybridize.   In control
experiments with adult liver chromatin no hybridization to globin
cDNA was seen.   Similarly in the absence of polymerase no
hybridization occurs with RNA isolated from incubations
containing foetal liver chromatin and carrier RNA.   These results
indicate that the appearance of globin specific mRNA sequences in
tissue specific, enzyme dependent and not due to carry over of
endogenous RNA by an artefact of the method.   In an additional
control experiment (Table 1) adult liver chromatin was
contaminated with purified globin mRNA.   Globin sequences
subsequently appeared in the thiol agarose purified Hg-transcript.
If the transcript is heated prior to the thiol agarose step or
if the mRNA is added during the phenol extraction very little
cross-contamination occurs.   While this agrees with the findings
of Zasloff and Felsenfeld that globin mRNA can be partially
transcribed by the polymerase and then co-isolated with the
transcript, it does not appear that endogenous RNA (i.e. chromatin
bound RNA) behaves in the same way.   This was demonstrated
further in experiments where Friend cells were grown in $[5 - {}^3H]$
uridine for 48 hrs. to achieve uniform labelling of endogenous
RNA.   The isolated chromatin was transcribed in the presence of
Hg - UTP and the purified transcripts applied to thiol agarose.

Table 1

Conditions	% cDNA hybridised
(1) No additions	0
(2) 20 ng globin mRNA at 0 time	89
(3) As above, with heat treatment	2
(4) 20 ng globin mRNA after incubation	3

Legend to Table 1:   Transcription reactions (2 ml) containing
adult mouse liver chromatin were incubated with Hg-UTP.   In
three cases globin mRNA was also added either before incubation
or after 1 hr incubation and cooling to 4°C.   Hg-transcripts
were purified on thiol agarose without heat denaturation, except
for sample (3) which was denatured at 100°C for 7 mins in 10 mM
Tris : HCl pH 7.5.

The results show (Table 2) that while substantial OD amounts of Hg - RNA were bound to the thiol agarose it contained virtually none of the ^{3}H-label. All of the applied radioactivity was recovered in the unbound fraction suggesting that insignificant amounts of complementary copying or chain elongation employing Hg - UTP had occurred. This result is not in accord with the findings of Shih et al. (1977) who found that up to 20% of the endogenous RNA in a mouse cell line chromatin is elongated in the presence of Hg - UTP. These authors did not carry out a control for aggregation artefacts, already mentioned previously. It is also apparent that after labelling cells for 18 hrs most of the label is in completed RNA chains and very little in nascent chains. That 20% of the RNA label is still in nascent chains after this time seems remarkable and may be explained more simply by non-specific aggregation.

It is concluded that in the mouse foetal liver system globin mRNA sequences are transcribed de novo and do not arise from endogenous RNA sequences. It is clear that the effect of exogenous RNA polymerase on added globin mRNA and on endogenous RNA in this chromatin is totally different. Indeed the results of Zasloff and Felsenfeld (1977a,b) could equally well be explained by a sequestration of added polymerase on cytoplasmic mRNA contaminating the reticulocyte chromatin. Indeed the concentration of globin mRNA in the reticulocyte cytoplasm is extremely high compared with that of foetal mouse liver; if this artefact is specific for globin mRNA then even very small levels of cytoplasmic contamination in reticulocyte chromatin could explain the findings.

Table 2

Thiol Agarose fraction	Total [^{3}H] cpm	% of original ^{3}H RNA in chromatin
Bound	424	0.002
Unbound	1.1×10^5	61

Legend to Table 2:   Friend cells were grown for three generations (48 hrs) in Ham's F12 medium containing 0.5 μCi/ml ^{3}H-uridine.   Isolated chromatin containing ^{3}H-endogenous RNA was transcribed with E. Coli RNA polymerase in the presence of Hg-UTP.   Transcripts were not heat-denatured prior to thiol agarose purification.   The [^{3}H] RNA in bound and unbound fractions was recovered by ethanol precipitation and counted in Triton-Toluene scintillation fluor.

Another factor which affects the ability of E. coli polymerase to copy added globin mRNA is the ionic conditions of the incubation.   Table III shows the fate of ^{125}I-labelled globin mRNA after addition to transcription incubation containing Hg-UTP and a variety of divalent cation conditions.

Zasloff and Felsenfeld (1977a,b) found that the amount of globin mRNA copied by the polymerase during the transcription of duck reticulocyte chromatin was unaffected by changing the divalent cation.   Giesecke (1977) also found that the amount of mRNA copying was the same whether 4 mM Mg^{2+} or 4 mM Mg^{2+}/1 mM Mn^{2+} was used.   Table 3 suggests a different effect with the foetal liver system.   It would appear that replacement of Mn^{2+} ions with Mg^{2+} ions reduces the degree of mRNA c opying.   This observation has also been made by O'Malley and co-workers (personal communication).

Clearly chromatin transcription systems vary considerably and not all findings for a given system can be generalised. Each system must therefore be analysed individually for its endogenous RNA levels, tendency for mRNA copying and sensitivity to divalent cations.   Perhaps the single most useful experiment for testing a system is shown in Table 4.   Foetal liver transcripts containing highly labelled with ^{32}P-ATP were isolated as already described.   Endogenous RNA is present but unlabelled.   The RNA

Table 3

		% Globin mRNA		Bound	RNA synthesised
		Unheated		Heated	(μg)
1.	2.5 mM $MnCl_2$ (Gilmour et al. 1975)	18.3		7.1	20
2.	1 mM $MnCl_2$ (Axel et al. 1973)	50.1		19.2	11
3.	5 mM $MgCl_2$ (Astrin, 1973)	8.6		7.2	20
4.	1 mM $MnCl_2$ 5 mM $MgCl_2$ (Biessmann et al. 1976)	8.5		1.4	15

Legend to Table 3:   Foetal liver chromatin (400 μg) was incubated with 400 μg E. coli polymerase and 10 ng [125] Iodinated globin mRNA under four different sets of incubation conditions.   Purified Hg-transcripts were purified on thiol agarose one half with a heat denaturation step, the remainder without denaturation.   The fraction of the total 125I counts bound to the thiol agarose indicates the degree to which mRNA transcription has taken place.

was hybridized to an excess of $[^3H]$-globin cDNA which had been
chemically mercurated according to the method of Dale et al.(1975).
This modification did not alter its hybridization properties and
conferred complete retention on thiol agarose columns.   The
presence of excess cDNA ensured that all globin mRNA sequences
whether ^{32}P labelled or not were hybridized.   After ribonuclease
treatment to remove unhybridized RNA as described in legend,
the Hg-cDNA was bound to thiol agarose and the amount of ^{32}P-RNA
as hybrid determined.   The background for the experiment was
determined by repeating the procedure with non-mercurated globin
cDNA.   The results indicate the presence of de novo synthesised
globin mRNA sequences in the transcription product in amounts
comparable to that found for transcripts purified by the Hg-UTP
technique.

(c)   Fidelity of in vitro transcription

     The investigation of specific gene expression by in vitro
transcription has been criticised on the grounds that bacterial
polymerases may transcribe chromatin less accurately than
homologous polymerases.   Various findings have been cited as
demonstrating this.

I.   Rat liver RNA polymerase II transcribes chromatin at a
greater rate than M. luteus RNA polymerase (Butterworth et al.
(1971).

Table 4

Reactants	^{32}P-RNA in Hybrid	% globin sequence
40 μg ^{32}P-RNA (3.7 x 10⁷cts.) + 10 ng. Hg-cDNA	1850 cts = 2 ng	0.005
40 μg ^{32}P-RNA (3.7 x 10⁷ cts) + 10 ng. cDNA	306 cts = 0.34 ng	0.0008

Legend to Table 4:   Foetal liver transcripts were synthesised in
the presence of ^{32}P-ATP to give RNA 5.A approx 10^6 cpm/μg.
cDNA or Hg cDNA (both ^3H-labelled) was hybridized in an excess of
at least 20-fold over the estimated globin mRNA sequences present.
Hybridizations were carried out in 10 μl 0.24 m sodium phosphate
buffer pH 6.8, 0.1% SDS; 1 mM EDTA at 60°C for 5 hrs.   The
incubations were diluted to 0.5 ml with 0.2 M NaCl treated with
20 μg pancreatic ribonuclease (heated 100°C for 5 mins) and
passed over Sephadex G50 in CB.   The voided material was then run
on to thiol agarose (3 mls) and washed with 100 mls CB.   A
correction was made for a small loss of Hg-cDNA (10%) binding
capacity as judged by $[^3H]$ recovery.   Retained material was
eluted with CB + 0.3 M mercaptoethanol, precipitated with
1.5 mls of ethanol and radioactivity estimated.

II.   Eukaryotic and bacterial RNA polymerases bind to different
sites on the template as judged by the lack of competition
between the enzymes for the same sites (Butterworth et a 1.1971).
In the case of the animal polymerase these sites are more
specific on the basis of a comparison of initiating triphosphates
(Keshgegian et al. 1973) or the kinetics of RNA synthesis
(Keshgegian and Furth, 1972).

III.   RNA transcribed from rat liver chromatin by the homologous
polymerase II has a larger molecular weight than E. Coli polymerase
transcripts (Maryanka and Gould, 1973),

     While these experiments demonstrate that differences between
the enzyme can be detected they fail to demonstrate whether one
polymerase is more accurate in selecting for the transcription
of tissue specific sequences than the other.   Accurate
comparisons of animal and bacterial polymerases have been made in
the particular cases of ribosomal and 5s RNA transcription.
Reeder (1973) found that the bacterial enzyme transcribed both
genes aberrantly in Xenopus liver chromatin;   both strands and
spacer regions of each gene are expressed and in the case of
X. mulleri. x laevis hybrids where virtually none of the
mulleri r-RNA genes are expressed in vivo,   the polymerase
transcribes actively from this gene in the hybrid chromatin.
In further studies Honjo and Reeder (1974) demonstrated that both
X. laevis polymerases I and II also transcribe aberrantly r-RNA
and 5s genes from chromatin.   Both strands are transcribed
although in vivo transcription is asymmetric and carried out by
polymerase III.   In a more recent study Parker and Roeder (1977)
examined the transcription of 5s genes in Xenopus oocyte chromatin
by exogenous polymerase I and III from X. laevis ovaries,
polymerase II from mouse plasmacytoma cells and E. Coli polymerase.
In this system only polymerase I II is capable of stimulating
specifically the asymmetric synthesis of 5s RNA;   the other
polymerases stimulate total RNA synthesis but do not effect a
specific stimulation of 5s RNA transcription.   When polymerases I
and III and Coli polymerases are compared on a X. laevis DNA
template not only is the synthesis of 5s RNA symmetric but the
specificity shown by polymerase III for 5s genes in chromatin
is lost.   Taken together these findings suggest that chromatin
associated proteins are required for the selective and asymmetric
transcription of 5s genes in amphibians and that a specific RNA
polymerase is required to do this.   Clearly the 5s gene represents
a special case.   It is subject to fine controls which are not
even recognised by the homologous type I and II animal polymerases
as well as E. Coli RNA polymerase.   As yet no analogous
specificity has been shown between m-RNA genes and specific RNA
polymerase II species and hence to extrapolate from the aberrant

transcription of 5s genes by Coli polymerase to an analogous
situation with m-RNA genes is not judtified.

At present it is difficult to make an assessment of the
overall fidelity of chromatin transcription of chromatin by
bacterial polymerase, especially in light of the findings of
Zasloff and Felsenfeld (1977,a,b).   In the case of globin gene
transcription in mouse foetal liver (Gilmour et al., 1975) and
rabbit marrow (Wilson et al., 1975a) chromatins where de novo
synthesis of globin mRNA sequences is demonstrated, it can be
concluded that there is some degree of tissue specific transcription
by the bacterial enzyme.   However the sequences analysed represent
a minute fraction of the total RNA.   The question of fidelity
can only be resolved by carrying out a more extensive analysis of
the sequences transcribed in vitro from chromatin and comparing
them with the sequences present in vivo.   Batchelor and Smith
(1976) compared mouse liver nuclear RNA and E. Coli polymerase
transcripts from mouse liver chromatin by hybridizing in RNA
excess to labelled mouse DNA fractions of high, intermediate
and low reiteration frequencies.   The results suggest that all
the sequences transcribed in vivo are also present in chromatin
transcripts;   however the latter contained additional sequences
transcribed from specific regions of high and low reiteration
sequences of the DNA.   Unfortunately in this study no attempt was
made to isolate in vitro synthesised RNA from endogenous sequences
and in the absence of the appropriate controls it is not possible
to ascertain how many of the in vivo sequences detected were
actually synthesised in vitro.

This question has also been studied by Biessmann et al.(1976)
by transcribing Drosophila chromatin with E. Coli polymerase and
Hg-UTP.   The purified Hg-transcripts were analysed by
hybridisation to cDNA made to the polyadenylated nuclear RNA of
the tissue.   The results indicate that while most of the
sequences represented in the cDNA probe were also found in the
transcript   some aberrant, symmetrical transcription also occurred.
In these experiments no precautions were taken to exclude carry
over of endogenous RNA in the event of RNA copying by the
polymerase.   In addition, if this occurred, it would also give
rise to mercurated 'anti sense' strands.   It is therefore
difficult to say whether the results are due to symmetrical
de novo transcription of DNA in the chromatin or due to the
co-purification of endogenous RNA and its de novo synthesised
complementary strand.   It is clear that if copying of
endogenous RNA is generally prevalent it will also take place
where unsubstituted nucleotides are present.   This possibility
suggests an alternative explanation for reports of poor strand
selection in in vitro transcripts (Reeder, 1973;   Honjo and
Reeder, 1974;   Astrin, 1973;   Wilson et al., 1975b).

In conclusion, it has to be admitted that the possibility of transcriptional artefacts seriously complicates the assessment of the overall fidelity of chromatin transcription. Future studies will have to consider the possible presence of artefacts in the particular system under investigation, especially where strand selectivity is considered. While the incorporation of mercurated nucleotides into chromatin transcripts can largely overcome the problem of endogenous RNA contamination providing heat denaturation is carried out, this technique will not eliminate mercurated copies of endogenous RNA in the event of RNA copying by the polymerase.

Conclusions

The presence of endogenous RNA sequences in chromatin has undoubtedly been responsible for a good deal of spurious data. The recent discovery revealed through the use of mercurated nucleotides that bacterial polymerase can copy m-RNA sequences present as a contaminant in chromatin preparations provides an additional caveat. The results of Zasloff and Felsenfeld (1977a,b) are particularly disconcerting in this respect since they fail to show significant transcription from chromatin in a system where the globin gene is known to be active (Fodor and Doty, 1977). We have presented data to show that this is not the case with the globin gene in mouse foetal liver chromatin. However it is not clear how many of the other systems under investigation are prone to this criticism. Transcriptional studies with native chromatin require a more careful characterisation to eliminate gross artefactual effects. Also it is still not possible to assess how accurate an approximation to the in vivo state the present methods give. It is quite possible that bacterial polymerase does not respond to the fine transcriptional controls in chromatin but rather transcribes randomly from the specific regions of chromatin normally active in vivo because they are structually more accessible. Indeed it has proved difficult to construct a satisfactory in vitro transcription system which is not open to criticism because the mode of action of the homologous eukaryotic RNA polymerase II is so poorly understood. If transcription with exogenous bacterial polymerase simply reflects the accessibility of DNA sequences then this may put a limitation on the information these systems can provide. For example, if chromosomal proteins bind to active DNA sequences causing a gross structural alteration then bacterial polymerase may be able to distinguish active chromatin on this basis. However, if the primary regulatory event takes place round the promotor regions and then requires correct initiation by the polymerase before transcription can occur, the question of polymerase source becomes crucial.

It is also important to bear in mind that transcriptional

control in chromatin in vivo may not involve a simple on-off
mechanism.   Hybridization studies on total messenger
populations shows that there exists a wide variation in the
numbers of copies of different messenger RNAs.   This
distribution approximates three frequency classes, one
representing a few genes in high abundance, a second representing
several hundreds of genes in moderate abundance and a third class
where several thousand genes are represented by only a few copies
each. (Birnie et al., 1974; Bishop et al., 1974; Young et al.
1976).   The inference from these findings is that the phenotype
of eukaryotic cells may be determined more by the relative
abundances of m-RNA sequences than by their absolute presence or
absence.   The question of whether the low abundance, high
complexity class of mRNA seen in most tissues is functionally
significant or whether it represents 'leaky' repressed genes
has been considered by Galau et al. (1976).

An important generalisation which can be drawn from these
experiments is that the total fraction of the genome expressed
as mRNA in the polysomes of a particular tissue is relatively
small, probably only a few percent.   Only a small portion of the
total coding potential of the genome is utilised; however, it
is not clear how many of the remaining sequences actually represent
structural genes.   Transcriptional modulation could involve
therefore either activation of previously silent genes or rate
limiting controls on genes which are more or less permanently
active.   Clearly it may not be possible to make this
distinction in all cases.   However both must involve selection
mechanisms which influence the activity of specific genes.

References

Allfrey, V.G. and Mirsky, A.E. (1962) Proc. Nat, Acad. Sci. U.S.
      48, 1590.
Allfrey, V.G., Littau, V.C. and Mirsky, A.E. (1963) Proc. Nat.
      Acad. Sci. U.S. 49, 414.
Ashburner, M. (1972) In "Developmental studies on giant
      Chromosomes" (W. Beermann, ed.) Vol. 4, p.101. Springer
      Verlag, Berlin and New York.
Astrin, S.M. (1973) Proc. Nat. Acad. Sci. U.S. 70, 2304.
Axel, R., Cedar, H. and Felsenfeld, G. (1973) Proc. Nat. Acad.
      Sci. U.S. 70, 2029.
Barrett, T., Maryanka, D., Hamlyn, P.H. and Gould, H.J. (1974)
      Proc. Nat. Acad. Sci. U.S. 71, 5057.
Batcheler, L.T. and Smith, K.D. (1976) Biochemistry 15, 3281.
Beebee, T.J.C. and Butterworth, P.H.W. (1976) Eur. J. Biochem.
      66, 543.
Bekhor, I., Kung, G.M. and Bonner, J. (1969) J. Mol. Biol. 39,351.
Biessmann, H., Gjerset, R.A., Levy, W.B. and McCarthy, B.J.(1976)
      Biochemistry 15, 4356.

Birnie, G.D., MacPhail, E., Young, B.D., Getz, M.J. and Paul, J. (1974) Cell Differentiation 3, 221.

Bishop, J.O. and Freeman, R.B. (1973) Cold Spring Harb. Symp. Quant. Biol. 38, 707.

Bishop, J.O. and Rosbash, M. (1973) Nature New Biol. 241, 204.

Bishop, J.O., Morton, J.G., Rosbash, M. and Richardson, M. (1974) Nature 250, 199.

Brown, I.R. and Church, R.B. (1972) Develop. Biol. 29, 73.

Butterworth, P.H.W., Cox, R.F. and Chesterton, C.J. (1971) Eur. J. Biochem. 23, 229.

Campo, M.S. and Bishop, J.O. (1974) J. Mol. Biol. 90, 649.

Church, R.B. and McCarthy, B.J. (1970) Biochim. Biophys. Acta. 199, 103.

Church, R.B. and Brown, I.R. (1972) In "Results and Problems in Differentiation" Vol. 3 (H. Urspring, ed.) p.11 Springer Verlag, New York.

Crouse, G.F., Fodor, E.J. and Doty, P. (1976) Proc. Nat. Acad. Sci. U.S. 70, 1564.

Cuatrecasas, P. (1970) J. Biol. Chem. 245, 3059.

Dale, R.M.R., Martin, E., Livingston, D.C. and Ward, D.C. (1975) Biochemistry 14, 2447.

Dale, R.M.R. and Ward, D.C. (1975) Biochemistry 14, 2458.

Davidson, E.H., Hough, B.R., Klein, W.H. and Britten, R.J. (1975) Cell 4, 217.

Firtel, R.A. (1972) J. Mol. Biol. 66, 363.

Fodor, E.J.B. and Doty, P. (1977) Biochim. Biophys. Res. Comm., 77, 1478.

Fox, C.F., Robinson, W.S., Haselkorn, R. and Weiss, S.B. (1974) J. Biol. Chem. 239, 186.

Frenster, J.H., Allfrey, V.G. and Mirsky, A.E. (1963) Proc. Nat. Acad. Sci. U.S. 50, 1026.

Galau, G.A., Klein, W.H., Davis, M.M., Wold, B.J., Brutten, R.J. and Davidson, E.H. (1976) Cell 7, 487.

Giesecke, K., Sippel, A.E., Nguyen-Hu, N.C., Groner, B., Hynes, N.E., Wurtz, T. and Schutz, G. (1977) Nuc. Acids Res. 4, 3943.

Gilmour, R.S. and Paul, J. (1973) Proc. Nat. Acad. Sci. U.S. 70, 3440.

Gilmour, R.S., Harrison, P.R., Windass, J.D., Affara, N.A. and Paul, J. (1974) Cell Differentiation 3, 9.

Glisen, U.R., Glisen, M.V. and Doty, P. (1966) Proc. Nat. Acad. Sci. U.S. 56, 285.

Goldberg, R.B., Gallau, G.A., Britten, R.J. and Davidson, E.H. (1973) Proc. Nat. Acad. Sci. U.S. 70, 3516.

Greenberg, J.R. and Perry, R.P. (1971) J. Cell Biol. 50, 774.

Grouse, L., Chilton, M.D., McCarthy, B.J. (1972) Biochemistry 11, 798.

Harris, S.E., Means, A.R., Mitchell, W.M. and O'Malley, B.W. (1973) Proc. Nat. Acad. Sci. U.S. 70, 3776.

Harris, S.E., Schwartz, R.J., Isai, M.J., Roy, A.K. and O'Malley, B.W. (1975) J. Biol. Chem. 251, 524.

Harrison, P.R., Birnie, G.D., Hell, A,, Humphries, S., Young, B.D.
    and Paul, J. (1974) J. Mol. Biol. 84, 539.
Honjo, T. and Reeder, R.H. (2974) Biochemistry 13, 1896.
Huang, R.C. and Huang, P.C. (1969) J. Mol. Biol. 39, 365.
Humphries, S., Windass, J. and Williamson, R. (1974) Cell 7, 267.
Keshgegian, A.A. and Furth, J.T. (1972) Biochem. Biophys. Res.
    Commun. 48, 757.
Keshgegian, A.A., Garibian, G.S. and Furth, J. (1973) Biochemistry
    12, 4337.
Klein, W.H., Murphy, W., Attardi, G., Britten, R.J. and Davidson,
    E.H. (1974) Proc. Nat. Acad. Sci. U.S. 71, 785.
Konkel, D.A. and Ingram, U.M. (1977) Nuc. Acids. Res. 4, 1979.
Krakow, J.S. and Ochoa, S. (1963) Proc. Nat. Acad. Sci. U.S.
    49, 88.
Leder, P., Ross, J., Gielen, J., Packman, S., Ikawa, Y., Aviv, H.,
    and Swan, D. (1973) Cold Spring Harbor Symp. Quant. Biol.
    38, 753.
Lewin, B. (1975) Cell 4, 77.
Liarakos, C.D., Rosen, J.M. and O'Malley, B.W. (1973) Biochemistry
    12, 2809.
Lodish, H.F., Firtel, R.A. and Jacobson, A. (1973) Cold Spring
    Harbor Symp. Quant. Biol. 38, 899.
Maitra, U., Nakata, Y. and Hurwitz, J. (1967) J. Biol. Chem.
    242, 4908.
McCarthy, B.J. and Hoyer, B.H. (2964) Proc. Nat. Acad. Sci. U.S.
    52, 915.
MacGillivray, A.J. and Rickwood, D. (1974) In Biochemistry of
    Differentiation and Development, p.301, Ed. J. Paul. Oxford
    Medical and Technical Publishing Co.
Maryanka, D. and Gould, H. (1973) Proc. Nat. Acad. Sci. U.S. 70,
    1161.
Melli, M. and Pemberton, R.E. (1972) Nature 236, 172.
Panitz, R. (1972) In "Developmental Studies on Giant Chromosomes"
    (W. Beermann, ed.) Vol. 4, p.209, Springer Verlag, Berlin and
    New York.
Parker, C.S. and Roeder, R.G. (1977) Proc. Nat. Acad. Sci. U.S.
    74, 44.
Paul, J. and Gilmour, R.S. (1966) J. Mol.Biol. 16, 242.
Paul, J. and Gilmour, R.S. (2968) J. Mol. Biol. 34, 305.
Reeder, R.H. (1973) J. Mol. Biol. 80, 299.
Sawada, H., Crain, W.R. and Sanders, G.F. (1972) Biochim.
    Biophys. Acta 291, 643.
Steggles, A.W., Wilson, G.N., Kantor, J.A., Picciano, D.J.,
    Falvey, A.K. and Anderson, W.F. (1974) Proc. Nat. Acad. Sci.
    U.S. 71, 1219.
Stein, G., Stein, J., Thrall, C. and Park, W. (1975) In
    "Chromosomal Proteins and their role in the Regulation of
    Gene Expression" (Stein and Kleinsmith, eds.) Academic Press,
    New York and London, p.1.

Shih, T.Y., Young, H.A., Parks, W.P. and Scolnick, E.M. (1977) Biochemistry 16, 1795.

Smith, K.D., Church, R.B. and McCarthy, B.J. (1969) Biochemistry 8, 4271.

Smith, M.M. and Huang, R.C. (1976) Proc. Nat. Acad. Sci. U.S. 73, 775.

Spradling, A., Penman, S., Campo, M.S. and Bishop, J.O. (1974) Cell 3, 23.

Spelsberg, T.C. and Hnilica, L.S. (1969) Biochim. Biophys. Acta 195, 63.

Sutton, W.D. (1971) Biochim. Biophys. Acta 240, 522.

Suzuki, Y., Gage, L.P. and Brown, D.B. (1972) J. Mol. Biol. 70, 637.

Szybalski, W. (1968) In "Methods in Enzymology", (Colowick S.P. and Kaplan, N.O. eds.) Academic Press, New York and London, vol. 12B, p.330.

Tan, C.S. and Miyagi, M. (1970) J. Mol. Biol. 50, 641.

Towle, H.C. Tsai, M.J., Tsai, S.Y. and O'Malley, B.W. (1976) J. Biol. Chem. 251, 4713.

Weinberg, E.S., Birnstiel, M.L., Purdom, I.F. and Williamson, R. (1972) Nature 240, 225.

Wilson, G.N., Steggles, A.W., Kantor, J.A., Nienhuis, A.W. and Anderson, W.F. (1975a) Proc. Nat. Acad. Sci. U.S. 25, 8604.

Wilson, G.N., Steggles, A.M. and Neinhuis, A.W. (1975b) Proc. Nat. Acad. Sci. U.S. 72, 4815.

Young, B.D., Harrison, P.R., Gilmour, R.S., Birnie, G.D., Hell, A., Humphries, S. and Paul, J. (1974) J. Mol. Biol. 84, 558.

Young, B.D., Birnie, G.D. and Paul, J. (1976) Biochemistry 15, 2823.

Zasloff, M. and Felsenfeld, G. (1977a) Biochim. Biophys. Res. Commun. 75, 598.

Zasloff, M. and Felsenfeld, G. (1977b) Biochemistry 16, 5135.

CIRCULAR DICHROISM OF DNA, PROTEINS AND CHROMATIN

Gerald D. Fasman

Grad. Dept. Biochemistry, Brandeis University

415 South St., Waltham, Massachusetts 02154 U.S.A.

I. INTRODUCTION

Although great strides have been made towards the elucidation of the structure of chromatin, many details remain unanswered as to the manner in which DNA is folded and condensed in the nucleus. The role played by the histones in the condensation of DNA in the nucleosome and the higher order tertiary structure still awaits clarification. The state of condensation of DNA plays an important role in genetic regulation as this conformational flexibility determines which regions of the DNA are available for transcription.

The structure of chromatin has been probed by both biochemical and biophysical techniques. Examination of nuclease digestion products has yielded new insights into chromatin organization (see recent reviews (1,2)). Physical-chemical methods have also been widely used and one of these, circular dichroism spectroscopy, has contributed significantly to our understanding of chromatin structure. Circular dichroism (CD) refers to the phenomenon of light interaction with an asymmetric structure, when the wavelength of incident light is within an absorption band of the molecule. The asymmetry of proteins and nucleic acids is strongly dependent upon the conformation of these structures, as well as their individual asymmetric centers. Thus CD is an extremely sensitive probe to follow conformational transitions in both components of chromatin, the nucleic acid as well as the protein components. CD is capable of delineating between the α-helical, β-sheet and random structure of proteins, alterations in the DNA secondary conformation due to variations in base stacking arrangements in the double strand, as well as any tertiary organization (e.g., super helix) of the DNA. Extensive reviews are available concerning the application of CD to conformational studies (3-6).

        This review is concerned with the application of CD to study
the DNA conformation in chromatin, as well as the conformation of
the proteins found in the nucleoprotein complex. The optical activity of DNA arises from several sources. The nucleotides are
inherently optically active due to the presence of the asymmetric
pentose moiety. The major CD contribution arises from the asymmetry of the base-base interactions due to the double helix formation. Any alterations in the base stacking which may be caused by
association with proteins (secondary structure) or due to asymmetric
condensation of the DNA in chromatin (tertiary structure) may be
detected by CD. Thus CD spectroscopy can serve as a sensitive
monitor of changes of either secondary or tertiary structure which
are concomitant with the condensation of DNA in chromatin.

                    II.   CIRCULAR DICHROISM OF DNA

    A. Conformation of DNA: Theoretical and Experimental Spectra

        The chromophoric groups in nucleic acids with absorption
bands which contribute most significantly, in the instrumentally
accessible spectral region, are the pyrimidine and purine bases.
These bases exhibit $\pi-\pi^*$ and $n-\pi^*$ electronic transitions in the
ultraviolet region. The major $\pi-\pi^*$ absorption band of DNA is
located near 260 nm. In the double helix, the stacking of the
bases leads to a strong interaction between the $\pi-\pi^*$ transition
dipole moments. According to the theory for CD of polynucleotides,
developed by Tinoco and coworkers (7-9), the interaction between
identical electronic transitions of the stacked bases gives rise
to a CD spectrum which resembles the first derivative of the absorption band. The crossover (zero ellipticity) point of this band
occurs at the absorption maximum. The spectrum has positive and
negative bands of equal magnitude, and is termed conservative.
In addition, there will be coupling between different electronic
transitions (e.g., $\pi-\pi^*$ transitions in the near ultraviolet
coupling with $\pi-\pi^*$ transitions in the far ultraviolet); this leads
to a normal gaussian CD band (nonconservative) centered at the
absorption maximum. The observed CD band contains contributions
from both of these interactions. The signs and magnitudes of the
separate contributions depend strongly on the relative orientations
and separation distances of the base transition moments. Thus, the
arrangement of the bases in a particular nucleic acid conformation
gives rise to a specific CD spectrum.

        The above-described theory has met with considerable success
in prediction of nucleic acid CD spectra, because base-base interactions provide the largest spectral contribution under normal
circumstances. It is not a simple matter, however, to resolve

observed CD spectra into those two types of contributions. There
are additional minor contributions from the n-π* transitions and
the intrinsic nucleotide optical activity. The determination of
DNA conformation therefore usually rests on comparison of observed
spectra with spectra recorded for known conformations.

X-ray diffraction analysis of DNA fibers formed in the presence of various salts and at differing relative humidities has
detected three canonical forms termed A, B, and C (10-12). These
forms of DNA differ with respect to the angle between the base
planes and a perpendicular to the helix axis; the tilt is small
for B and C forms (2° and 6°, respectively) but large for A form
(20°). The B form of DNA has exactly 10 base pairs per helical
turn, A form has 11, and C form has a non-integral 9-1/3 residues
per turn. The B form exhibits a 3.37 to 3.46 Å rise per base
pair; the C form rise is 3.32 Å and the A form rise is 2.55 Å.
It is clear that there should be differing base-base electronic
interactions for these structures, and consequently differing
CD spectra.

Tunis-Schneider and Maestre (13) studied films of DNA under
conditions essentially identical to those employed in the X-ray
studies, and observed characteristic CD spectra for the A, B, and
C forms of DNA (Figure 1) (147).

The B form spectrum observed was essentially identical to the
CD spectrum found for DNA in low ionic strength aqueous solution
(14). The positive ellipticity band maximum occurred at 275 nm,
the negative band at 245 nm, and the crossover point at 257 nm.
In low ionic strength solution, the positive band magnitude is
approximately 8400 deg. cm^2dmole^{-1}. The spectral shape compares
favorably with that calculated by Johnson and Tinoco (9) for B form
DNA, although the measured ellipticity is much smaller than the
calculated value (15). Wide angle X-ray scattering data has provided some evidence that the structure of DNA in low ionic strength
solution, although still of the B-family, differs slightly from the
solid-state structure. The winding angle between base pairs differs,
so that there are closer to eleven base pairs per turn (16). This
conformational change will have a small effect on the CD spectrum.

In consideration of the above findings, it is now well established that DNA structures of the B family exhibit nearly conservative CD spectra due to the small tilt of the bases relative
to a line perpendicular to the helix axis.

The CD spectrum assigned by Tunis-Schneider and Maestre (13)
to the A form of DNA is nonconservative, with a maximum near 260
nm. According to the Johnson and Tinoco theory (9), the large tilt
of the bases in A form DNA (and also in RNA) results in this non-

conservative CD characteristic. Experimental spectra for RNA in
aqueous solution and DNA in alcoholic solutions show the A type CD
spectrum (14).

Greater controversy exists concerning the spectral properties
of C form DNA. Utilizing a salt concentration and relative humidity
conditions similar to those employed in the original C form X-ray
structure determination (11), Tunis-Schneider and Maestre (13) obtained the CD spectrum shown in Figure 1. The ellipticity above
270 nm is very low, while the negative 245 nm band is comparable
to that observed for B form DNA. Similar spectra may be obtained
for DNA in aqueous salt solutions (17,18). There are difficulties
with this assignment on both theoretical and experimental grounds.
Theoretically, the small conformational difference between B and C
form DNA (C form DNA may be regarded as part of the B family by

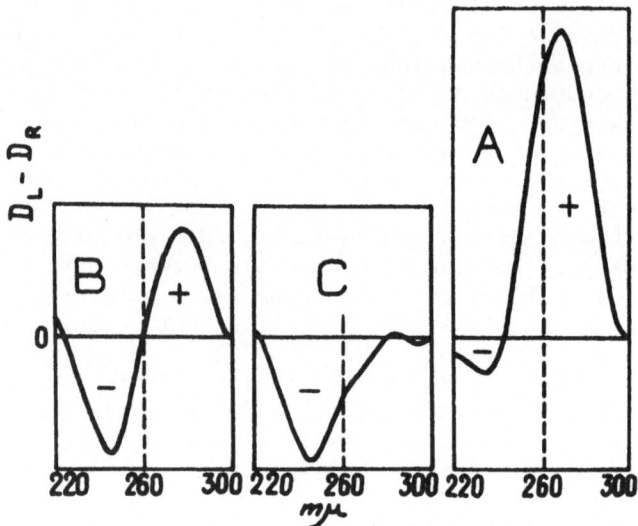

Figure 1. Schematized representation of the CD spectra for the A,
B, and C forms. The dotted lines are drawn through the absorption
maxima. Ivanov, V.I., et al. (147). Reprinted with permission of
John Wiley and Sons, Inc.

many criteria) should not result in the observation of a nonconservative CD spectrum, but rather should result in a low magnitude
conservative spectrum (9). Experimentally, Brunner and Maestre (19)
found difficulty in reproducing the original C form spectrum for
LiDNA films at low relative humidities. Since the original assignment of the CD spectrum for the C-form DNA depended on the similarity
in the conditions of film preparation for the X-ray studies and the
CD studies, that assignment is now open to question.

B. Ψ DNA

In the presence of a neutral polymer, such as polyethylene
oxide, and salt, the structure of DNA is altered so that a unique
CD spectrum of large magnitude is observed. Lerman and coworkers
(20,21) have shown that under these conditions high molecular weight
DNA becomes intramolecularly condensed to an ordered tertiary structure, as a result of excluded volume effects. As a critical concentration of polymer and salt are required for this induced transition, they termed the structure PSI or Ψ-DNA. A series of CD spectra of Ψ-DNA are seen in Figure 2, produced by various concentrations of polyethylene oxide, after various times of mixing. Extremely
large negative ellipticity bands are observed. The Johnson and
Tinoco theory (9) for polynucleotide CD cannot account for this
spectrum. Lerman (20-22) believes this spectrum derives from a
superfolding of the DNA and may be comparable to the condensed form
of DNA in the chromosome. X-ray scattering studies (22) have demonstrated that Ψ-DNA is still in the B-form. Thus, it is possible
to condense DNA into an asymmetric tertiary structure, without de-
stroying the basic B-form structure of the Watson-Crick double
helix, and obtain a new type of CD spectrum. By analogy to studies
of cholesteric liquid crystals (23-27), the unusual CD spectrum
may be attributed to long range anisotropy in the arrangement of
the stacked DNA bases within the condensed structure. Destruction
of this ordered structure, by heating to a temperature below that
at which DNA denatures, results in a reversion of the CD spectrum
to the typical B type (28).

C. Salt Effects

The CD spectrum of DNA in aqueous solution can be altered by
the addition of salts. Permogorov et al. (29) and Fric and Sponar
(30) first demonstrated that in 2M NaCl the CD spectrum of DNA
shows a decrease and slight red shift in the positive band near
275 nm. Extensive investigations by other authors (17,31,32,147)
have shown that a number of monovalent and divalent salts produce
this effect (Figure 3). In all cases, the negative band near 245
nm shows relatively little change, while the positive band is

greatly reduced and may become negative. The greatest effect documented is that in 13 molal LiCl; DNA exhibits two large negative
CD bands centered near 285 nm and 240 nm (18,33).

One explanation of the salt effects relies upon the published
reference spectrum of Tunis-Schneider and Maestre (13) for C form
DNA. The DNA spectrum in 5.5M NH_4Cl, for example, is nearly identical to the C form reference spectrum (Figure 3). Salt addition is
therefore postulated to cause a B→C conformational change in DNA.
The increase in winding angle between adjacent base pairs, which
would accompany this change, may be considered to arise from effective cation neutralization of the DNA phosphates (17,31,34,147

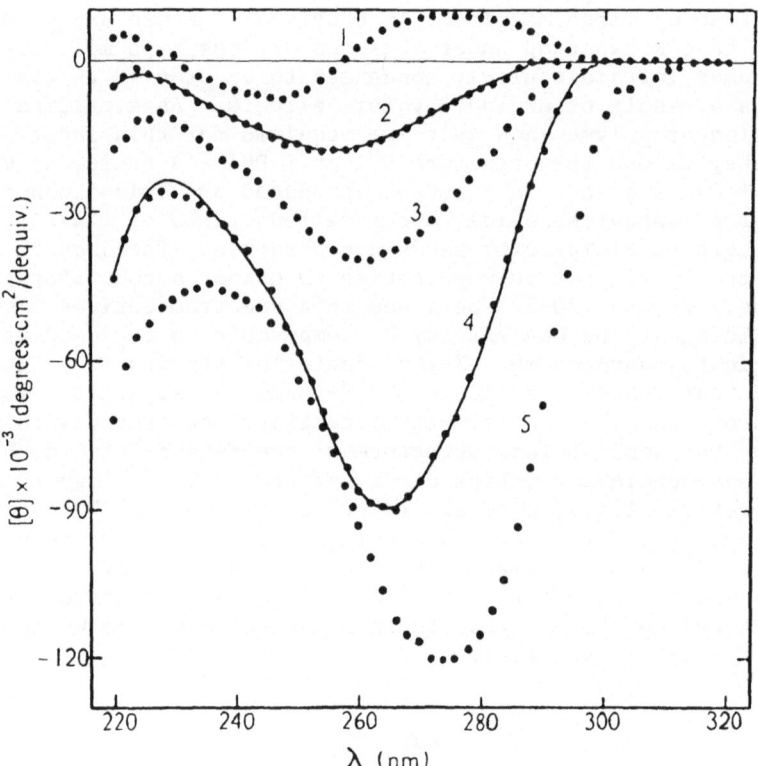

Figure 2. Representative circular dichroism spectra of T7 DNA in
different concentrations of polyethylene oxide (PEO) and at different
times after mixing. The circles represent measured spectra in PEO
concentrations, and at the time after mixing as follows: 1, 80.9
mg/ml, 0.7 hr; 2, 98.1 mg/ml, 0.7 hr; 3, 126.5 mg/ml, 0.7 hr; 4,
90.1 mg/ml, 0.7 hr; 5, 126.5 mg/ml, 48 hr. DNA = 22 x 10^{-6}M, Na^+
= 0.26M. Jordan, CF., et al. (21). Reprinted with permission of
Macmillian Journals, Ltd.

35). The increase in winding angle need not stop at the canonical C form, and thus conformations more tightly wound could occur and give rise to further CD changes as was the case for high LiCl concentrations.

However, there is some evidence which is not consistent with the above explanation. Wide angle X-ray scattering data obtained for DNA in 6.0M LiCl (22), which exhibits the C-type CD spectrum, have been interpreted as being consistent only with the B-form DNA structure. As previously described (Section II, A), the reference spectrum for C-DNA may be questioned on several grounds. There is, however, some evidence that the winding angle between bases in

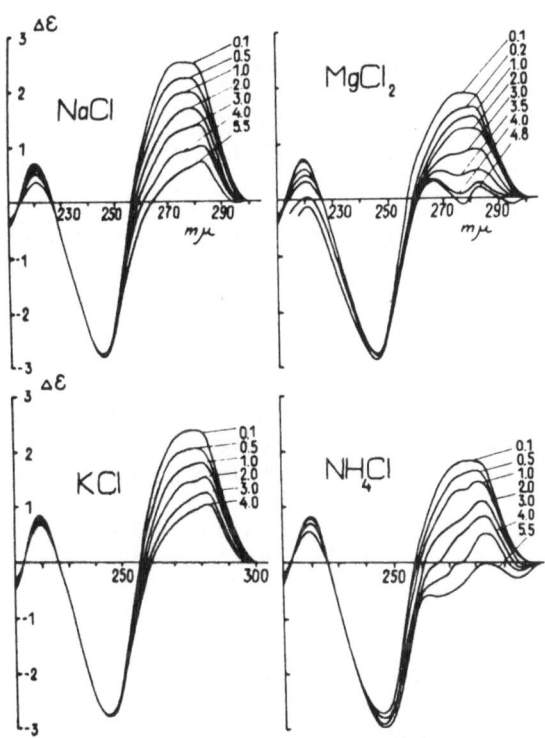

Figure 3. Influence of the cations upon the CD spectrum of the water solution of DNA. The DNA in concentration of 10^{-4}M (PO$_4$) was in the form of Na salt and initially contained 2.5 x 10^{-4}M NaCl. The curves are labelled with the molar concentration of the cations. Ivanov, V.I., et al. (147). Reprinted with permission of John Wiley and Sons, Inc.

DNA can vary with ionic strength (36) and yet remain in the B
family of conformations.

An alternate explanation for the effect of salt on the DNA CD
spectrum may be offered by analogy to the Ψ-DNA studies. Cation
binding by DNA in solution may allow an intramolecular condensation
into an ordered structure. Wolf et al. (33) have shown that upon
increasing the LiCl concentration, the sedimentation coefficient of
DNA becomes larger. At very high LiCl concentrations intermolecular
aggregation occurs, as well as intramolecular condensation. The
existence of a compact ordered form could give rise to a liquid
crystal-type CD contribution. In this regard, it is interesting
to note that the CD spectra for Ψ-DNA (21), DNA in 13 molal LiCl
solution (33), and LiDNA films at low relative humidities (19) are
quite similar in band shape and position, despite some light scattering contributions. At lower salt concentrations, the CD spectra
of DNA may be resolved into contributions from the B-form DNA secondary structure and the Ψ-DNA tertiary ordering.

D. Solvent Effects

The conformation of DNA can be perturbed by organic solvents
causing either an increase or decrease in the ellipticity, depending on solvent. Ethylene glycol or 95% methanol induce changes in
the CD spectral character, similar to those observed in aqueous
salt solutions, namely, a decrease in the ellipticity at 275 nm,
$[\theta]_{275}$ (37-39). The effects of salt and these organic solvents are
additive (40,147). Upon increasing the concentrations of perturbing
solvent, the CD spectra tend to approach the C-form DNA reference
spectrum of Tunis-Schneider and Maestre (13). It is not known
whether these spectra contain any Ψ-like contributions.

Ethanol at low concentrations also produces the same effect on
the CD spectrum of DNA. As the ethanol content of the solvent is
increased from 0 to approximately 66% (v/v with water), the spectrum
changes from the B-type to the C-type (39), the magnitude being dependent on the base composition of the DNA. Between 66% and 78%
ethanol a significant increase in the 275 nm ellipticity occurs in
the CD spectrum, producing spectra similar to the A-form DNA reference spectrum (147), which reaches a maximum value at 80% ethanol
(14). These effects are reversible upon addition of water. Once
again, however, the interpretation of such CD data, in terms of
DNA conformations, is not straightforward. Girod et al. (39) have
demonstrated that the DNA in 80% ethanol solution is condensed.
Furthermore, DNA has been crystallized from ethanol, and has been
found to be in the B conformation (41). The observed CD spectra
are thus likely to be complicated by light scattering and tertiary
ordering (liquid crystal) contributions, so that the A type spectra

obtained in this manner may not be simply due to the A-form conformation.

E. Temperature Effects

The secondary structure of DNA can be destroyed by heating, and this helix-to-coil transition causes a loss of hypochromism in the absorption band at 260 nm, as the bases unstack [Marmur and Doty (42)]. Similarly, base unstacking is reflected in the CD spectrum as a change in band intensities. Both the absorption and CD changes are cooperative. The melting temperature of DNA is a function of solvent and ionic strength, the helical form being stabilized at higher salt concentrations (42).

The CD changes of DNA during heating are complex. Brahms and Mommaerts (14) demonstrated that below the melting temperature, the intensity of the 275 nm peak increases. This effect has been termed a "premelt", and may correspond to an unwinding and/or base tilting in the DNA conformation prior to denaturation (43). At the melting temperature, the CD band intensity is decreased as base-base interactions are destroyed (14). Denatured DNA retains the CD spectral properties of the intrinsically optically active nucleotides, as well as additional contributions from local interactions.

Destruction of the DNA tertiary structure (e.g., Ψ-DNA) by heating can also lead to cooperative CD changes. The condensed form of T2 phage DNA in 80% ethanol exhibits two CD melting bands (39). The lower temperature band corresponds to destruction of tertiary order, and the higher temperature band to the helix→coil secondary structural change.

F. Interpretation of CD Spectra for Unknown DNA Conformations

It is evident from the discussion presented in the preceding sections that the interpretation of a CD spectrum obtained for DNA in an unusual conformation is not straightforward, as the spectrum may contain both secondary and tertiary structure contributions. In general, the latter contributions are to be suspected when band magnitudes are extremely large and nonconservative, DNA is near its solubility limit, or the thermal denaturation profile (CD) exhibits more than one cooperative transition. The absence of these indications, however, does not preclude a smaller tertiary structure contribution, as was discussed for the CD of DNA in salt solutions.

Light scattering artifacts provide an additional complication to interpretation of spectra. Differential scattering of left and right circularly polarized light from optically active particles can result in band position and intensity changes (44-47). The differential scattering contribution can, however, be estimated and

at least partially corrected by both theoretical (45, 148) and experimental means (48-50, 149). In scattering solutions there is also
an absorption flattening effect, not correctable by experimental
means, which generally results in reduced band ellipticities. The
sum of these two effects can provide a significant perturbation of
the CD spectrum for both DNA and proteins (47) ans should therefore
be investigated whenever absorption spectra of such solutions indicate light scattering to be present.

The greatest utility of CD in conformational studies lies in
its extreme sensitivity to conformational changes. Thus when a
small change occurs in the base-base interaction geometry of DNA,
there can be a significant CD spectral alteration. CD may therefore be employed to monitor small secondary and tertiary conformational changes in DNA which may be observed with great difficulty
by X-ray scattering or other physical techniques.

III.   CIRCULAR DICHROISM OF HISTONES

A.   CD of Known Protein Conformations

The ultraviolet CD spectra of proteins arise from the $\pi-\pi^*$
and $n-\pi^*$ electronic transitions of the peptide bond chromophores.
As in the case for polynucleotides, interactions between the transition dipole moments of these chromophores, in ordered conformations,
can lead to unique spectral effects (7).

Reference spectra for various protein backbone conformations
have been obtained from studies of synthetic polypeptides under
conditions shown to yield particular homogeneous conformations; e.g.,
100% α-helix. The three main conformations to be considered are
the α-helix, the β-pleated sheet, and the random coil (or irregular
conformation). The CD reference spectra for these three conformations are seen in Figure 4 [Greenfield and Fasman (150)]. Curve
1, that observed for the α-helix, has two negative bands, at 222 and
208 nm (molar ellipticities are -35,700 and -32,600, respectively)
and a positive band at 191 nm (76,900). Curve 2 is that for the
β-pleated sheet, which has a negative band at 217 nm (-18,400) and
a positive band at 195 nm (31,900). The random coil conformation,
curve 3, shows a small peak at 217 nm (4,600) and a large trough
at 197 nm (-41,900). Thus, the basic conformations of proteins
are easily distinguished by their CD spectra. Other standard
curves (e.g., derived from proteins whose X-ray structure has been
determined) have been employed without significant improvement (51).

The β-turn conformation, which often occurs at the point of
chain reversal at the surface of globular proteins, will also have
a CD contribution similar to that of the β-pleated sheet. The

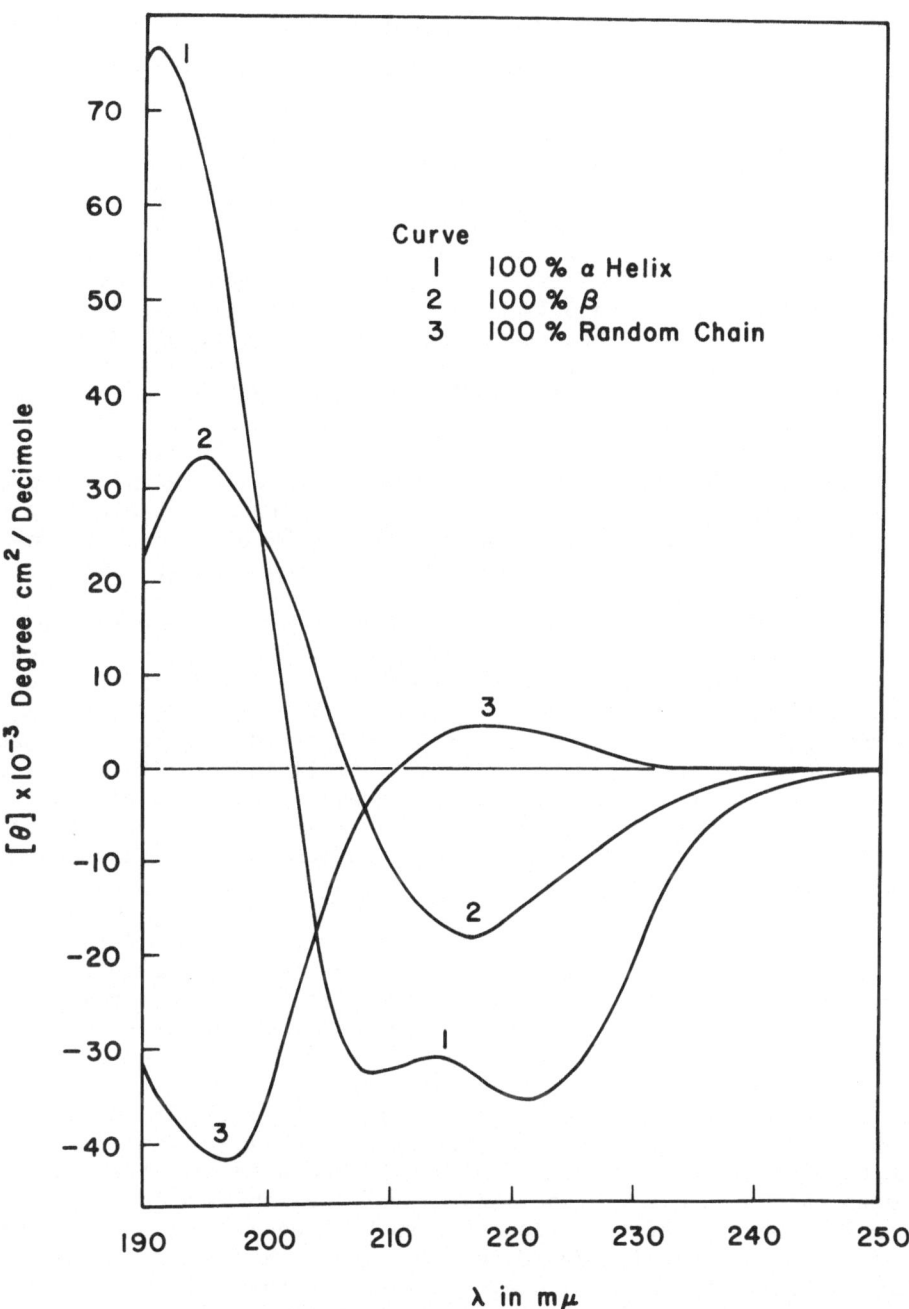

Figure 4. Circular dichroism spectra of poly-L-lysine in the α, β and random conformation. Greenfield, N., and Fasman, G.D. (150). Reprinted with permission of the American Chemical Society.

exact parameters of the β-turn CD spectrum have not yet been determined.

B.   Salt Effects on Histone Conformation

The CD spectra of histones dissolved in water or very low
ionic strength solution are essentially identical to the reference
spectrum for a random coil conformation (52-58,65). The spectrum observed for H4 in 0.01 M Tris, pH 7.0, is that associated with the
random conformation.   In 0.14 M NaF, the conformation of H4 is
markedly changed such that the CD spectrum can be resolved into
contributions of 24% α-helix, 36% β, and 40% random coil (53).
Similar changes due to variation in ionic strength can be observed
for the other histones.   Thus in the presence of salts the histones
can adopt highly ordered secondary structures.   This conclusion has
been confirmed by measurement of tyrosine fluorescence anisotropy,
which also indicates substantially increased structural rigidity
with increasing salt content (54-58).   A review of literature on
this topic has recently been published by Fasman et al. (59).

Isenberg and coworkers (54-58) have thoroughly investigated
the salt-dependent histone conformational changes.   Upon addition
of salt to solutions of low protein concentration, H3 and H4 undergo
both fast and slow conformational alterations.   Only the fast conformational change can be observed for H2A and H2B.   The fast change
was shown by CD studies to consist of a small increase in α-helical
structure.   The slow step (the rate of which depends on histone
concentration, salt concentration, type of salt, and temperature)
consists of β-pleated sheet formation, concomitant with histone
aggregation.

When mixtures of the histones were similarly studied [D'Anna
and Isenberg (60-62)], significant histone-histone interactions
were observed.   As noted above, H4 undergoes both fast and slow
conformational changes upon addition of salt, whereas H2B does not
exhibit a slow change.   However, a mixture of these two histones
did not yield intermediate CD changes, thus indicating interaction
between species.   The CD curve for a 1:1 mixture of H2B and H4
shows no time-dependent slow conformational change.   Similarly,
the CD spectrum of mixtures of H3 and H4 shows no slow changes, although each of these histones individually exhibits slow conformational changes.   Sedimentation studies have demonstrated that these
different effects arise from specific histone-histone complexing
which inhibits histone self-aggregation.   In the H3-H4 case, the
complex is a tetramer; for H2A-H2B and H2B-H4 the complexes are
dimers (61,63).

Such CD studies have demonstrated that specific histone-histone
complexes can form in aqueous salt solutions.   Moreover, it has been

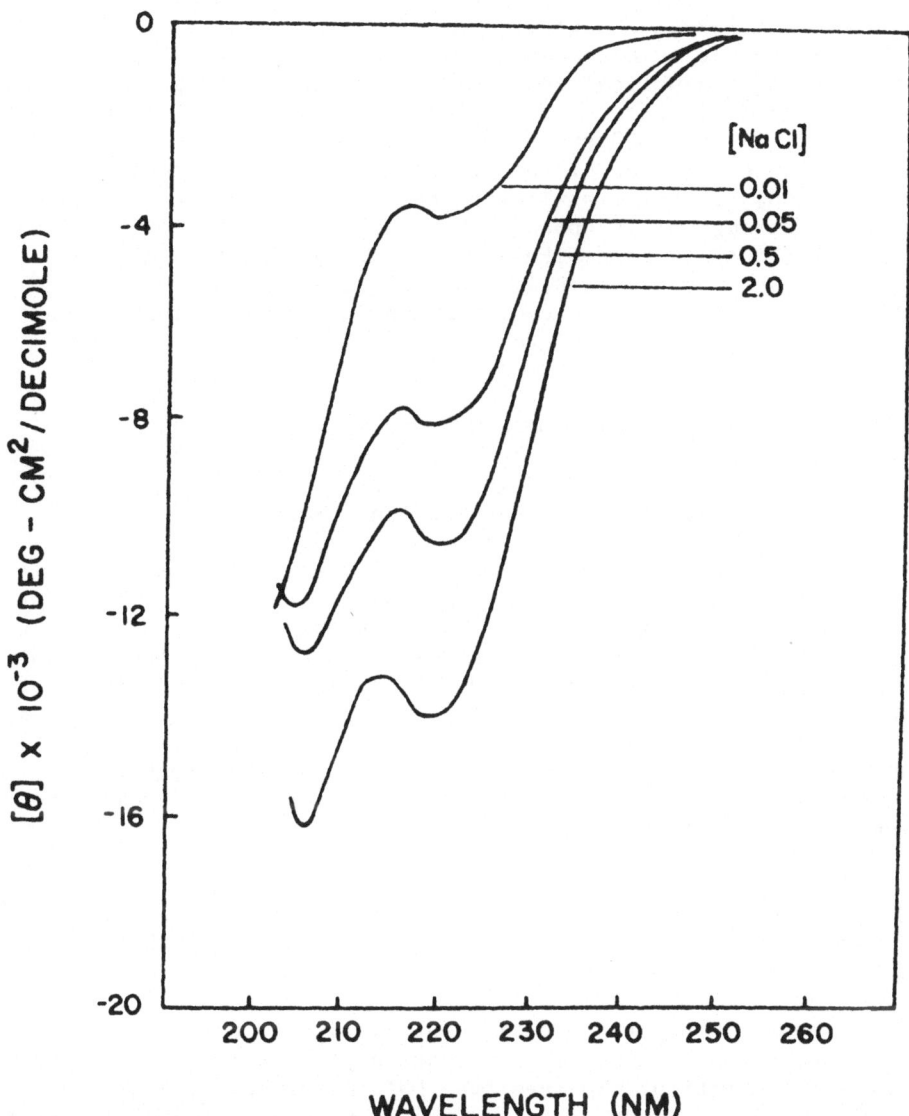

Figure 5. Circular dichroism spectra of chicken erythrocyte histones in solutions of 1 mM Tris-HCl (pH 7.5) and several concentrations of NaCl. Bidney, D.L., and Reeck, G.R. (64). Reprinted with
permission of the American Chemical Society.

shown that the fast conformational change results in complexes with substantially larger negative ellipticities at 220 nm than found present for the individual histones (60). The CD spectra indicate that upon complexation a greater α-helical content resulted. The α-helical content is dependent on salt concentration, higher salt yielding higher α-helical contents. This pattern [Figure 5, Bidney and Reeck (64)] is also observed for mixtures of total histone extracts (30,64). The CD spectrum, below 240 nm, in 2.0 M NaCl for the solution of all the histones, [as well as the histone core tetramer (66)] is quite similar to that obtained for chromatin at low ionic strength. It appears that the DNA-histone complex interaction results in a salt-like neutralization effect on histones so that a compact conformation is adopted. Independent evidence (67,68) indicates that the histone conformation in chromatin is like that adopted by the four core histone mixture in 2M NaCl.

IV.   CIRCULAR DICHROISM OF HIGH MOLECULAR WEIGHT CHROMATIN

A.   Chromatin at Low Ionic Strength

It is now well established [recently reviewed by Kornberg (1)] that chromatin exists as a repeating unit, in which specific histone complexes, containing two each of the histones H4, H3, H2A, and H2B, bind and condense a discrete length (about 140 base pairs) of DNA. The repeating units are connected by a variable shorted length of DNA which may be condensed or extended depending on solvent conditions and interaction with H1 (H5 in avian erythrocytes) (69,70). The interaction of DNA with histones could possibly result in a conformational change in the DNA and/or histones which should be amenable to investigation by CD.

The CD spectrum of calf thymus chromatin is shown in Figure 6 (71). Below 230 nm, chromosomal proteins provide the greatest spectral contribution. The band shapes and magnitudes in this region indicate substantial (∿40%) α-helical character in the proteins. Recent estimates of conformation gave 50% α, 5% β and 51% random (66). Above 250 nm, the protein contribution is negligible and consequently the CD bands in this region provide information about the DNA conformation. Whereas free DNA in solution exhibits a positive band at 275 nm with $[\theta]_{275}$ = 8,500 deg. cm^2 dmole^{-1}, that band is red-shifted and reduced to 4,000 for the protein-bound DNA of chromatin. The altered DNA spectrum is a direct consequence of protein binding, since dissociation of the proteins by addition of sodium dodecylsulfate results in the observation of a normal B-form DNA spectrum (71). This CD spectrum of chromatin has been amply confirmed (29,30,72-74). It is thus apparent that the secondary or tertiary ordering (or both) of DNA in chromatin is altered relative to B-DNA.

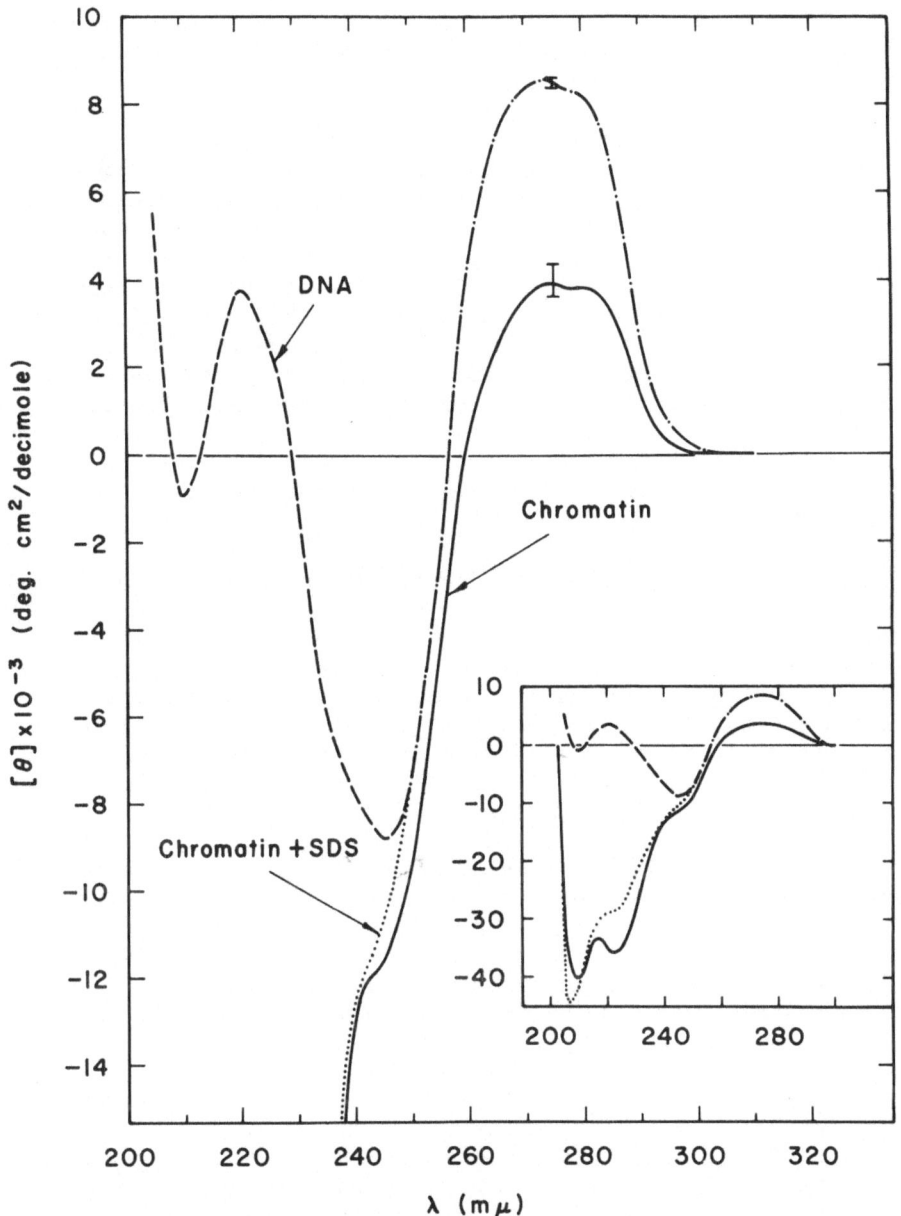

Figure 6.   CD spectra of calf thymus chromatin, chromatin in 0.1%
sodium dodecylsulfate, and DNA.   (——) calf thymus chromatin; (···)
chromatin in the presence of 0.1% (w/w) sodium dodecylsulfate; (---)
pure calf thymus DNA.   Solutions in 0.14M NaF, 0.01M Tris-HCl, pH
8.0.   Conc. DNA = 1.6 x 10^{-4}M (PO_4).   Mean residue ellipticity is
based on DNA residue concentration.   Shih, T.Y., and Fasman, G.D.
(71).   Reprinted with permission of Academic Press.

Calf thymus chromatin is a relatively simple system for CD
studies because it contains very little nonhistone protein or RNA.
Histones have few aromatic amino acids which could provide a CD
contribution above 230 nm. By contrast, the nonhistone proteins,
as a group, have a significant aromatic amino acid content. RNA,
as previously described (Section II, A), has a CD spectrum similar
to that of A-DNA. Thus the presence of nonhistone proteins and
RNA must be considered for the interpretation of the CD spectra obtained for various chromatin samples from active tissues (75-78).
Thus it has been found that the CD spectra of chromatin samples
isolated from cells at different stages in the cell cycle are distinguishable; these differences are in part due to changes in the
nonhistone protein and RNA contents [reviewed by Baserga and Nicolini (79) and Baserga in a Chapter in this volume]. The discussion
herein will be limited to studies of chromatin samples for which
the non-histone protein and RNA contributions are negligible, thus
allowing a more detailed understanding of the DNA and histone conformations.

What can be said about the conformation of DNA in chromatin?
Hanlon and coworkers (18, 80) have stated that a portion of the DNA
adopts the C canonical conformation, and the remainder is of the B
form. The CD spectra of chromatin isolated by two different procedures have been examined by Hanlon et al. (80) and compared to
the B and C-DNA reference spectra. It is, in fact, possible to
match the observed spectra for chromatin by a combination of the B
and C-DNA spectra. The C-DNA reference spectrum, however, corresponds to no adequately established DNA structure (see Sections II,
A and II, C). In addition, the existence of C-DNA in chromatin is
not supported by X-ray scattering studies (16) which show that DNA
retains the B conformation.

If the secondary structure of DNA is not altered significantly
by histone complexation in chromatin, the CD change may be due to
the existence of an ordered tertiary arrangement of the condensed
DNA. Support for this hypothesis may be found in the calculated
difference spectrum between free DNA and chromatin DNA [Figure 7,
Shih and Lake (81)]. The difference spectrum shows a negative
band centered at 275 nm, similar to the spectrum of Ψ-DNA. The
position of this band accounts for the lower $[\theta]_{275}/[\theta]_{284}$ ratio
in chromatin relative to free DNA. If this explanation is pursued,
any change in the extent of DNA condensation in chromatin will be
reflected in a change of the 275 nm ellipticity. Furthermore, the
changes at nearby wavelengths (e.g., the peak at 284 nm) will be
less extensive. Data which supports this explanation is discussed
in the following section.

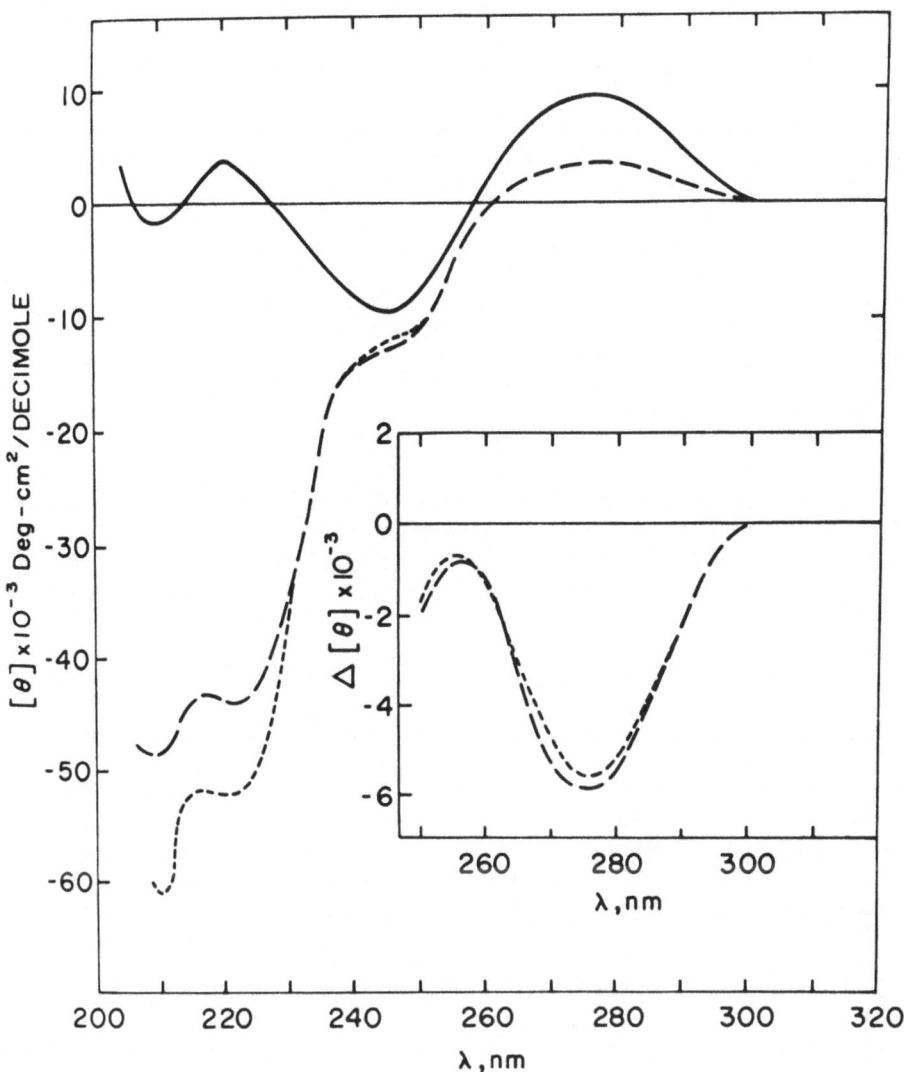

Figure 7. Circular dichroism spectra of metaphase and interphase
chromatin. (— — —), metaphase chromatin; (---), interphase
chromatin. Ellipticity, [θ], calculated on the basis of DNA residue
concentration. Difference spectra shown in the inset (Δ[θ] = [θ]^{chr}–
[θ]^{DNA}) were calculated by subtracting CD of pure DNA (——) measured
under similar conditions. Solvent was 0.01M NaCl, 0.001M Tris·HCl,
pH 7.6. Shih, T.Y., and Lake, R.S. (81). Reprinted with permission
of the American Chemical Society.

B. Tertiary Order in Chromatin

1. <u>Thermal Denaturation of Chromatin</u>. When DNA is condensed
into an ordered tertiary structure (see Section II, E), heating to
a temperature below the melting temperature of the DNA double helix
can sometimes lead to a reversion of the CD spectrum to that characteristic of the secondary structure alone. Thus, the tertiary
structure can be destroyed at a lower temperature than the secondary
structure. An analogous situation is found to occur in chromatin
temperature melts. The DNA in chromatin melts out at a much higher
temperature than free DNA indicating that the double helix is
stabilized by bound protein. Chromatin appears to have two temperature transitions. At the first transition, the ellipticity at 280
nm shows a large increase (Figure 8), followed by a decrease at the
second temperature transition (73,82,83). The CD ellipticity at
227 nm, $[\theta]_{227}$, observed at the first transition shows a reduction
in the histone α-helical content, simultaneously with an apparent
meltout of the tertiary order of the DNA (82). This is followed by
the meltout of the B form DNA (associated with histones) seen in
the second transition. Thus it would appear that the lower the
ellipticity at 275 nm, the higher the degree of tertiary structure
exists.

2. <u>Shearing Effects on Chromatin</u>. This topic will only be
briefly discussed. Very high molecular weight chromatin is not
easily solubilized for spectroscopic study. Since light scattering
contributions make interpretation of CD spectra difficult, many
laboratories have employed shearing methods to reduce the molecular
weight of DNA in chromatin and thus obtain soluble preparations.
If, however, the degree of DNA condensation in chromatin has an effect on the observed CD spectrum, then methods of isolation that
minimize the disruptive effects of shear forces should yield more
highly condensed chromatin, with a new characteristic CD spectrum
of lower magnitude. Thus if the negative Ψ-like spectral contribution centered at 275 nm plays a significant role, its contribution
should be larger for a more highly condensed and ordered chromatin
preparation. This explanation is consistent with data obtained by
several groups (49,75,80,84). When the CD light scattering contribution is corrected for in a condensed chromatin sample, the CD
spectrum shows a positive band centered near 280 nm, but with a
nearly two-fold reduction in ellipticity relative to sheared
chromatin (49).

3. <u>Specific Salt Effects on Chromatin</u>. At concentrations
far below those necessary to cause histone dissociation, salts can
effect the CD spectrum of chromatin. It has been well established
that the extent of chromatin condensation as viewed by the electron
microscope is dependent on the concentration and type of salt present [see for example Brasch (85); Finch and Klug (86)]. In general,
lower salt concentrations result in more extended chromatin. If
the positive CD ellipticity at 275 nm is a sensitive measure of

tertiary order in DNA, then a higher ellipticity is expected at
lower ionic strength. This has found to be true by a CD study which
was correlated with an electron microscope study to reveal the actual state of condensation (151). The electron micrographs, at
various concentrations of ammonium acetate, showed that the structure of chromatin is considerably more condensed at the higher salt
concentration. A parallel study of the CD properties, under the
same conditions, showed that the ellipticity at 278 nm was reduced
approximately 28% at the highest salt concentration, relative to
the lowest. Other reports, in which hydrodynamic measurements were
made to determine the extent of chromatin condensation, confirmed
the lowered CD ellipticity observed in solvents favoring a compact
structure for the chromatin (84,87).

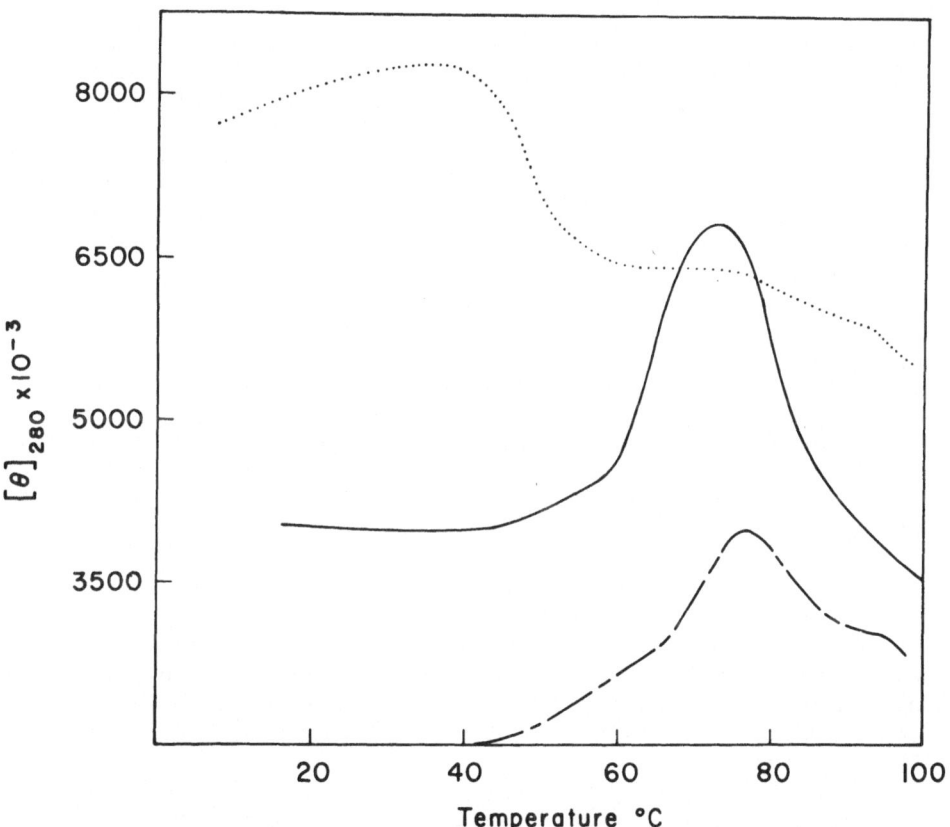

Figure 8. Circular dichroism ellipticity, $[\theta]_{280}$, versus temperature for chromatin (——); mononucleosomes (— — — -); and DNA (···)
in 2.5 x 10^{-4}M EDTA, pH 7.0. Mandel, R., and Fasman, G.D. (83).
Reprinted with permission of Information Retrieval Ltd.

    4. <u>Supercoiling</u>. A supercoiled arrangement of DNA need not lead to a negative Ψ-like spectral contribution. Maestre and Wang (88) observed a slight increase of ellipticity at 280 nm, relative to B-DNA, in closed circular supercoiled DNA from several sources. The folded chromosome of E. coli has been reported (89) to exhibit CD properties consistent with contributions only from RNA and simple B-DNA. By contrast, intact chromosomal fibers from equine spermatozoa do show a Ψ DNA CD spectrum (90).

C. Perturbation of Chromatin Structure

    1. <u>Effect of Trypsin</u>. Digestion of chromatin with trypsin results in the cleavage of peptide bonds adjacent to lysyl and arginyl residues in the histones which are exposed to the enzyme. Limited tryptic digests result in cleavage of bonds adjacent to basic residues which are not involved in ion-pair bonds with the DNA phosphates (91). Thus protein segments not bound or only weakly associated with the DNA backbone are digested. This results in disruption of histone-histone interactions and an unfolding of the DNA to an extended conformation (91). The CD changes with varying levels of trypsin digestion (92) are shown in Figure 9. This data illustrates the dependence of the depression of the positive CD band of DNA on the maintenance of histone-histone interactions. Unfolding of the condensed DNA structure results in a higher positive ellipticity.

    2. <u>Effect of Urea</u>. Histone-histone interactions are disrupted by the presence of urea. This disruption is a direct result of histone denaturation, and is reflected in a loss of the α-helix contribution as determined by the CD spectrum. There is a concomitant increase in the ellipticity of the positive DNA CD band (73, 74,81,87,93). Hydrodynamic studies show that chromatin DNA is more extended in urea solutions (87) in agreement with the above observations.

    3. <u>Effect of Chemical Crosslinking</u>. Formaldehyde, which can form both histone-histone and histone-DNA linkages, causes a marked decrease in the positive CD above 260 nm as a function of formaldehyde concentration (94). This reduction is most noticeable at 275 nm. Chemical crosslinking thus appears to enhance the Ψ-like CD contribution in chromatin, probably by tightening the condensed DNA structure.

    4. <u>Effect of Histone Chemical Modification</u>. The lysine residues in the histones may be chemically altered by specific reagents which may modify the DNA-histone interactions, leading to changes in the CD spectrum. Mild treatment of chromatin with acetic anhydride results in the acetylation of approximately 25% of all protein lysine residues (95). The extent of histone acetylation

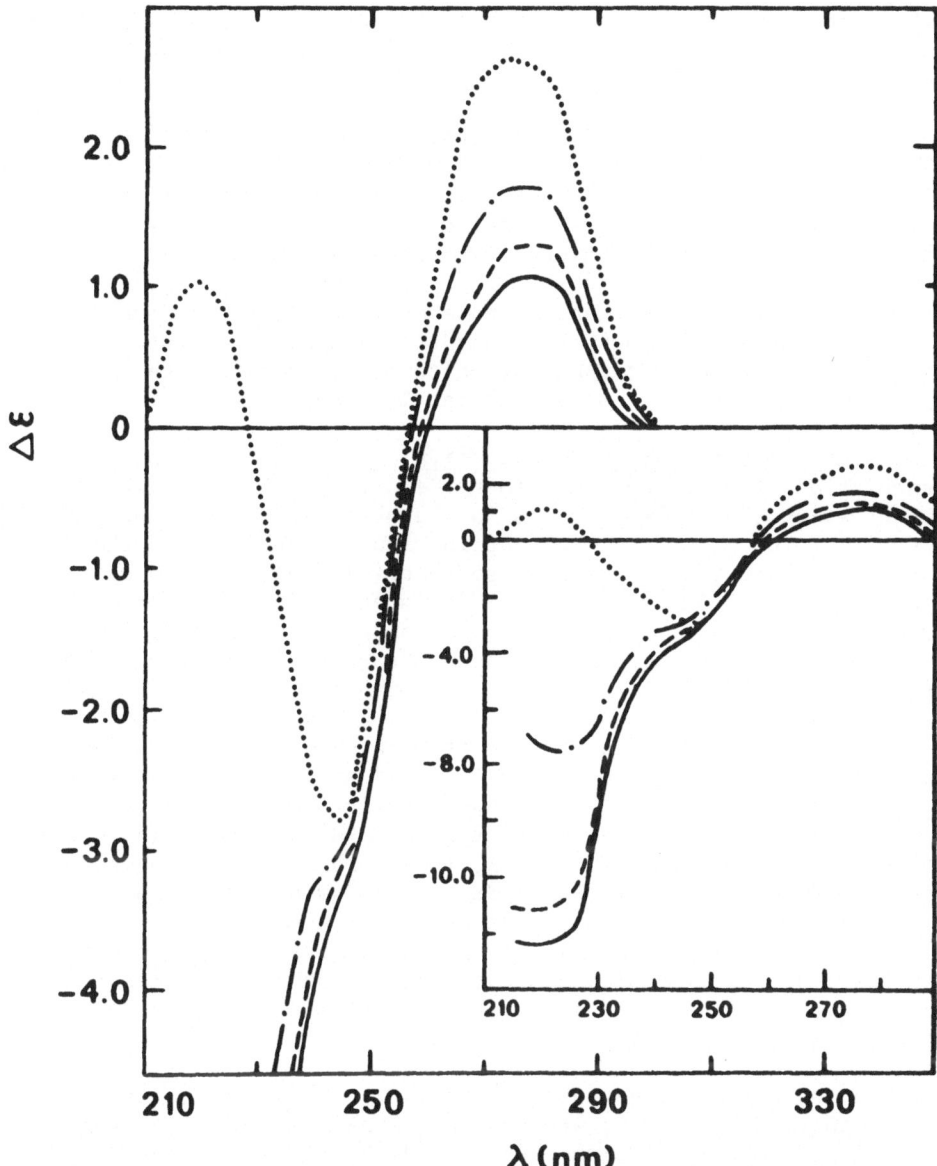

Figure 9.   CD spectra of trypsin-treated nucleohistones.   Trypsin
concentration in µg/ml is 0 (——), 10 (— — —), and 40 (-·-);
DNA (···).   Contribution of trypsin to CD below 240 nm has been
corrected.   It is noted that the total histone content, digested or
not, is the same for these samples with varied levels of trypsin.
Li, H.J. et al. (92).   Reprinted with permission of John Wiley and
Sons, Inc.

is much lower than the value for total chromosomal protein, as nonhistone proteins react more strongly. However, the CD spectrum above 250 nm was reported to be unchanged by this low level of modification (95).

Extensive modification of the histone lysine residues may be achieved with ethyl acetimidate (96). The altered amino acids retain a positive charge, and so may still interact with the DNA phosphate groups. Up to 90% of the histone lysines may be so altered without resulting in significant CD changes (96).

In both of the cases described above, the basic condensed structure of DNA was shown, by other techniques, to be retained in the chemically treated chromatin, which is consistent with the fact that the CD spectra are not significantly changed.

5. Effect of Histone Dissociation. Essentially complete dissociation of histones from DNA by detergents (53) or high salt concentration (29,30,74) results in the observation of a chromatin CD spectrum for which the DNA contribution above 250 nm is the same as that for free DNA in the same solvent. At intermediate salt concentrations (0.6M to 1.6M NaCl) it is possible to selectively dissociate particular histones from chromatin, and thus determine the extent to which those histones are responsible for the condensation of DNA (97).

H1 is effectively dissociated from chromatin with 0.6M NaCl (73). This histone is thought to bind and at least partially condense the spacer region of DNA between nucleosomes (69,98). As shown in Figure 10, removal of H1 from chromatin results in an increased ellipticity due to the DNA, in accord with the expected results for DNA unfolding (92). Several reports confirm this observation (72,87,99,100), but a decrease or lack of change in the positive ellipticity of chromatin upon H1 removal have also been reported (73,101). Results obtained in the author's laboratory support the observation of increased ellipticity in H1 and H5-depleted chicken erythrocyte chromatin.

H2A and H2B are removed from chromatin at NaCl concentrations of 0.6M to 1.2M NaCl (97), whereas H3 and H4 are removed between 1.2M and 2.0M NaCl. Published reports differ in conclusions regarding the properties of H1, H2A, H2B-depleted chromatin. Thus it has been observed that chromatin depleted of all histones except H3 and H4 exhibits a positive CD band of DNA similar to free DNA (72,73); or that these histones do exert some conformation-perturbing effect on DNA (87,92) as shown in Figure 10.

Removal of the different histone classes from chromatin thus destroys the difference in CD properties between chromatin DNA and free DNA. Thermal denaturation studies show a concomitant loss of

the higher temperature stabilization of the DNA (82). With increasing temperature, free DNA shows one downward CD transition, centered
at about 44°C, whereas intact chromatin DNA shows an increase in
ellipticity around 60°C and a reduction in ellipticity transition
centered near 72°C. As the histones are progressively removed, by
extraction with higher NaCl concentrations, the higher temperature
transitions decrease in size and after the 2.0M NaCl extraction
the CD plot is similar to that of native DNA. Thus, the importance
of histone:DNA and histone:histone interactions is demonstrated for
maintenance of the chromatin structural stability.

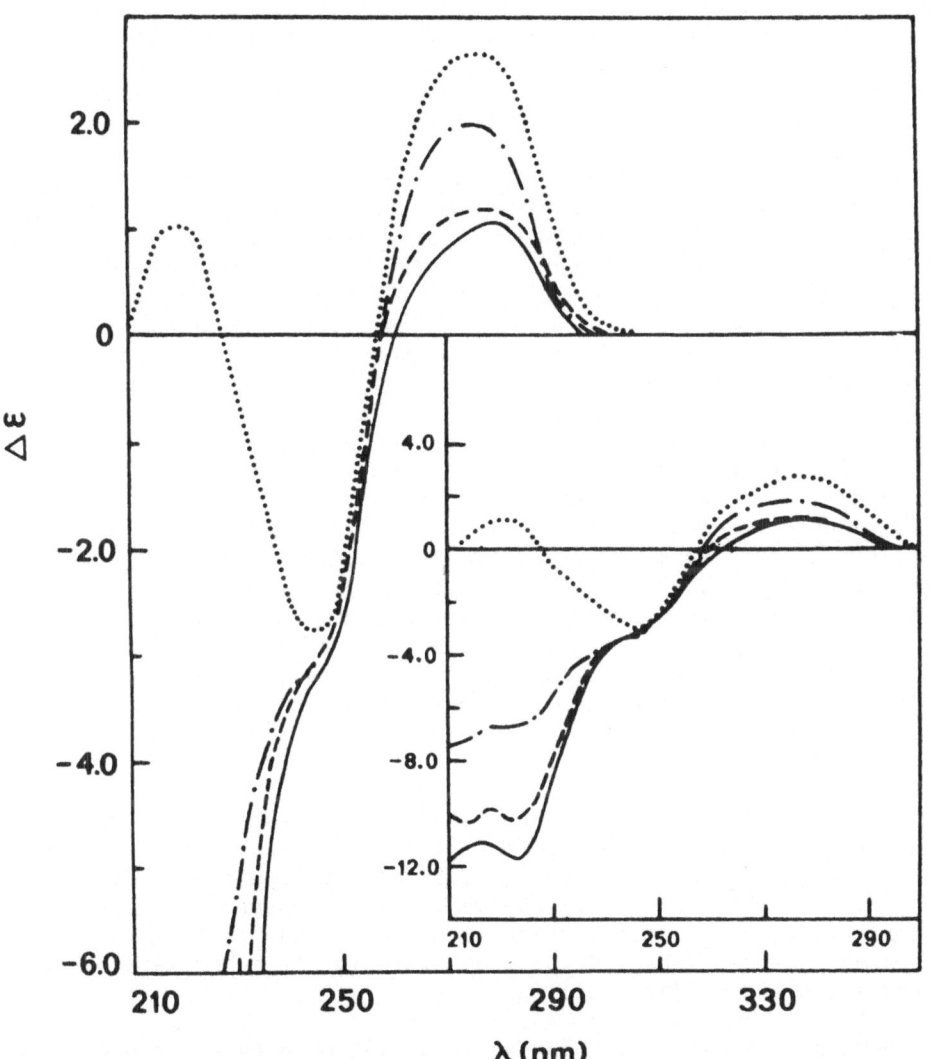

Figure 10. CD spectra of NaCl-treated nucleohistones. 0 M NaCl
(——); 0.6M NaCl (---); 1.6M NaCl (-·-), DNA (···). Li, H.J. et al.
(92). Reprinted with permission of John Wiley and Sons, Inc.

V.   CIRCULAR DICHROISM OF CHROMATIN SUBUNITS (NUCLEOSOMES)

A.   CD of Isolated Nucleosomes

    The repeating subunit of chromatin structure is termed the
nucleosome, with a spacer of DNA between nucleosomes. The nucleosome core contains about 140 base pairs of DNA and 8 histone molecules, two each of H2A, H2B, H3, and H4 [recently reviewed by
Kornberg (1), Felsenfeld (2)]. Nucleosomes (mono, di, tri, etc.)
are isolated from micrococcal nuclease digests of nuclei or chromatin. Depending on the extent of nuclease digestion, the mononucleosome may also contain varying size "tails" of DNA to which H1 is
bound (69,102,103); the "tails" originate from the approximately
60 base pair spacer DNA between nucleosome cores in chromatin and
are usually shorter than their original length in most large-scale
nucleosome preparations. Finch et al. (104) have recently proposed
a detailed model for the conformation of DNA within nucleosome
cores. In this model, the 140 base pairs of DNA form 1-3/4 superhelical turns around a histone core. The pitch of the superhelix
is 28 Å. Thus the adjacent coils of DNA are spatially close
enough for cation or histone salt bridges between phosphates.
Little information exists about the conformation of DNA "tails"
attached to the nucleosome cores.

    The condensation of DNA into supercoils around a histone core
may provide the electronic interactions necessary to reduce the
ellipticity of chromatin compared to free DNA. The type of CD
spectrum observed above 250 nm for isolated PS particles which contain in the order of 100 base pairs of DNA, i.e., subnucleosome
particles, in low ionic strength solution is shown in Figure 11
(105). The positive band exhibits a maximum at 284 nm, and a
shoulder at 275 nm. A new negative band is observed at 295 nm.
The positive CD band ellipticity is approximately one-half that of
chromatin, and the ratio of ellipticity at 275 nm to that at 284
nm is lower relative to chromatin. These changes are consistent
with the addition of a large negative CD band centered at 275 nm
to the CD band of B-form DNA; that is, more Ψ character in the
spectrum. Shih and Lake (81) have shown that the calculated CD
difference spectrum (above 250 nm) between B-DNA and chromatin is
a single negative band centered at 275 nm. It is apparent from
the spectra in Figure 11 that the mononucleosome spectrum contains
a greater percentage contribution from the Ψ band, than in whole
chromatin, which may be interpreted as arising from tertiary order
in DNA arrangement (supercoiling).

    Mononucleosomes containing at least 140 base pairs of DNA
yield similar CD spectra (83,103,106-108). Ellipticity values reported for the CD maximum near 284 nm range from approximately
1400 to 2200 deg. cm^2 $dmole^{-1}$.

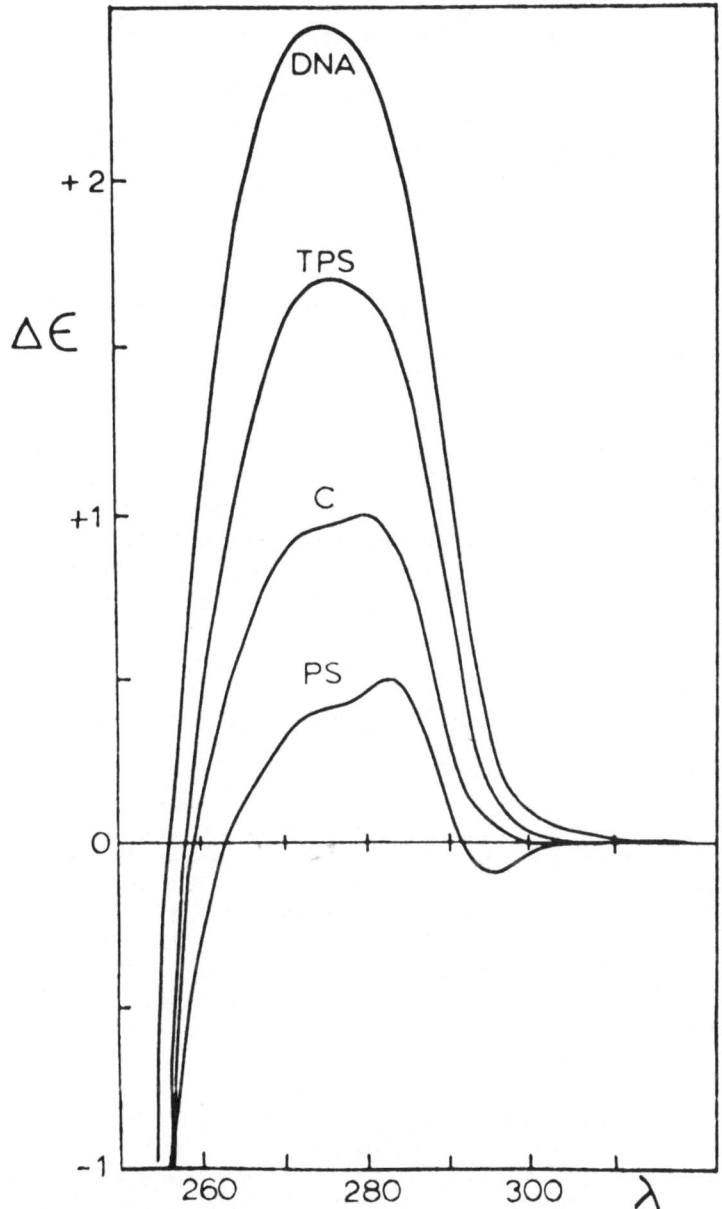

Figure 11. Circular dichroism spectra of whole calf thymus chromatin (C), nuclease-resistant fragments (PS), trypsin-digested PS fragments (TPS), and DNA. Solvent is 10 mM Tris-HCl, pH 8.0, room temperature. Sahasrabuddhe, C.G., and Van Holde, K.E. (105). Reprinted with permission of the American Society of Biological Chemists.

There is additional evidence that the CD spectrum of mononucleosomes contains a strong contribution from an ordered tertiary
structure of DNA. The temperature dependence of the 280 nm ellipticity for DNA, chromatin, and mononucleosomes (83) is compared in
Figure 8. The mononucleosome meltout is very similar to chromatin:
there is a strong increase in ellipticity around 70°C followed by
a decrease near 80°C. The first transition may be interpreted as
destruction of tertiary order; the second transition corresponds
to denaturation of the double helix.

B.   Perturbation of Mononucleosomes

1. Effect of Salts. Mononucleosomes may be precipitated
partially by KCl, and completely by a number of salts containing
divalent cations (109). Finch and Klug (86) demonstrated by electron microscopic evidence that aggregated mononucleosomes in water
form a superhelical fiber; this aggregate can be destroyed by
chelating agents which remove residual cations. Olins et al. (110)
reported a fibrous appearance for aggregated nucleosomes in 5 mM
$MgCl_2$. Moreover, these authors found nucleosomes precipitated by
$MgCl_2$ to exhibit low-angle X-ray scattering with the same maxima
as chromatin.

It has not yet been possible to reproduce chromatin-like CD
spectra from associated mononucleosomes. Sahasrabuddhe and Saunders (108) have studied mononucleosome aggregates (nucleosomes
precipitated with 10 mM $MgCl_2$, then resolubilized by addition of
$[NH_4]_2SO_4$) and found that the CD spectrum of the initially solubilized mononucleosome preparation exhibited an ellipticity at 282 nm
only one-half that for untreated mononucleosomes in EDTA at low
ionic strength. Additional features of the CD spectrum observed
were a lower $\theta_{275}/\theta_{282}$ ratio and a larger negative CD band at 294
nm relative to the untreated nucleosomes. In the presence of additional $(NH_4)_2SO_4$, beyond that necessary for solubilization of
the precipitated nucleosomes, the CD spectra showed a trend toward
the spectrum of chromatin, but did not yield a true chromatin-like
spectrum.

2. Effect of Urea. The effect of urea on mononucleosomes is
analogous to the effect on chromatin (see Section IV, C, 2). In
6M urea, nucleosomes exhibit a CD spectrum having a DNA contribution similar to that of free DNA in the same solvent (103,111). A
thorough study of the effect of increasing urea concentration on
mononucleosomes has been made by Olins and coworkers (112). The
CD data from that study is shown in Figure 12. The positive CD
band above 250 nm linearly increased in ellipticity with increasing
urea concentrations up to 8M. Parallel hydrodynamic studies found
that the viscosity and sedimentation properties of the mononucleosomes showed similar linear changes indicative of DNA uncoiling.
By contrast, the histone CD bands showed a cooperative loss of α-

helical structure between 4 and 7M urea. This data suggests that
the graduate unfolding of the nucleosome leads to a loss of the
spectral perturbation of protein-bound DNA relative to free DNA
and that the condensation of DNA is not directly related only to
the secondary structure of the histones, although histone-histone
interactions cannot be evaluated in this study.

    3. Effect of Trypsin. The digestion of the histones, within
the mononucleosomes, with trypsin results in a change in the nucleosome shape from globular to extended (105,113). As a consequence
of the DNA unfolding, the CD contribution for DNA within the nucleosome changes towards that of free DNA (Figure 11). The CD spectral
change found for these subnucleosomal PS particles (105) is in
agreement with results found for mononucleosomes containing 140
and 160 base pairs (113). Thus the histone segments responsible
for DNA supercoiling are readily attacked by trypsin.

    4. Effect of Histone Crosslinking. Linkages formed between
histones in isolated mononucleosomes by dimethylsuberimidate (114)
result in only a small reduction in the DNA CD band at 275 nm; the
CD ellipticity decrease is much less at 284 nm. A slight tightening of the histone-histone interactions by chemical crosslinking
thus may cause a concomitant small increase in condensation of the
DNA. This effect is analogous to the observed changes in chromatin
CD after formaldehyde treatment (94) discussed in Section IV, C, 3.

    VI. CIRCULAR DICHROISM OF MODEL CHROMATIN SYSTEMS

    Significant information concerning the nature of protein:DNA
interactions in the complex structure of chromatin can be obtained
from the study of model protein:DNA systems. Firstly, it can be
established whether or not any single histone or model histone can
induce a conformational change in DNA similar to that found in
chromatin. Secondly, the stabilizing forces involved in histone:
DNA interactions can be elucidated by studying synthetic polypeptides
of different amino acid compositions. The relative importance of
electrostatic and hydrophobic interactions in the DNA:histone complex may be evaluated from investigation of these different model
protein structures.

    A. Polypeptide:DNA Complexes

    The use of polypeptide models for histones in the investiga-
tion of histone:DNA interactions allows the determination of the
degree to which electrostatic and hydrophobic forces determine

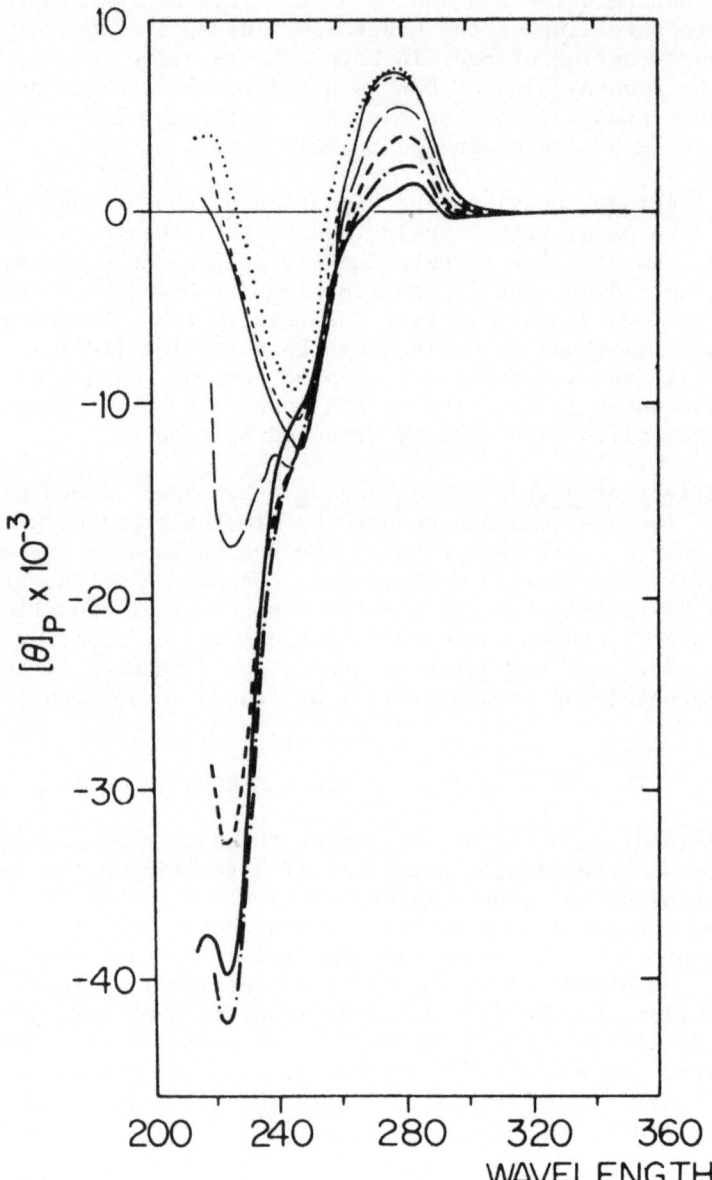

Figure 12. Circular dichroism of mononucleosomes at different urea concentrations. Mononucleosomes in 0 (——), 2 (-·-), 4 (---), 6 (— —) M urea (thick lines), or 8 (----), 10 (——) M urea (thin lines). Also shown is DNA in 0.1M NaCl (····). Ellipticity calculated on basis of DNA residue concentration. Olins, D.E. et al. (112). Reprinted with permission of Information Retrieval Ltd.

specific aspects of the complexation. A model polypeptide for
investigation of predominately electrostatic protein:DNA binding
is poly-L-lysine. When poly-L-lysine is complexed to DNA, by dialysis to 0.85-1.0M NaCl, the CD spectrum of DNA shows large changes
(115-117). A large negative ellipticity band centered at approximately 270 nm appears, the ellipticity of which is dependent on the
ratio of lysine:PO_4, and can reach 25 times the normal range of
ellipticity values for DNA (Figure 13). X-ray diffraction studies
of the complexes (117) show that the DNA retains the B conformation.
Electron microscopic (117) and light scattering (116) investigations
found that the complexes are aggregated into highly solvated particles
of nearly uniform size with a doughnut-like shape. Lerman and coworkers (21) attributed the CD spectrum observed for these poly-L-

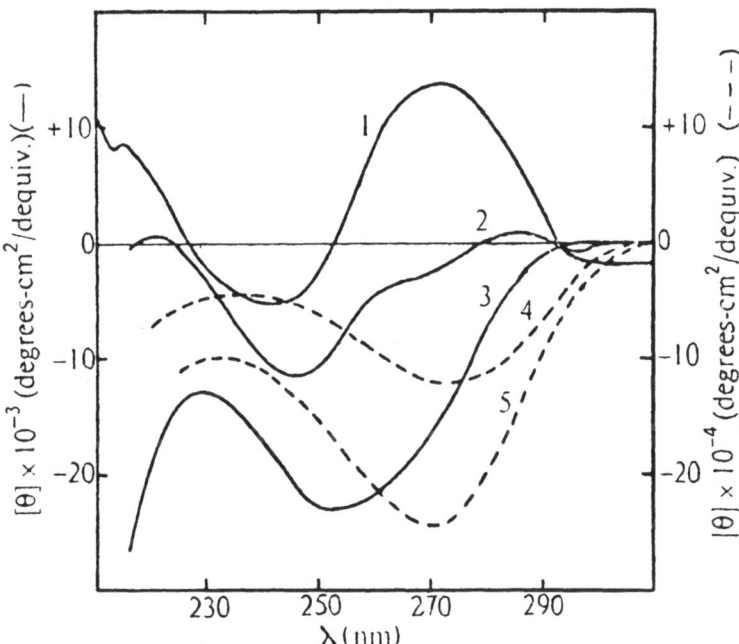

Figure 13. The circular dichroism spectra of DNA in various complexes or perturbing solvents. Curves are redrawn from published
data. 1, H4:DNA complex, r = 1.5; 2, DNA in ethylene glycol; 3,
H1:DNA complex, r = 1.0; 4, T7 DNA in 0.2M NaCl and 126.5 mg/ml
PEO, 48 hr. after mixing; 5, poly-L-lysine:DNA complex, r = 1.1.
All data except curve 4 with calf thymus DNA. The ordinate on the
left applies to solid curves, while the ordinate on the right,
corresponding to a 10-fold larger value, applies to the dashed
curves. Jordan, C.F. et al. (21). Reprinted with permission of
Macmillian Journals, Ltd.

lysine:DNA complexes to electronic interaction between adjacent
DNA helices in a condensed DNA structure; that is, similar to Ψ-
DNA. Thus, electrostatic interactions alone can play a significant
role in the condensation of DNA by proteins.

Synthetic polypeptides containing hydrophobic residues as well
as basic residues have been interacted with DNA, and their complexes
studied. The polypeptides studied include the random copolymers of
L-Lys and L-Ala (118-120), L-Lys and L-Leu (121-122), L-Lys and L-
Val (123-124), and L-Lys, L-Ala, and Gly (125). Sequential polypeptides which have been studied are: $(L-Lys-L-Ala-Gly)_n$, (L-Ala-
$L-Lys-L-Lys-L-Pro-L-Lys)_n$, $(L-Ala-L-Lys-L-Pro)_n$ (119,126); (L-
$Lys-Gly)_n$ (127), $(L-Lys-L-Ala)_n$ (128), $(L-Lys-L-Ala-Gly)_n$ (125),
and $(L-Lys-L-Ala-L-Pro)_n$ (125).

The inclusion of hydrophobic residues into the polypeptide
structure affects the manner in which the basic groups can change
the CD spectrum of DNA. Long runs or a large fraction of leucine
(hydrophobic groups) with lysine cause aggregation and a positive
contribution at 280 nm. Large runs of lysine with neutral amino
acids (Gly, Pro) cause Ψ-type DNA spectra. Lysine and valine random
copolymers cause little DNA CD change. Thus the nature of the hydrophobic side chain and its sequential order is important (124).

B. Histone:DNA Complexes

Histone H4, molecular weight ≈11,000 has a high concentration
of basic amino acids near the N-terminal end of the molecule (129).
Upon binding H4 to DNA, by salt gradient dialysis, it is possible
to obtain two types of CD spectra (53). It is necessary to add
urea to the dialysis solutions to prevent H4 aggregation, and
these two spectra depend upon the salt concentration at which urea
is removed in the dialysis procedure. When the urea is removed at
a salt concentration of 0.14M NaCl, then altered CD spectra are
obtained. As the ratio of histone amino acid residues to DNA phosphates (r) is increased from 0 to 0.5, the CD spectra are essentially
unchanged; from r = 1.0 -1.5, a gradual blue shift and an increase
in amplitude of the 275 nm band is observed and a maximum value of
$[\theta]_{270}$ = 13,800 is obtained at r = 1.5. The 245 nm negative ellipticity band is concurrently decreased in amplitude and blue
shifted. A new negative band centered at 305 nm is generated.
When r becomes larger than 1.5, the magnitude of the CD change is
decreased, and at r = 2.5 the CD spectrum is close to that of B
form DNA. The spectra observed for H4:DNA complexes show no concentration dependence. The CD spectrum obtained at r = 1.5 (53)
looks similar to the CD spectra of RNA, the theoretically calculated
CD spectrum for A-form DNA (9), and the measured CD spectrum for
A-form DNA (13).

H4:DNA complexes give different CD spectra if urea is removed
at 0.015M NaCl (53). No changes in the CD spectra of these complexes relative to B-DNA are observed. Thus, the exact environmental media are important in formation of specific complexes.
Spectra of this latter type, where there is no DNA CD change, can
be utilized to evaluate the conformation of the bound histone. By
subtracting out the CD contribution of native DNA, the CD spectrum
of bound histone may be obtained. It was thus found that, upon
interaction with DNA, H4 assumes an ordered conformation with considerable β structure. The sensitivity of H4 conformation to ionic
strength has already been discussed (Section III, B). Perhaps this
change of conformation with ionic strength is the reason for the
differences in the CD spectra when binding H4 to DNA under different
ionic conditions. These experiments illustrate the interdependence
of the conformations of both the DNA and protein.

Complexes of the other core histones (H2A, H2B, H3) with DNA
have been studied in a similar fashion. These studies will be
described briefly. H2B, a slightly lysine-rich histone containing
125 amino acid residues, has a molecular weight of 13,800 and has
a lysine:arginine ratio of 2.5 (130). In 0.14M NaF, this histone
shows a CD spectrum indicative of about 30% α-helix (131). Upon
H2B complexation with DNA, at different histone:DNA ratios, the
CD spectra obtained (131) are similar to those observed for H4:DNA
complexes. At a ratio of r = 3.0, the H2B:DNA complex shows a
positive CD band in the 280 nm region with an ellipticity greater
than 40,000.

Histone H2A, a slightly lysine-rich protein, has 129 amino
acids, with the more basic region residing near the N-terminal end
(132,133). Complexes reconstituted from DNA and H2A show similar
CD spectra to those for H2B:DNA and H4:DNA complexes (134). Thus,
the positive CD band of the DNA is increased in magnitude and
slightly blue-shifted.

The arginine-rich histone H3 contains 135 residues, and has a
typical histone amino acid distribution with many basic residues
clustered in the N-terminal part of the protein (135). When H3 is
complexed with DNA, only slight alterations in the DNA-CD spectrum
are observed (134).

The perturbation of DNA structure caused by the core histones
is apparently due primarily to interaction between DNA and the more
basic sections of the histone structure (131). This was demonstrated
in a study of H2B cleaved into two fragments of nearly equal size
but different charge densities. The more basic fragment caused a
conformational change in DNA similar to that described for intact
H2B (but lower in magnitude), whereas the less basic fragment did
not significantly change the DNA CD spectrum (131).

Further evidence that electrostatic interactions play a primary role in the histone:DNA complexation can be obtained from studies of modified histones. Histone H4 is enzymatically acetylated (136) in vivo to form ε-N-acetyl lysine residues at specific sites. The CD spectra for complexes of DNA with H4 at various specific levels of acetylation (137) show that modification of even one lysine residue in the basic region of the histone moderates the histone effect on DNA conformation. As it has been shown that H4 is acetylated at the time of extensive gene activation in sea urchin embryos (138), a possible mechanism is suggested by this CD study.

CD studies of complexes formed between DNA and any of the core histones show that each of these histones can perturb the conformation of DNA to some extent. This effect is dependent on strong electrostatic interactions. Individual histones, however, do not perturb the DNA structure in the same manner as the histone-histone complex in nucleosomes. This is evident from the CD spectra, which show a reduction in ellipticity and a red shift of the band for DNA in chromatin relative to free DNA, but an increase in ellipticity and a blue shift of the band position for DNA complexed by individual core histones. These model studies illustrate the importance of histone-histone interactions in chromatin.

The histones H1 and H5 (present in avian erythrocytes) are thought to interact primarily with the spacer regions of DNA between nucleosomes (69,98,102). These histones are very lysine-rich, and higher in molecular weight than the core histones.

Histone H1 has a molecular weight of 21,000. It contains 214 amino acids, 61 of which are lysine, 3 are arginine, and 16 are acidic amino acids. The basic groups are not evenly distributed, and 39 are found in the carboxyl end of the molecule (139). Upon complexing H1 with DNA, by gradient dialysis, an altered CD spectrum is obtained, as seen in Figure 15 (52). As increasing amounts of histone are added, the peak at 275 nm decreases and red shifts, being completely eliminated at a ratio (r, amino acid residues per DNA phosphate) of 1.0. The extent of the CD change for any particular H1:DNA ratio is concentration dependent, being greater at higher concentrations. Thus aggregation between complexes plays a role in the CD changes.

Histone H5, which largely replaces H1 in avian erythrocyte chromatin, has a molecular weight of 21,450. There are 197 residues, of which 49 are lysine, and 22-23 are arginine (140). H5 exerts a conformational effect on DNA similar to that of H1 (141), in that the positive CD band of DNA is reduced and shifted to longer wavelengths.

H1 and H5 are thus seen to perturb the DNA structure in a

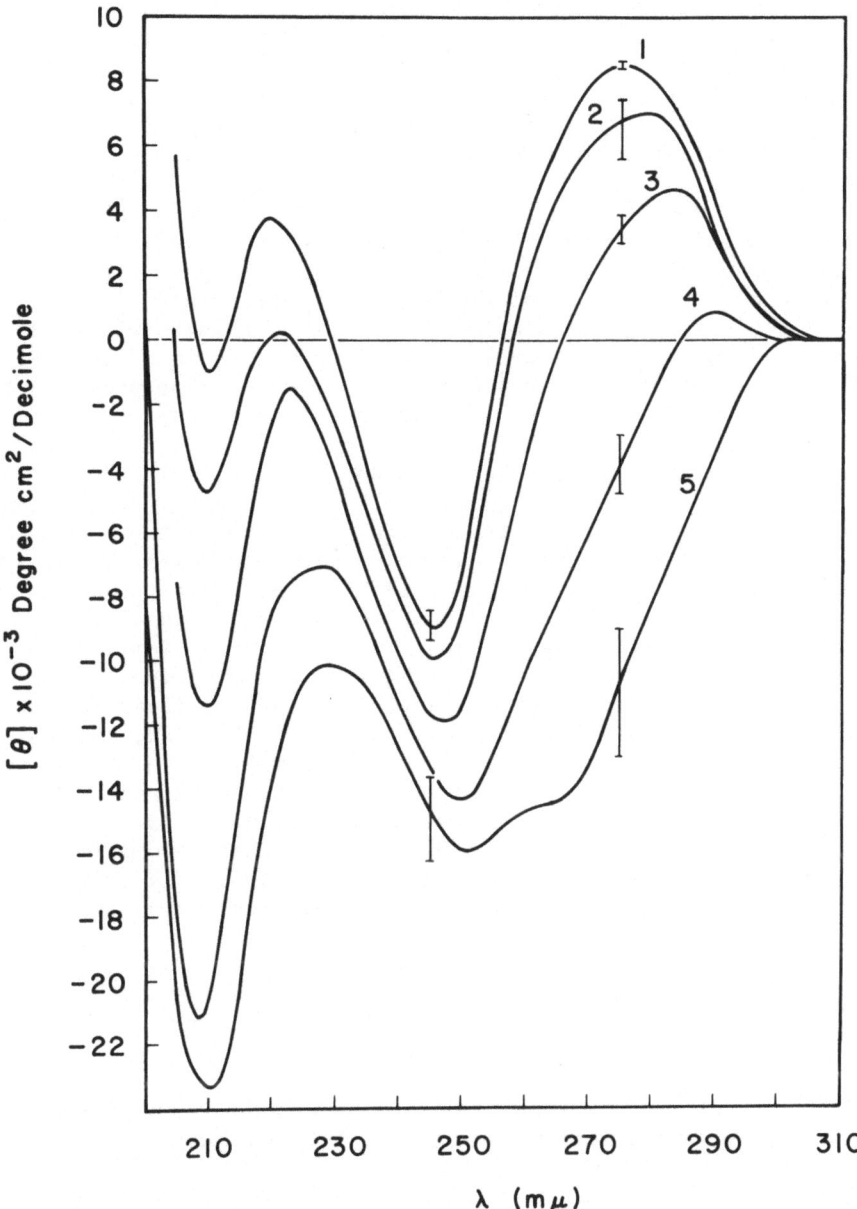

Figure 14. Circular dichroism spectra of H1:DNA complexes as a function of r, the histone residue : PO_4 ratio, DNA concentration = 10^{-3}M (PO_4); solvent 0.14 M NaF, pH 7.0. 1, DNA (calf thymus); 2, r = 0.25; 3, r = 0.50; 4, r = 0.75; 5, r = 1.0. Error bars represent reproducibility and noise dependence. Uncorrected for histone contribution. Fasman, G.D. et al. (52). Reprinted with permission of the American Chemical Society.

manner similar to that found in chromatin, or in Ψ-DNA. This is
of great interest, since these histones are postulated to condense
the spacer regions of DNA (69,70) between nucleosomes. The individual core histones, by contrast, do not perturb DNA in this same
manner, and must act as a specific histone-histone complex to
condense DNA into nucleosomes.

Enzymatic modification of H1 (142) producing mono and di phosphorylated species, occurs in response to the pancreatic hormone
glucagon, upon initiation of protein synthesis. Thus it is postulated that phosphorylation of H1 results in derepression of the
template activity of the associated DNA. When phosphorylated H1
is complexed to DNA, the change in the CD spectrum found for H1:DNA
complexes is completely reversed (143). Perhaps these phosphoryla-
tions cause a modification of the histone:histone and/or histone:
DNA interaction, reducing the state of aggregation in chromatin,
thus making DNA more accessible for transcription.

DNA may also be reconstituted with a mixture of the four core
histones. The extent to which such a complex resembles native
chromatin (depleted of H1) provides information about the mechanism
of chromatin assembly. That is, can the core histones alone organize
DNA into an ordered condensed state? Circular dichroism is an excellent sensitive probe for evaluating such condensation. By a
salt step gradient dialysis procedure, Garel et al. (144) showed
that an equimolar mixture of the four histones H2A, H2B, H3, and
H4 could complex and condense DNA into a chromatin-like structure.
Both the CD spectra and CD melting profiles of the reconstituted
material were identical to H1-depleted chromatin. This report confirmed the observations of Stein et al. (145) that chromatin reconstituted by a gradient dialysis procedure exhibits a CD spectrum
similar to that for native chromatin. An earlier review of this
subject matter should be consulted for further references (146).

VII.   SUMMARY

The utilization of circular dichroism can be of great assistance
in defining the conformational states of both DNA and the proteins
found in chromatin.

The conformation of the histones can be evaluated with a high
degree of confidence, especially when isolated from the nucleoprotein complex. A reasonable estimate of their conformation can be
made in chromatin using the CD bands below 240 nm, as the DNA contribution in this region is minimal, compared to that of the protein
contribution. The CD spectrum has been interpreted to indicate
that the conformation of the histones in chromatin are composed of
40-50% α-helix, 0-5% β-sheet and 51-60% random coil. A mixture of

all the histones exhibits similar CD spectra in 2M NaCl as found
in chromatin, which suggests that the conformation of the histones
in chromatin is dependent in part on the electrostatic interaction
with DNA.

The conformation of DNA may be evaluated by examination of
the CD bands above 240 nm, where the histones have a negligible
contribution. Upon binding of the histones (and other proteins),
the CD spectrum of DNA becomes significantly altered, with the
largest change centered around 275 nm. This observation has been
interpreted as either 1) a change in the secondary structure of
DNA (e.g., B→C form) or 2) an ordering of DNA into a tertiary
structure. Concominant changes in other physical parameters are
known to be caused by the condensation of DNA, without necessarily
altering the secondary structure. Therefore, the formation of a
tertiary structure would be strongly implicated. Thus CD can be
used both to evaluate the degree of condensation, ie.., tertiary
structure, as well as the secondary structure. CD spectra temperature profiles can be used to study the stabilization of both
the secondary and tertiary structure of chromatin.

Great caution must be exercised in interpreting the CD spectra
of chromatin. Factors which must be evaluated are the composition
of the complex (proteins, DNA, RNA) and optical artifacts such as
scattering and absorption phenomena due to particle size.

As a probe for conformational changes in the structural organization of chromatin, circular dichroism offers a tool of great
sensitivity and ease of use.

ACKNOWLEDGMENTS

The writing of this review was generously supported in part
by grants from the U. S. Public Health GM 17533 and the National
Science Foundation (PCM76-21856). This is Publication No. 1190
from the Graduate Department of Biochemistry, Brandeis University,
Waltham, Massachusetts 02154. The author wishes to thank Dr. M.
Cowman for her major contribution to the writing of this review.

REFERENCES

1.   Kornberg, R.D. (1977). Ann. Rev. Biochem. 46, 931-954.
2.   Felsenfeld, G. (1978). Nature 271, 115-122.
3.   Tinoco, I., Jr., and Cantor, C.R. (1970). In "Methods of Bio-
     chemical Analysis" (D. Glick, ed.), Vol. 18, pp. 81-203. Wiley,
     New York.

4. Brahms, J., and Brahms, S. (1970). In "Fine Structure of Pro-
    teins and Nucleic Acids" (G.D. Fasman and S.N. Timasheff, eds.),
    pp. 191-270. Marcel Dekker, Inc., New York.
5. Sears, D.W., and Beychok, S. (1973). In "Physical Principles
    and Techniques of Protein Chemistry" (S.J. Leach, ed.), Part
    C, pp. 445-593. Academic Press, New York.
6. Adler, A.J., Greenfield, N.J., and Fasman, G.D. (1973). Meth.
    Enzymology 27, Part D, 675-735.
7. Tinoco, I., Jr. (1964). J. Am. Chem. Soc. 86, 297-298.
8. Tinoco, I., Jr. (1968). J. Chim. Phys. 65, 91-97.
9. Johnson, W.C., Jr., and Tinoco, I., Jr. (1969). Biopolymers 7,
    727-749.
10. Langridge, R., Marvin, D.A., Seeds, W.E., Wilson, H.R., Hooper,
    C.W., Wilkins, M.H.F., and Hamilton, L.D. (1960). J. Mol. Biol.
    2, 38-64.
11. Marvin, D.A., Spencer, M., Wilkins, M.H.F., and Hamilton, L.D.
    (1961). J. Mol. Biol. 3, 547-565.
12. Fuller, W., Wilkins, M.H.F., Wilson, H.R., Hamilton, L.D., and
    Arnott, S. (1965). J. Mol. Biol. 12, 60-80.
13. Tunis-Schneider, M.J.B., and Maestre, M.F. (1970). J. Mol.
    Biol. 52, 521-541.
14. Brahms, J., and Mommaerts, W.F.H.M. (1964). J. Mol. Biol. 10,
    73-88.
15. Studdert, D.S., and Davis, R.C. (1974). Biopolymers 13, 1377-
    1389.
16. Bram, S. (1971). J. Mol. Biol. 58, 277-288.
17. Studdert, D.S., Patroni, M., and Davis, R.C. (1972). Bio-
    polymers 11, 761-779.
18. Hanlon, S., Johnson, R.S., Wolf, B., and Chan, A. (1972).
    Proc. Nat. Acad. Sci. U.S.A. 69, 3263-3267.
19. Brunner, W.C., and Maestre, M.F. (1974). Biopolymers 13,
    345-357.
20. Lerman, L.S. (1971). Proc. Nat. Acad. Sci. U.S.A. 68, 1886-1890.
21. Jordan, C.F., Lerman, L.S., and Venable, J.H., Jr. (1972).
    Nature New Biol. 236, 67-70.
22. Maniatis, T., Venable, J.H., Jr., and Lerman, L.S. (1974). J.
    Mol. Biol. 84, 37-64.
23. Saeva, F.D., and Wysocki, J.J. (1971). J. Am. Chem. Soc. 93,
    5928-5929.
24. Saeva, F.D. (1972). J. Am. Chem. Soc. 94, 5135-5136.
25. Chabay, I. (1972). Chem. Phys. Lett. 17, 283-287.
26. Holzwarth, G., Chabay, I., and Holzwarth, N.A.W. (1973). J.
    Chem. Phys, 58, 4816-4819.
27. Holzwarth, G., and Holzwarth, N.A.W. (1973). J. Opt. Soc. Am.
    63, 324-331.
28. Cheng, S.M., and Mohr, S.C. (1975). Biopolymers 14, 663-674.
29. Permogorov, V.I., Debabov, V.G., Sladkova, I.A., and Rebentish,
    B.A. (1970). Biochim. Biophys. Acta 199, 556-558.
30. Fric, I., and Sponar, J. (1971). Biopolymers 10, 1525-1531.

31. Zimmer, C., and Luck, G. (1973). Biochim. Biophys. Acta 312,
     215-227.
32. Hanlon, S., Brudno, S., Wu, T.T., and Wolf, B. (1975).
     Biochemistry 14, 1648-1660.
33. Wolf, B., Berman, S., and Hanlon, S. (1977). Biochemistry 16,
     3655-3662.
34. Zimmer, C., and Luck, G. (1974). Biochim. Biophys. Acta 361,
     11-32.
35. Ivanov, V.I., Lysov, Yu.P., Malenkov, G.G., Minchenkova, L.E.,
     Minyat, E.E., Schyolkina, A.K., and Zhurkin, V.B. (1976).
     Studia Biophysica 55, 5-13.
36. Bram, S., and Tougard, P. (1972). Nature New Biol. 239, 128-131.
37. Green, G., and Mahler, H.R. (1968). Biopolymers 6, 1509-1514;
38. Nelson, R.G., and Johnson, W.C., Jr. (1970). Biochem. Biophys.
     Res. Comm. 41, 211-216.
39. Girod, J.C., Johnson, W.C., Jr., Huntington, S.K., and Maestre,
     M.F. (1973). Biochemistry 12, 5092-5096.
40. Bronner, M., and Pysh, E.S. (1976). Biopolymers 15, 589-590.
41. Giannoni, G., Padden, F.J., Jr., and Keith, H.D. (1969).
     Proc. Nat. Acad. Sci. U.S.A. 62, 964-971.
42. Marmur, J., and Doty, P. (1962). J. Mol. Biol. 5, 109-118.
43. Gennis, R.B., and Cantor, C.R. (1972). J. Mol. Biol. 65,
     381-399.
44. Urry, D.W. (1972). Biochim. Biophys. Acta 265, 115-168.
45. Schneider, A.S. (1973). Meth. Enzymology 27, Part D, 751-767.
46. Gordon, D.J., and Holzwarth, G. (1971). Proc. Nat. Acad. Sci.
     U.S.A. 68, 2365-2369.
47. Gordon, D.J. (1972). Biochemistry 11, 413-420.
48. Schneider, A.S., and Harmatz, D. (1976). Biochemistry 15,
     4158-4162.
49. Nicolini, C., Baserga, R., and Kendall, F. (1976). Science 192,
     796-798.
50. Bohren, C.F. (1977). J. Theor. Biol. 65, 755-767.
51. Chen, Y.-H., Yang, J.T., and Martinez, H.M. (1972). Biochemistry
     11, 4120-4131.
52. Fasman, G.D., Schaffhausen, B., Goldsmith, L., and Adler, A.
     (1970). Biochemistry 9, 2814-2822.
53. Shih, T.Y., and Fasman, G.D. (1971). Biochemistry 10, 1675-
     1683.
54. Li, H.J., Wickett, R., Craig, A.M., and Isenberg, I. (1972).
     Biopolymers 11, 375-397.
55. Wickett, R.R., Li, H.J., and Isenberg, I. (1972). Biochemistry
     11, 2952-2957.
56. D'Anna, J.A., Jr., and Isenberg, I. (1972). Biochemistry 11,
     4017-4025.
57. D'Anna, J.A., Jr., and Isenberg, I. (1974). Biochemistry 13,
     2093-2097.
58. D'Anna, J.A., Jr., and Isenberg, I. (1974). Biochemistry 13,
     4987-4992.

59. Fasman, G.D., Chou, P.Y., and Adler, A.J. (1977). In "The
     Molecular Biology of the Mammalian Genetic Apparatus" (P.O.P.
     Ts'o, ed.), Vol. 1, pp. 1-52. Elsevier/North-Holland Bio-
     medical Press, Amsterdam.
60. D'Anna, J.A., Jr., and Isenberg, I. (1973). Biochemistry 12,
     1035-1043.
61. D'Anna, J.A., Jr., and Isenberg, I. (1974). Biochemistry 13,
     2098-2104.
62. D'Anna, J.A., Jr., and Isenberg, I. (1974). Biochemistry 13,
     4992-4997.
63. D'Anna, J.A., Jr., and Isenberg, I. (1974). Biochem. Biophys.
     Res. Comm. 61, 343-347.
64. Bidney, D.L., and Reeck, G.R. (1977). Biochemistry 16, 1844-
     1849.
65. Bradbury, E.M., Cary, P.D., Crane-Robinson, C., Rattle, H.W.E.,
     Boublik, M., and Sautiere, P. (1975). Biochemistry 14, 1876-
     1885.
66. Thomas, G.J., Jr., Prescott, B., and Olins, D.E. (1977).
     Science 197, 385-388.
67. Weintraub, H., Palter, K., and Van Lente, F. (1975). Cell 6,
     85-110.
68. Pardon, J.F., Cotter, R.I., Lilley, D.M.J., Worcester, D.L.,
     Campbell, A.M., Wooley, J.C., and Richards, B.M. (1977).
     Cold Spring Harbor Symposium, in press.
69. Noll, M., and Kornberg, R.D. (1977). J. Mol. Biol. 109, 393-404.
70. Renz, M., Nehls, P., and Hozier, J. (1977). Proc. Nat. Acad.
     Sci. U.S.A. 74, 1879-1883.
71. Shih, T.Y., and Fasman, G.D. (1970). J. Mol. Biol. 52, 125-129.
72. Simpson, R.T., and Sober, H.A. (1970). Biochemistry 9, 3103-
     3109.
73. Henson, P., and Walker, I.O. (1970). Eur. J. Biochem. 16, 524-
     531.
74. Matsuyama, A., Tagashira, Y., and Nagata, C. (1971). Biochim.
     Biophys. Acta 240, 184-190.
75. Tashiro, T., and Kurokawa, M. (1975). FEBS Lett. 59, 250-253.
76. Hjelm, R.P., Jr., and Huang, R.C.C. (1975). Biochemistry 14,
     1682-1688.
77. Nicolini, C., and Baserga, R. (1975). Arch. Biochem. Biophys.
     169, 678-685.
78. Huang, C.-H., and Baserga, R. (1976). Biochemistry 15, 2829-
     2836.
79. Baserga, R., and Nicolini, C. (1976). Biochim. Biophys. Acta
     458, 109-134.
80. Johnson, R.S., Chan, A., and Hanlon, S. (1972). Biochemistry
     11, 4347-4358.
81. Shih, T.Y., and Lake, R.S. (1972). Biochemistry 11, 4811-4817.
82. Wilhelm, F.X., DeMurcia, G.M., Champagne, M.H., and Daune,
     M.P. (1974). Eur. J. Biochem. 45, 431-443.
83. Mandel, R., and Fasman, G.D. (1976). Nucl. Acids Res. 3, 1839-
     1855.

84. Rees, A.W., Dubuysere, M.S., and Lewis, E.A. (1974). Biochim. Biophys. Acta 361, 97-108.
85. Brasch, K. (1976). Exp. Cell Res. 101, 396-410.
86. Finch, J.T., and Klug, A. (1976). Proc. Nat. Acad. Sci. U.S.A. 73, 1897-1901.
87. Bartley, J., and Chalkley, R. (1973). Biochemistry 12, 468-474.
88. Maestre, M.F., and Wang, J.C. (1971). Biopolymers 10, 1021-1030.
89. Baase, W.A., and Johnson, W.C., Jr. (1976). Nucl. Acids Res. 3, 3123-3131.
90. Sipski, M.L., and Wagner, T.E. (1977). Biopolymers 16, 573-582.
91. Simpson, R.T. (1972). Biochemistry 11, 2003-2008,
92. Li, H.J., Chang, C., Evagelinou, Z., and Weiskopf, M. (1975). Biopolymers 14, 211-226.
93. Chang, C., and Li, H.J. (1974). Nucl. Acids Res. 1, 945-958.
94. Senior, M.B., and Olins, D.E. (1975). Biochemistry 14, 3332-3337.
95. Simpson, R.T. (1971). Biochemistry 10, 4466-4470.
96. Tack, L.O., and Simpson, R.T. (1977). Biochemistry 16, 3746-3753.
97. Ohlenbusch, H.H., Olivera, B.M., Tuan, D., and Davidson, N. (1967). J. Mol. Biol. 25, 299-315.
98. Whitlock, J.P., Jr., and Simpson, R.T. (1976). Biochemistry 15, 3307-3314.
99. Williams, R.E., Lurquin, P.F., and Seligy, V.L. (1972). Eur. J. Biochem. 29, 426-432.
100. Vengerov, Yu. Yu., and Popenko, V.I. (1977). Nucl. Acids Res. 4, 3017-3027.
101. Hjelm, R.P., Jr., and Huang, R.C.C. (1974). Biochemistry 13, 5275-5283.
102. Varshavsky, A.J., Bakayev, V.V., and Georgiev, G.P. (1976). Nucl. Acids Res. 3, 477-492.
103. Whitlock, J.P., Jr., and Simpson, R.T. (1976). Nucl. Acids Res. 3, 2255-2266.
104. Finch, J.T., Lutter, L.C., Rhodes, D., Brown, R.S., Rushton, B., Levitt, M., and Klug, A. (1977). Nature 269, 29-36.
105. Sahasrabuddhe, C.G., and Van Holde, K.E. (1974). J. Biol. Chem. 249, 152-156.
106. Ramsay Shaw, B., Corden, J.L., Sahasrabuddhe, C.G., and Van Holde, K.E. (1974). Biochem. Biophys. Res. Comm. 61, 1193-1198.
107. Lawrence, J.-J., Chan, D.C.F., and Piette, L.H. (1976). Nucl. Acids Res. 3, 2879-2893.
108. Sahasrabuddhe, C.G., and Saunders, G.F. (1977). Nucl. Acids Res. 4, 853-866.
109. Olins, A.L., Carlson, R.D., Wright, E.B., and Olins, D.E. (1976). Nucl. Acids Res. 3, 3271-3291.

110. Olins, A.L., Breillatt, J.P., Carlson, R.D., Senior, M.B., Wright, E.B., and Olins, D.E. (1977). In "The Molecular Biology of the Mammalian Genetic Apparatus" (P.O.P. Ts'o, ed.), Vol. 1, pp. 211-237. Elsevier/North Holland Biomedical Press, Amsterdam.

111. Rill, R., and Van Holde, K.E. (1973). J. Biol. Chem. 248, 1080-1083.

112. Olins, D.E., Bryan, P.N., Harrington, R.E., Hill, W.E., and Olins, A.L. (1977). Nucl. Acids Res. 4, 1911-1931.

113. Lilley, D.M.J., and Tatchell, K. (1977). Nucl. Acids Res. 4, 2039-2055.

114. Stein, A., Bina-Stein, M., and Simpson, R.T. (1977). Proc. Nat. Acad. Sci. U.S.A. 74, 2780-2784.

115. Cohen, P., and Kidson, C. (1968). J. Mol. Biol. 35, 241-245.

116. Shapiro, J.T., Leng, M., and Felsenfeld, G. (1969). Biochemistry 8, 3219-3232.

117. Haynes, M., Garrett, R.A., and Gratzer, W.B. (1970). Biochemistry 9, 4410-4416.

118. Stokrova, S., Sponar, J., Havranek, M., Sedlacek, B., and Blaha, K. (1975). Biopolymers 14, 1231-1244.

119. Sponar, J., Blaha, K., and Stokrova, S. (1973). Stud. Biophys. 40, 125-133.

120. Pinkston, M.F., and Li, H.J. (1974). Biochemistry 13, 5227-5234.

121. Ong, E.C., Snell, C. and Fasman, G.D. (1976). Biochemistry 15, 468-477.

122. Ong, E.C., and Fasman, G.D. (1976). Biochemistry 15, 477-486.

123. Mandel, R., and Fasman, G.D. (1974). Biochem. Biophys. Res. Comm. 59, 672-679.

124. Mandel, R., and Fasman, G.D. (1976). Biochemistry 15, 3122-3130.

125. Schwartz, A.M., and Fasman, G.D. (1977). Biochemistry 16, 2287-2299.

126. Sponar, J., Fric, I., and Blaha, K.B. (1975). Biophys. Chem. 3, 255-262.

127. Williams, R.F., and Kielland, S.L. (1975). Can. J. Chem. 53, 542-548.

128. Privat, J.-P., Spach, G., and Leng, M. (1972). Eur. J. Biochem. 26, 90-95.

129. DeLange, R.J., Smith, E.L., Fambrough, D.M., and Bonner, J. (1968). Proc. Nat. Acad. Sci. U.S.A. 61, 1145-1146.

130. Iwai, K., Ishikawa, K., and Hayashi, H. (1970). Nature 226, 1056-1058.

131. Adler, A.J., Ross, D.G., Chen, K., Stafford, P.A., Woiszwillo, M.J., and Fasman, G.D. (1974). Biochemistry 13, 616-623.

132. Yeoman, L.C., Olson, M.O.J., Sugano, N., Jordan, J.J., Taylor, C.W., Starbuck, W.C., and Busch, H. (1972). J. Biol. Chem. 247, 6018-6023.

133. Sautiere, P., Tyrou, D., Laine, B., Mizon, J., Ruffin, P., and Biserte, G. (1974). Eur. J. Biochem. 41, 563-576.

134. Adler, A.J., Moran, E.C., and Fasman, G.D. (1975). Biochemistry 14, 4179-4185.
135. DeLange, R.J., Hooper, J.A., and Smith, E.L. (1972). Proc. Nat. Acad. Sci. U.S.A. 69, 882-884.
136. Pogo, B.G.T., Pogo, A.O., Allfrey, V.G., and Mirsky, A.E. (1968). Proc. Nat. Acad. Sci. U.S.A. 59, 1337-1344.
137. Adler, A.J., Fasman, G.D., Wangh, L.J., and Allfrey, V.G. (1974). J. Biol. Chem. 249, 2911-2914.
138. Wangh, L., Ruiz-Carrillo, A., and Allfrey, V.G. (1972). Arch. Biochem. Biophys. 150, 44-56.
139. Rall, S.C., and Cole, R.D. (1971). J. Biol. Chem. 246, 7175-7190.
140. Sautiere, P., Briand, G., Kmiecik, D., Loy, O., Biserte, G. (1976). FEBS Lett. 63, 164-166.
141. Wagner, T.E., Hartford, J.B., Serra, M., Vandegrift, V., and Sung, M.T. (1977). Biochemistry 16, 289-290.
142. Langan, T.A. (1969). J. Biol. Chem. 244, 5763-5765.
143. Adler, A.J., Langan, T.A., and Fasman, G.D. (1972). Arch. Biochem. Biophys. 153, 769-777.
144. Garel, A., Kovacs, A.M., Champagne, M., and Daune, M. (1976). Nucl. Acids Res. 3, 2507-2519.
145. Stein, G.S., Mans, R.J., Gabbay, E.J., Stein, J.L., Davis, J., and Adawadkar, P.D. (1975). Biochemistry 14, 1859-1866.
146. Fasman, G.D. (1977). in "Chromatin and Chromosome Structure" (Eds. Li, H.J. and Eckhardt, R.A.). Academic Press, pp. 71-142.
147. Ivanov, V.I., Minchenkova, L.E., Schyolkina, A.K., and Poletayev, A.I. (1973). Biopolymers 12, 89-110.
148. Holzwarth, G., Gordon, D.G., McGinness, J.E., Dorman, B.P., and Maestre, M.F. (1974). Biochemistry 13, 126-132.
149. Dorman, B.P., and Maestre, M.F. (1973). Proc. Nat. Acad. Sci. U.S.A. 70, 255-259.
150. Greenfield, N., and Fasman, G.D. (1969). Biochemistry 8, 4108-4116.
151. Slayter, H.S., Shih, T.Y., Adler, A.J., and Fasman, G.D. (1972). Biochemistry 11, 3044-3054.

IMPORTANT HYDRODYNAMIC AND SPECTROSCOPIC TECHNIQUES IN THE FIELD OF CHROMATIN STRUCTURE

Donald E. Olins

University of Tennessee-Oak Ridge Graduate School of
Biomedical Sciences, Biology Division, Oak Ridge
National Laboratory, Oak Ridge, Tennessee 37830

INTRODUCTION

If chromatin were an enzyme or, at least if it bound reversibly and tightly to a readily measurable ligand, we could directly and meaningfully assay for chromatin functional states. We could then search for conformational effector molecules, cooperative structural transitions, and allosteric interactions. There are reasons to anticipate such conformational properties in chromatin. In all probability, the nucleosome has several properties reminiscent of hemoglobin, the prototype of allosteric multisubunit proteins. The nucleosome probably possesses a pairwise stoichiometry of the constituent inner histones, an internal dyad axis with the heterotypic tetramers acting as protamers, and a close-packing of highly α-helical, globular polypeptide chains[1,2]. Unfortunately, simple functional assays are not yet the province of students of chromatin structure. Save for complex transcriptional assays, conformational studies of chromatin and nucleosomes are largely confined to enzymatic and biophysical probes.

The use of enzymatic probes has underscored the necessity of identifying chromatin conformational states. The employment of DNAse I, or of trypsin followed by micrococcal nuclease, has correlated altered chromatin conformational states with transcriptional activation[1,2]; but these assays are destructive and clearly not reversible. Many biophysical techniques, on the other hand, are nondestructive and conformational dynamics can be readily studied; but the properties of molecules are necessarily averaged, and these techniques are not easily focused upon a subset of nucleosomes with conformations different than those of the bulk particles. Most biophysical studies to date are confined to "inactive" nucleosomes or chromatin; amplification of the material, or the signal, is essential for comparable studies on "active" states.

The purpose of this chapter is to illustrate the usefulness of two broad classes of biophysical techniques - hydrodynamic and spectroscopic - emphasizing those currently in use in association with this author's laboratory. Results of studies on the response of mononucleosomes (ν_1) to various solvent parameters and perturbants will document the contribution of the different techniques. Despite the organization of this chapter around particular techniques, it is important to recognize that different techniques must be used in concert in order to develop a comprehensive view of nucleosome conformational states.

Hydrodynamic Techniques

Conventional hydrodynamic techniques (transport processes) have the capability of yielding measurements of molecular weight, size and shape, under varying solvent conditions[3,4]. Analytical ultracentrifugation, combined with determinations of partial specific volume, have permitted accurate estimation of the molecular weight of ν_1 - by equilibrium sedimentation, and by use of the Svedberg equation (Table I).

Solvent-induced conformational changes that do not involve changes in molecular weight are readily detected by monitoring sedimentation (S), diffusion (D) or intrinsic viscosity [η]. Measurements of S and molecular weight, or of D alone, permit estimation of the frictional coefficient (f) which is a function of the shape (asymmetry) and volume (hydration) of the macromolecules. The intrinsic viscosity of a molecule is also a function of particle volume and asymmetry. Assumptions of particle volume (hydration) are necessary to extract estimates of particle shape. It is important to remember that changes in the S, D or [η] of ν_1 with varying solvent parameters can equally well be interpreted as arising from changes in particle shape or volume, or a combination of both. The Scheraga-Mandelkern factor (β) can be obtained by combining measurements of S and [η], or of D and [η]. β is a function only of particle shape (i.e., the axial ratio of an equivalent ellipsoid of revolution). But, unfortunately, β does not effectively distinguish between oblate, spherical or slightly prolate objects. Low-angle neutron scattering combined with the technique of "contrast-variation" (i.e., employing buffers of varying H_2O/D_2O) undoubtedly yields the most complete description of particle size and shape in solution. Few of us, however, have ready access to a high flux neutron source. Neutron scattering experiments are, therefore, probably best employed after the conformational states are defined by conventional hydrodynamic techniques.

Examples will be presented on the influence of simple solvent parameters on the size and shape of ν_1. The measurement of S, employing the analytical ultracentrifuge equipped with scanner optics, is well-described in textbooks[3,4], and needs no further explanation. Determination of D in the ultracentrifuge is based upon boundary-spreading analysis of the solvent/solution interphase generated in a synthetic boundary centerpiece, a technique of moderate accuracy and lengthy computation. The advent of laser light scattering techniques has permitted a rapid and accurate determination of D by analysis of the time-decay of the intensity of the "twinkles" of light scattered from groups of molecules, or the frequency-broadening of scattered light due to Doppler shifts.[5]

Urea Effects. Exposure of ν_1 to urea (0-10 M) in buffers containing 0.2 mM EDTA, pH 7.0, results in size and shape changes without detectable dissociation of histones and DNA[6]. Figure 1 presents the results of S and [η] measurements as a function of urea. Combining S, [η] and partial specific volume ($\bar{v}$) following the theory of Scheraga and Mandelkern, permitted an estimation of the dependence of β upon urea concentration. β was calculated to increase monotonically between 0-8 M urea, indicating a transition of ν_1 toward a more prolate-type structure, from its slightly oblate shape in the absence of urea.

TABLE 1. Hydrodynamic parameters of ν_1 in 0.1 M KCl

$S^\circ_{20,w}$ =	11.11 ± 0.30
$D^\circ_{20,w}$ =	3.90 ± 0.13
$\bar{v}$ =	0.670 ± 0.005
$\bar{M}_W$ (S & D) =	$209{,}547 \pm 9{,}634$
$\bar{M}_W$ (Equil Sed) =	$221{,}515 \pm 2{,}143$
[η] =	5.628 ± 0.003
β =	$2.119 \pm 0.058 \times 10^6$

<u>Ionic Strength Effects</u>. Isolated chicken erythrocyte ν_1 has
been observed to undergo two sharp conformational transitions
between 1.0 and 100 mM ionic strengths[7]. Molecular weight determinations over the same range of ionic strength indicate that
dissociation is not taking place. Figure 2 contains a summary of
the hydrodynamic data (i.e., $S_{20,w}$ and $D_{20,w}$) over the ionic
strength range. Transitions are observed at approximately 1 and
10 mM, and are reversible.

In the absence of additional data it is not possible to ascribe
these changes in S, D and f to changes in particle shape and/or
particle volume (and hydration). As is usually done, one can make
certain reasonable assumptions and examine the consequences. We

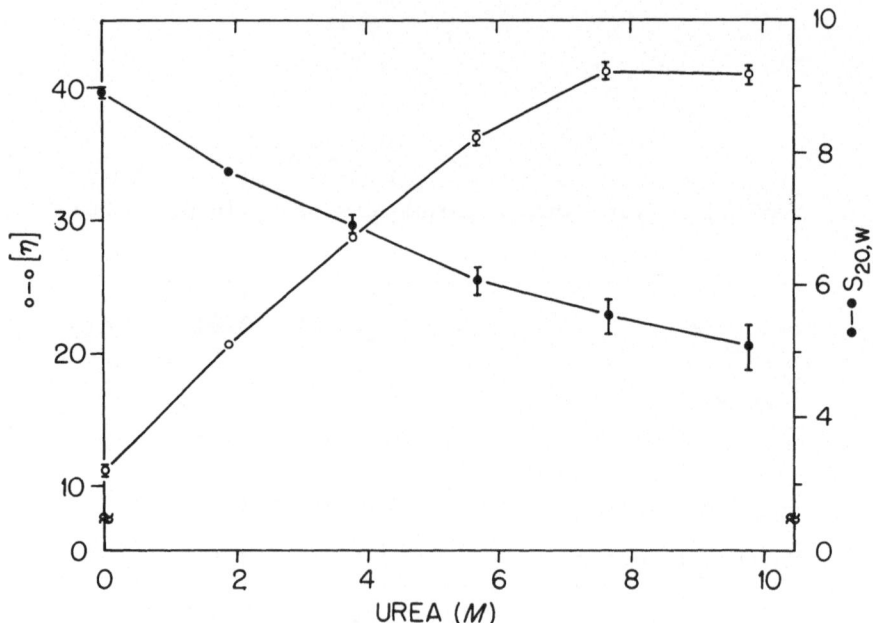

Figure 1. Hydrodynamic properties of ν_1 as a function of urea molarity. [η] units are cm^3/g; sedimentation coefficients corrected
to $S_{20,w}$.

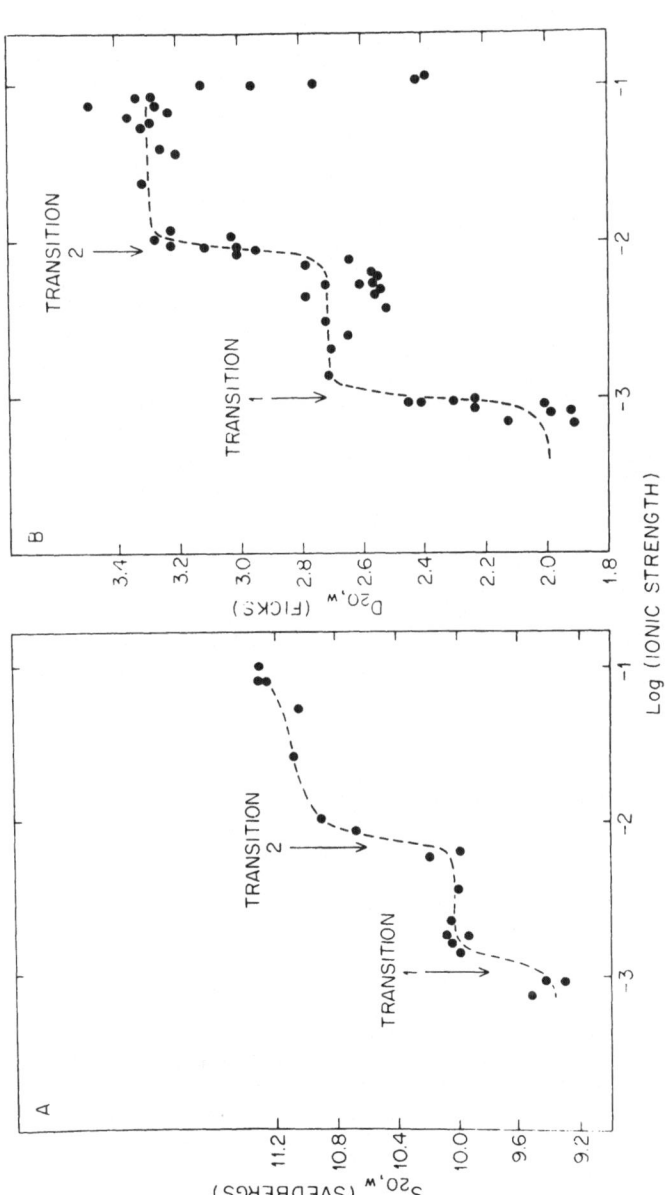

Figure 2. A. $S_{20,w}$ plotted as a function of the logarithm of the ionic strength. Initially the monomer particles were prepared in 0.2 mM EDTA (pH = 7.0), at an ionic strength of 1.0 mM. Then the ionic strength was increased by additions of 0.25 M KCl in 0.2 mM EDTA. The temperature was regulated close to 20°C, and the speed was 48,000 rpm. B. $D_{20,w}$ plotted as a function of the logarithm of the ionic strength. The monomer particles were prepared in 0.2 mM EDTA (pH = 7.0) at an ionic strength 1.0 mM. Then the ionic strength was increased by additions of o.25 M KCl in 0.2 mM EDTA. The autocorrelation function was measured and fit to a single exponential plus baseline to obtain D.

TABLE 2.   Estimated Shape Changes of ν_1 at Different Ionic
           Strengths

$\underline{mM}$ Ionic Strength	$\times 10^{13}$ $S_{20,w}$	$\times 10^{8}$ $f_{20,w}$[†]	f/f_0[‡]	Oblate a/b	Prolate a/b
≤ 1.0	9.40	12.6	1.22	4.9	4.5
$1.0 - 10.0$	10.05	11.8	1.15	3.8	3.5
$10.0 - 100.0$	11.05	10.7	1.04	2.0	1.9

[†]Calculated from: $f = \dfrac{M (1 - \bar{v}\rho)}{NS}$

Assuming: $M = 2.1 \times 10^{5}$ g/mole

$\bar{v} = 0.661$ ml/g

$\rho = 0.998$ g/ml

$N = 6.02 \times 10^{23}$ particles/mole

[‡]Calculated from: $f_0 = 6\,\pi\eta_0 \left(\dfrac{3\,v}{4\,\pi}\right)^{1/3} = 6\,\pi\eta_0 r_0$

Where: $v = \dfrac{M}{N}(\bar{v} + \delta_1 v^0)$

Assuming: $\delta_1 = 1.3$ g H_2O/g ν_1

$v^0 = 1.00$ ml/g

$\eta_0 = 10^{-2}$ g/sec-cm

Yields: $r_0 = 54.6$ Å (radius of an
                      equivalent hydrated
                      sphere)

$f_0 = 10.29 \times 10^{-8}$ g/sec

will assume that ν_1 is hydrated to an extent similar to ribosomes
(i.e., $\delta_1 = 1.3$ g H_2O/g ν_1) and that particle hydration does not
significantly change over the ionic strength range[8]. Table 2 is
a compliation of computations, interpreting the ionic-strength
dependence of sedimentation.

An oblate ellipsoid with axial ratio 2/1 would be in reasonable
agreement with current data on the shape of ν_1 by low-angle neutron
scattering and x-ray crystallography[6]. The particle could become
more oblate or more prolate with decreased ionic strength. Collabo-
rative studies are in progress with J. P. Baldwin and E. M. Bradbury, employing low-angle neutron scattering on ν_1 at different
ionic strength, to permit a clearer selection of the many alternative models of particle conformational change.

pH Effects at Different Ionic Strengths. Spectroscopic studies
of ν_1 as a function of pH (discussed below) have suggested that its
conformation is reasonably stable over a wide pH range ($\sim$5-9), with
a drastic and irreversible transition $\leq$ pH 4.5. Parallel hydrodynamic studies[9] indicate that to some extent, decreased pH can
stabilize ν_1 against the conformational effects of low ionic strength. $S_{20,w}$ was measured on ν_1 over the pH range (4.5-9) at three
different ionic strengths corresponding to the stable conformational
states at neutral pH (i.e., 1, 5.5 and 11 mM; see Figure 2). At
11 mM salt, $S_{20,w}$ remains constant at $\sim$ 11.1 over the entire pH
range. At 5.5 mM salt, $S_{20,w} \stackrel{\sim}{\sim} 10.2$ down to pH 5.4; a transition
toward a more compact conformation occurs between pH 5.4 - 5.0.
A similar sharp increase in $S_{20,w}$ between pH 5.4 - 5.0 is observed
in 1 mM ionic strength. Neither the exact conformational states
nor the mechanism of stabilization are understood. Weintraub[10]
has reported that in 2 M NaCl buffers, the heterotypic tetramer
(H4, H3, H2A, H2B) converts to a mixture of the homotypic tetramer
$(H3, H4)_2$, H2A and H2B. It is conceivable that similar pH effects
occur within the nucleosome; stabilization of H4, H3 contacts at
low pH, tending to oppose the low-ionic strength unfolding of ν_1.

Spectroscopic Techniques

Our laboratory has utilized a number of spectroscopic techniques
to examine changes in conformational states of ν_1 under varying solvent conditions or perturbing treatments. A combination of circular
dichroism (CD) and laser Raman spectroscopy has been especially useful in examining the secondary structure of the histones and the
DNA[11]. The value of these methods arises from the considerable
background of reference spectra on model and on natural macromolecules. A detailed understanding of the theory of each method
is beyond the scope of this chapter, and unnecessary for the present
empirical level of data interpretation. A brief comparison of CD
and laser Raman is sufficient for the present purposes.

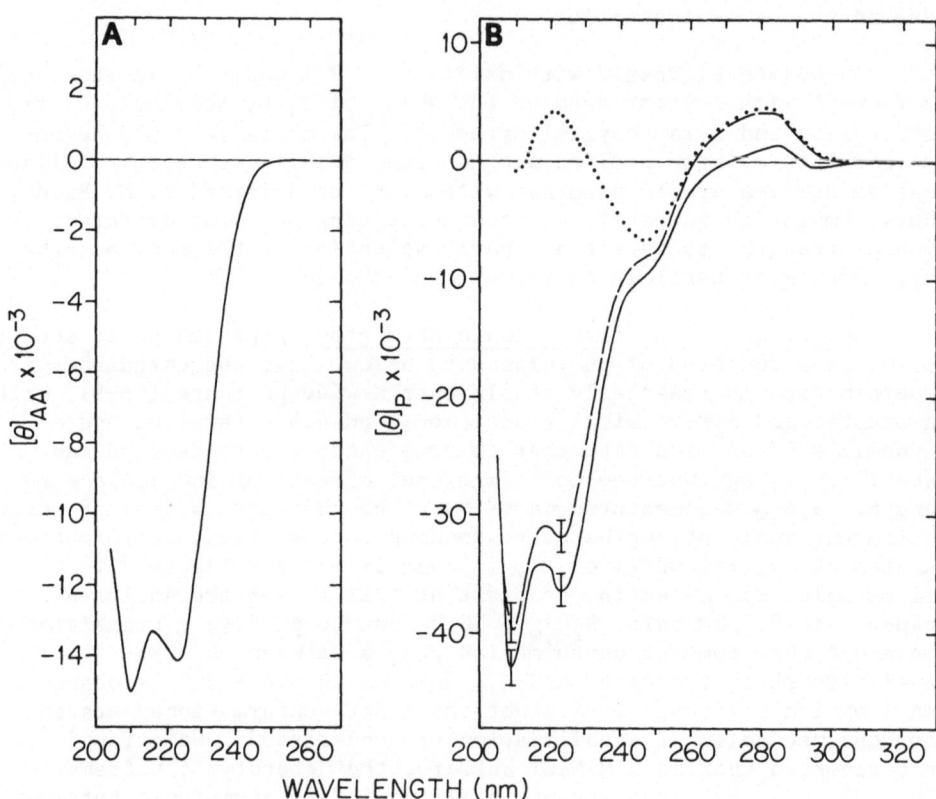

Figure 3. (A) Circular dichroic spectrum at 25°C of inner histone tetramer in 2.0 M NaCl, 10 mM Tris (pH 7), and 0.1 mM DTT. The total concentration of amino acid residues was estimated by quantitative amino acid analysis of an HCl hydrolyzate. Data are expressed as molecular ellipticity per mole of amino acid residues, $[\theta]_{AA}$. (B) Circular dichroic spectra at 25°C of ν_1 in 0.2 mM EDTA (——) and in 2.0 M NaCl, 10 mM Tris (pH 7), and 0.1 mM DTT (– – – –) compared to the spectrum of chicken DNA (····) in the same 2.0 M NaCl buffer. Data are expressed as molecular ellipticity per mole of DNA phosphate $[\theta]_p$, assuming the absorptivity per mole of DNA phosphate $(\varepsilon_{p,260})$ is equal to 6500 for DNA and for ν_1.

Circular dichroism measures the difference in absorbance between left and right-handed circularly polarized light. Difference in absorbance arises from asymmetry within molecules, especially from the helical structures of DNA and protein. In chromatin and nucleosomes, the spectral region 240 - 300 nm is dominated by DNA helix conformation; between 200 - 240 nm, by the considerable proportion of α-helix within the histones (Figure 3).

Laser Raman spectroscopy consists of measuring the vibrational spectrum of inelastically scattered light. The Raman frequencies or lines have been assigned to particular subgroups. For example, a strong and prominent Amide I ($\sim$1650 cm^{-1}; arising primarily from carbonyl stretch) associated with a weak Amide III ($\sim$1230 cm^{-1}; arising from C-N-H in plane bending and C-H stretching modes), indicates the presence of considerable protein α-helix. A more quantitative estimate of the amount of α-helix can be obtained by comparing spectra in H_2O and in D_2O. Laser Raman spectra can also distinguish effectively between the B and A genus of DNA structure. B is characterized by a strong dioxy symmetric stretch of the PO_2^- group ($\sim$1094 cm^{-1}) and a weak phosphodiester backbone stretch of O-P-O ($\sim$830 cm^{-1}). Laser Raman spectroscopy requires small volumes ($<10\mu l$) of material at concentration 10^2-10^3 times greater than those required for CD studies, and clear spectra can be obtained on precipitates. Laser Raman and CD complement each other when different solvents or additives are employed[12]. For example, due to the high ultraviolet absorbance of dimethylsulfoxide, CD measurements are impossible; whereas, useful "windows" are available to laser Raman. Vibrational spectra of DNA, inner histones, ν_1 and chromatin are presented in Figure 4.

From these CD and laser Raman studies we were able to estimate the amount and nature of protein and DNA secondary structure[11]. We concluded that nucleosomal DNA is in the B-genus, and that associated inner histones consist of $\sim$50% α-helix and negligible β-sheet structure. We further suggested that the α-helix is primarily confined to the apolar globular regions of the histones, which would consist of a localized α-helix content of 70-80%, comparable to myoglobin and hemoglobin. Trypsin studies of isolated nucleosomes, described below, are consistent with this conclusion.

The introduction of site-specific extrinsic fluorescent chromophore into a macromolecule is a widely used technique for probing a localized chemical environment. It is necessary, of course, that the probe does not damage the native molecular structure. The existence of a single cysteine residue at position 110 in chicken erythrocyte H3 afforded the opportunity[13] to place a thiol-specific fluorescent probe [i.e., NPM or N-(3-pyrene) maleimide] in a well-defined site. Due to the lack of reactivity of the H3 thiol within the intact nucleosome, NPM was allowed to react with ν_1 in high

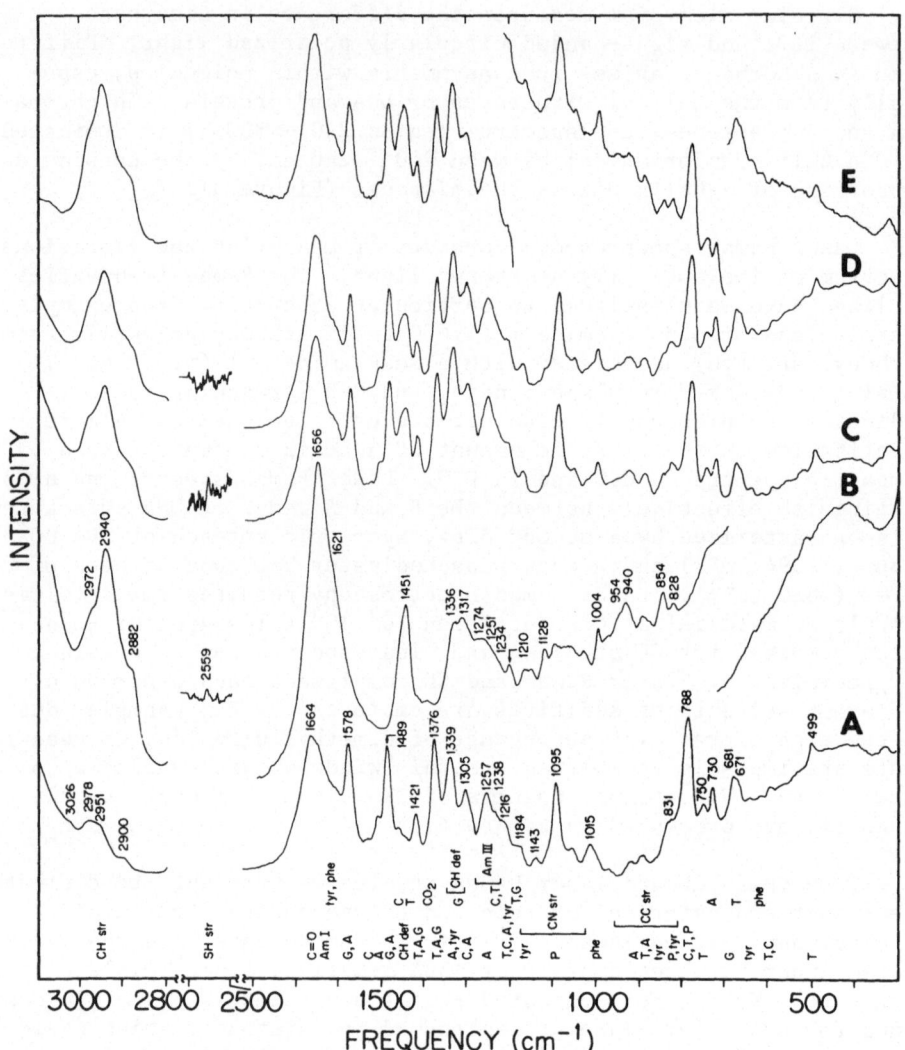

Figure 4. Laser Raman spectra at 32°C of (A) chicken DNA (4 percent by weight) in 2.0 M NaCl; (B) inner histones (9 percent) in 2.0 M NaCl, 10 mM Tris (pH 7), and 0.1 mM dithiothreitol (DTT); (C) ν_1 (10 percent) in 0.2 mM EDTA: (D) ν_1 (10 percent) in 2.0 M NaCl and 0.2 mM EDTA; and (E) chicken erythrocyte chromatin (26 percent) in 0.2 mM EDTA. Conditions: excitation wavelength, 488.0 nm; spectral slit width, 10 cm^{-1}; radiant power, 300 mw; amplification, A = 1 (300 to 1800 cm^{-1}), A = 3 (2500 to 2600 cm^{-1}). Abbreviations: str, stretching; def, deformation; A, T, C and G, adenine, thymidine, cytosine, and guanine; P, phosphate, phe, phenylalanine; tyr, tyrosine; and Am, amide.

NaCl, followed by reassociation and purification of the resulting NPM-v_1. The emission spectrum of NPM-v_1 exhibited peaks at 374 and 392 nm, and a long wavelength fluorescence at 460 nm (excimer). This "excimer" fluorescence, believed to arise from a pairwise association of pyrene groups attached to the two thiols of an intact nucleosome, has proven to be remarkably sensitive to conformation changes induced by various perturbants. Presumably this sensitivity of the probe arises from the stringent geometric constraints required for energy transfer within the excited state. Changes in the quantum yield of monomer fluorescence are also informative of the extent of exposure and quenching of the NPM group.

The utility of these three spectroscopic techniques to examine the conformational transitions of v_1 in response to various treatments and perturbants will be illustrated with the following examples.

Trypsin Digestion of v_1. Previous studies have indicated that trypsin rapidly degrades the N-terminal basic portions of the inner histones in chromatin or mononucleosomes[10]. Approximately 25% of the total histone sequences are digested. On the basis of CD and laser Raman studies we have suggested that very little α-helix is contained within the N-terminal basic tails. In a recent study[14] we have confirmed this suggestion by monitoring the CD of v_1 during the course of digestion with trypsin (Figure 5). It is evident that by 60-90 minutes all of the native inner histones have been degraded, with virtually no decrease in α-helix content. On the other hand, the "suppression" of the DNA spectrum is clearly dependent upon the integrity of the histone basic tails.

Urea Effects. Hydrodynamic studies, described earlier, have demonstrated that the addition of urea (0-10 M) to v_1 results in a progressive non-cooperative increase in the frictional coefficient (f), consistent with progressive swelling or increase in particle asymmetry. CD studies[6], on the other hand, demonstrate that the α-helix content of v_1 remains unchanged up to ~4M urea, followed by a cooperative destabilization of α-helix between 4-7 M. Spectrofluorometric studies of NPM-v_1 as a function of urea concentration substantiate the usefulness of this fluorescent probe (Figure 6). Excimer fluorescence disappears at low urea concentrations during moderate and reversible swelling of v_1. Monomer fluorescence is quenched simultaneously with the destruction of α-helix (i.e., between 4-6 M urea), presumably due to destruction of the solvent-protected environment around the pyrene chromophore.

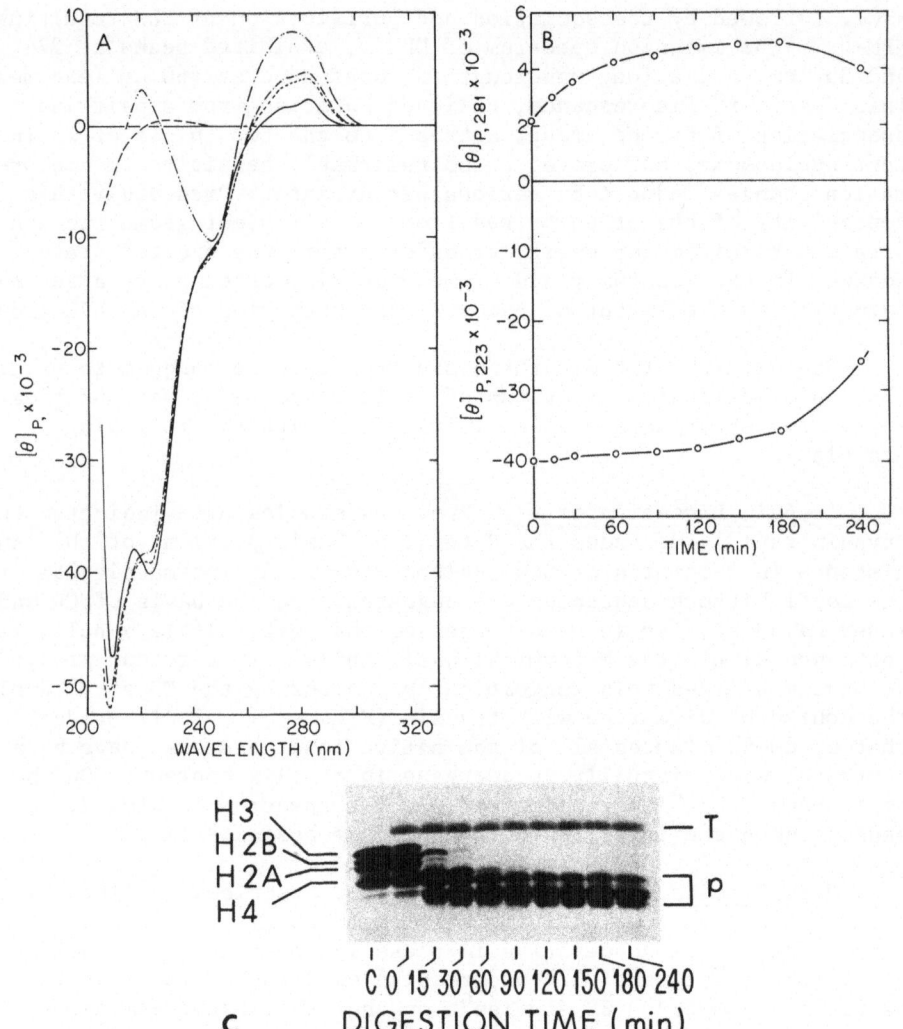

Figure 5. (A) Circular dichroic spectra of trypsin-digested ν_1.
Samples: chicken DNA (——·——); ν_1 (——); ν_1, digested 60 min (----);
ν_1, digested 120 min (——·——); trypsin + soybean trypsin inhibitor
(—— —— ——), at some concentration as present with digested ν_1.
Solvent systems: ν_1, 5 mM Tris (pH 8.0); DNA, 65 mM NaCl, 4.75 mM
Tris (pH 7.3). (B) Kinetics of changes in molecular ellipticities
at 281 nm (DNA conformation) and 223 nm (α-helix content), with time
of digestion by trypsin. (C) SDS gel electrophoresis of native and
trypsin digested ν_1. After varying periods of digestion with trypsin at room temperature, soybean trypsin inhibitor was added.
Abbreviations: C, control ν_1; 0-240 min, digested ν_1; T, doublet
band of trypsin and trypsin-inhibitor; p, limit tryptic fragments
of the inner histones.

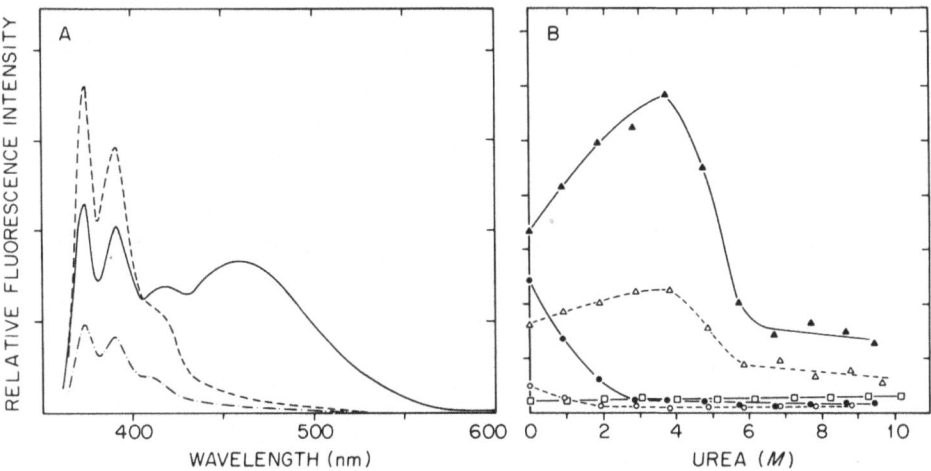

<u>Figure 6.</u>   (A) Fluorescence emission spectra of NPM-ν_1 (G or good) complexes at different urea concentrations.   Urea concentrations: 0 (———); 2.83 (----) and 6.71 (———·———) M urea, containing 0.2 mM EDTA (pH 7.0).   Spectra are presented at constant NMP-ν_1 concentrations.   Excitation wavelength: 340 nm.   (B) Fluorescence intensity changes as a function of urea molarity.   Samples: NPM-ν_1 (G) at 374 nm (▲——▲) and 460 nm (●——●); NPM-ν_1 (B or bad) at 374 nm (△---△) and 460 nm (O---O); NPM-β-mercaptoethanol at 374 nm (▢——▢).   G and B complexes are normalized to the same DNA concentration; NMP-β-mercaptoethanol at the same NPM concentration (based upon A_{340}nm) as the G fraction.   All solvents contained 0.2 mM EDTA (pH 7.0).   Excitation wavelength: 340 nm.

<u>Ionic Strength Effects</u>.   Spectrofluorometric and circular dichroism studies have been conducted in parallel to the hydrodynamic data described above[9,13].   NPM-ν_1 exhibits a decrease in excimer, and an increase in pyrene monomer fluorescence when the ionic strength is decreased in the vicinity of transition 1 (i.e., $\sim 10^{-3}$ M).   Apparently unstacking of the pyrene groups is promoted by the lowered ionic strength.   A less well-defined fluorescence changes is also observed in the vicinity of transitions 2 (i.e., $\sim 10^{-2}$ M), with a sharp decrease in both monomer and excimer fluorescence progressing from low to high ionic strength.   Thus, the local environment of the pyrene chromophores reflects the larger scale hydrodynamic conformational transitions.

Preliminary CD studies of ν_1 over the range 10^{-3}-10^{-1} M ionic strength indicate essentially negligible change in α-helix content, and small, but measureable effects on the DNA contribution to the spectrum.   The higher ionic strength conformations of ν_1 exhibit

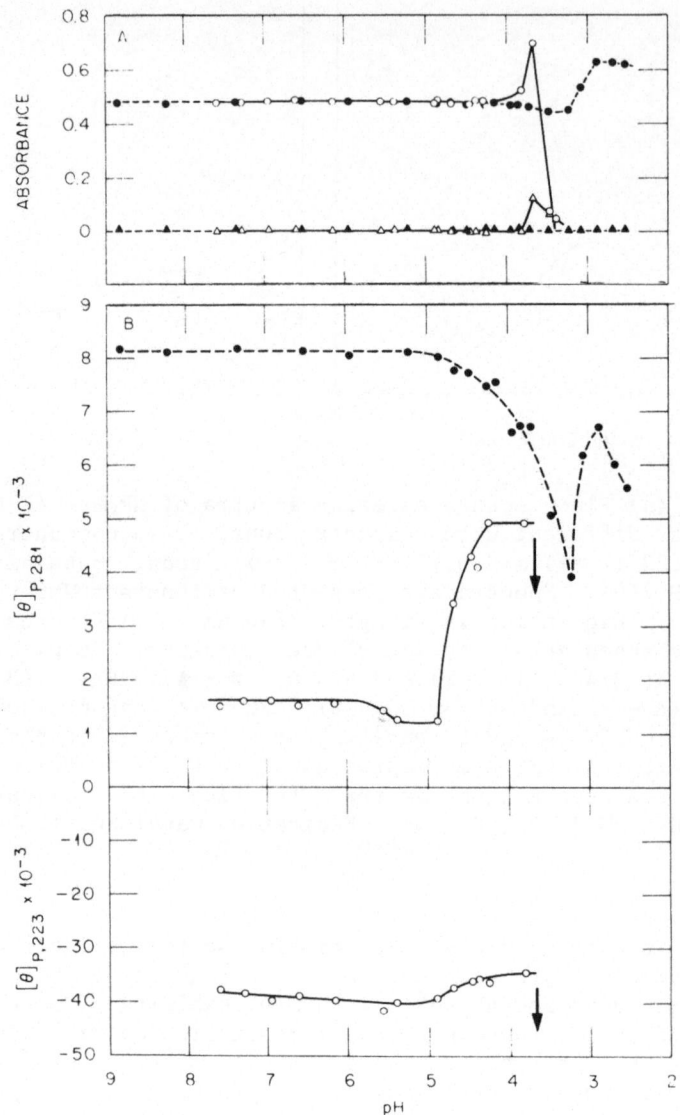

Figure 7. Absorption and circular dichroism of DNA and ν_1 as a function of pH. (A) A_{260} for DNA (●) and ν_1 (O), A_{325} for DNA (▲) and ν_1 (△). (B) $[\theta]_{p,281}$ for DNA (O) and ν_1 (O), $[\theta]_{p,223}$ for ν_1 (O). 4.75 and 0.5 mM buffers were used for DNA and ν_1, respectively. Buffer solutions in the pH range 2-9 were prepared by using the following systems: Tris-HCl; sodium cacodylate-cacodylic acid; sodium acetate-acetic acid and glycine-HCl.

slightly greater "suppression" of the DNA signal, than is observed
at lower salts.

pH Effects. Figure 7 illustrates a summary of recent spectroscopic studies of ν_1 conformation at different pH. Progressing from
neutral pH towards acid, several regions can be discerned: (A) pH
7.5 - 5.5, negligible changes in DNA conformation or α-helix content;
(B) 5.5 - 4.8, "suppression" of DNA ellipticity with no loss of
α-helix; (C) 4.6 - 4.2, sharp increase in DNA ellipticity towards
the conformation of naked DNA in solution, with slight loss of α-
helix; (D) pH 4-3.8, aggregation of ν_1. V. Gordon et al[9] has carefully compared these CD observations with hydrodynamic properties
of ν_1, and concluded that region (B), the "suppression" of DNA
ellipticity, parallels the pH-induced increase in S at low ionic
strength. Laser Raman studies[14] demonstrate that protonation of
the bases C and A in DNA or ν_1 does not occur significantly until
below pH $\sim$4. The mechanism and reversibility of these pH-induced
conformational transitions remains to be illucidated. There is,
however, a clear positive correlation between hydrodynamic "compactness" of ν_1, and "suppression" of DNA conformation as assayed
by circular dichroism.

CONCLUSION

Combining hydrodynamic and spectroscopic techniques in the
study of conformational states of ν_1 induced by a variety of
perturbants has led us to a general conception: the two structural
domains of ν_1 (i.e., the DNA-rich outer shell and the α-helix-rich
apolar histone core) exhibit differential responsiveness. In
general, the α-helical regions are more resistant, than DNA conformation or ν_1 size and shape, to the perturbing effects of urea,
decreased ionic strength and pH, trypsin treatment, or a variety
of water-miscible organic solvents. There are a number of reasonable conceptual models to explain this differential responsiveness
of the structural domains of ν_1. A few of these models can be denoted with self-descriptive names: e.g., "shell-swell", "open-
clam", or "unravelled". Distinction between the various models
will require data from many other biophysical techniques, such as
neutron scattering studies. Whatever the exact geometry of the
transitions, it is tempting to speculate that the α-helical histone
core constitutes the "restoring force" of the nucleosome--returning
the conformationally perturbed nucleosome of its "inactive" compact
form.

ACKNOWLEDGEMENTS

The author is grateful to his many colleagues and collabora-
tors. Their names can be found, in detail, in the Bibliography.
Research sponsored jointly by research grants GM19334 (DEO); PCM77-

21498 (ALO); AI11855 (GJT); and GM13914 (VNS) and the Division of Biological and Environmental Research, U.S. Department of Energy under contract W-7405-eng-26 with the Union Carbide Corporation.

BIBLIOGRAPHY

1. Cold Spring Harbor Symposium of Quantitative Biology (1977) XLII Cold Spring Harbor, NY.
2. Felsenfeld, G. (1978) Nature $\underline{271}$, 115-121.
3. Tanford, C. Physical Chemistry of Macromolecules (1959) John Wiley & Sons, Inc., NY.
4. Van Holde, K. E. Physical Biochemistry (1971) Prentice-Hall, Inc., Englewood, Cliff, NJ.
5. Ford, Jr., N. C. (1972) Chemica Scripta $\underline{2}$, 193-206.
6. Olins, D. E., Bryan, P. N., Harrington, R. E., Hill, W. E. and Olins, A. L. (1977) Nucleic Acids Res. $\underline{6}$, 1911-1932.
7. Gordon, V. C., Knobler, C. M., Olins, D. E., and Schumaker, V. N. (1978) Proc. Natl. Acad. Sci. USA, $\underline{75}$, 660.
8. Olins, A. L., Carlson, R. D., Wright, E. B., and Olins, D. E. (1976) Nucleic Acids Res. $\underline{3}$, 3271-3291.
9. Gordon, V. C., Knobler, C. M., Olins, D. E., and Schumaker, V. N. (1978) Biophys. J. $\underline{21}$, 66a.
10. Weintraub, H., Palter, K., and Van Lente, F. (1975) Cell $\underline{6}$, 85-110.
11. Thomas, Jr., G. J., Prescott, B. and Olins, D. E. (1977) Science $\underline{197}$, 385-388.
12. Zama, M., Olins, D. E., Prescott, B., and Thomas, Jr., G. J. (1978) Biophys. J. $\underline{21}$, 66a.
13. Zama, M., Bryan, P. N., Harrington, R. E., Olins, A. L., and Olins, D. E. (1977) Cold Spring Harbor Symp. Quant. Biol. XLII, in press.
14. Zama, M., Olins, D. E., Prescott, B. and Thomas, Jr., G. J. (1978) manuscript in preparation.

PREPARATION AND ANALYSIS OF CORE PARTICLES AND NUCLEOSOMES: A CONVENIENT METHOD FOR STUDYING THE PROTEIN COMPOSITION OF NUCLEOSOMES USING PROTAMINE-RELEASE INTO TRITON-ACID-UREA GELS

Barbara Ramsay Shaw and Randall G. Richards

Paul M. Gross Chemical Laboratory

Duke University, Durham, NC   27706

The nucleosomal repeat structure of chromatin consists of a core of 140 base pairs of DNA wrapped around an octamer of eight histones, and a spacer region which is variable in length and more susceptible to nuclease attack than the core[1-10].   Histone H1, which exhibits tissue specific variation, has been postulated to be associated with DNA in this spacer region[3,4,10-13].   The octameric histone complex in the core particle contains two copies each of histones H2A, H2B, H3 and H4.   It is known that these core histones undergo postsynthetic covalent modifications such as acetylation and phosphorylation at different times during the cell cycle[14].   Recently, Cohen et al.[15] have shown in the developing sea urchin and Franklin and Zweidler[16] have shown in the differentiated tissues of mammals that there are also variations within the primary sequence of histones H2A, H2B, and H3.   The molecular differences between variant histones arise from the conservative substitution of one or two amino acid residues in the polypeptide chain which can be detected on Triton/acid/urea gels. In the sea urchin system, different H2A and H2B variants are synthesized at different stages of embryonic development[15].   The existence of multiple forms of histones implies that there must be multiple forms of nucleosomes.

Our laboratory has investigated methods of separating and identifying different subsets of nucleosomes and 140 base pair core particles.   We report here the utility of polyacrylamide gel electrophoresis in carrying out systematic characterization of the homogeneity of such preparations.   In particular, samples of

125

nucleoprotein particles can be well resolved electrophoretically into component particles that differ by their histone content as well as the size of their DNA. Todd and Garrard[13] and Bakayev et al.[14] have reported that nucleosome particles having discrete sizes of DNA can be separated with polyacrylamide gel electrophoresis. We have used similar type gels to resolve the various structural types of nucleoprotein momomers, dimers and higher oligomers that result from digestion of chicken erythrocyte chromatin. We have observed that the resolution of monodisperse nucleoprotein particles obtained with nucleoprotein electrophoresis is comparable to that obtained by isolating and electrophoresing the component DNA. Such results suggest that nucleoprotein particles obtained from the digestion of chromatin with micrococcal nuclease are produced only in discrete packages, wherein the size of the DNA and the protein content of the products are found in only certain allowable combinations.

Presented here is a high resolution two-dimensional gel electrophoresis system which has been developed for rapid fractionation and analysis of histone variants and other acidic proteins found in the numerous products that result from digestion of chromatin or nuclei with micrococcal nuclease. The fundamental premise that underlies this procedure is that the complete removal of histone proteins from DNA can be accomplished by treating deoxyribonucleoprotein with a one-step protamine treatment discussed below. By combining gel electrophoresis of DNA-protein complexes in low ionic strength buffer in the first dimension with a protamine-released Triton X-100/acid/urea gel electrophoresis in the second dimension, this procedure can readily screen for variations in the histone and other acidic protein components that arise from discrete kinds of mononucleosomes, dinucleosomes and higher oligonucleosomes found in nuclei of many cells. The technique is easier and potentially more powerful than most of the current fractionation procedures, which employ columns or sucrose gradients to fractionate chromatin digests. The time-consuming collection, concentration and analysis steps required for determining the histone components in each nucleosome fraction can be avoided. More important, in one gel all the nucleoprotein products from one time point of digestion can be analyzed.

METHODS AND MATERIALS

(A) Isolation of Erythrocyte Nuclei. Blood was obtained from adult White Leghorn chickens by cardiac puncture in the presence of 0.15 M NaCl-0.015 M Na citrate, pH 7.2 (saline/citrate). After centrifugation at 3000 x g for 10 min at 4°C, the plasma and buffy coat were removed. The erythrocytes were washed twice with the isotonic saline solution and frozen at -60° until needed. The frozen erythrocytes were thawed at 37° in an equal volume of saline/citrate and centrifuged at 3000 x g for 10 min at 4°C;

the nuclear pellet was resuspended in 0.25% Nonidet P-40 in
saline/citrate. Throughout, 0.1 mM phenylmethane sulfonyl fluoride (PMSF) was added to inhibit protease action. The nuclei were
repelleted, washed with saline-citrate and repelleted. A more
detailed procedure can be found on pp. 78-79 of Ref. 23.

(B) Digestion of whole chromatin. Freshly prepared nuclei
from 6 ml of packed frozen red blood cells were, after the above
treatment, resuspended in about 200 ml of digestion buffer (0.3 M
sucrose-0.75 mM $CaCl_2$-10 mM Tris•HCL, pH. 7.2) at a concentration
of 2×10^8 nuclei per ml. Digestion of the nuclei by micrococcal
nuclease (Worthington) was carried out at 37° with 125 units of
nuclease per ml of nuclei suspension. The digestion reaction was
terminated by making the solution 2 mM in EDTA.

(C) Isolation and digestion of H1/H5-depleted chromatin.
Following the method of Tatchell and Van Holde[24], nuclei from 10
ml of packed red blood cells from part (A) were lysed with stirring in 250 ml of 10 mM Tris-cacodylic acid - 0.1 mM EDTA - 0.1 mM
PMSF (pH 7.2) (lysis buffer). After several hours, the solution
was made 0.6 M NaCl with solid NaCl and gentle stirring, and
allowed to swell overnight at 4°C. The chromatin gel was centrifuged at 18,000 x g for 45 min; the pellet was resuspended with
gentle stirring in 400 ml of lysis buffer containing 0.65 M NaCl,
allowed to stand 24 hr. and pelleted at 18,000 x g for 15 min.
This last step was repeated one time. The pelleted chromatin
($\sim$3000 A_{260} units) was brought up in 400 ml of lysis buffer and
stirred for 2 hrs. The chromatin, now in low ionic strength
buffer, was centrifuged at 18,000 x g for 30 min, brought up with
a Dounce homogenizer in 20-25 ml of lysis buffer containing 1 mM
$CaCl_2$, and the 3000 A_{260} units digested at 37° with 125 units/ml
of micrococcal nuclease (Worthington). The reaction was terminated after about 40 minutes by making the solution 10 mM EDTA and
cooling on ice.

(D) Fractionation of nucleosomes. The digested chromatin
from part (C) was pelleted at 10,000 x g for 10 min. Monomer and
dimer nucleosomes in the supernatant were isolated with gel chromatography on Bio-Gel A-5m by eluting with 10 mM Tris, 0.7 mM
EDTA, pH 7.2.[22,23] The core particles were obtained in a well
resolved peak in the included volume; however, some contamination
with H1/H5-depleted dimer nucleosomes occurred. The peak fractions
containing about 300-400 A_{260} units of monomer were pooled, concentrated to 100 OD/ml, dialyzed in an Amicon concentrator and
fractionated on 10 mM Tris, 0.7 mM EDTA, pH 7.2 sucrose gradients
isopycnic for particle density 1.51[25] to yield mainly purified
monomer and some dimer nucleosomes.

(E) Electrophoresis of nucleoproteins. Nucleoprotein electrophoresis was carried out at 25°C using 4.0% acrylamide
(acrylamide: N, N'methylene-bisacrylamide ratio, 25:1) tube gels
(12 x 0.3 cm) or slab gels (13 x 0.15 cm). We have found that
this acrylamide:bis ratio gives excellent resolution of nucleosomes in the monomer/dimer region. The buffer used to make the
gel and circulate in the electrophoresis apparatus was that used
by Todd and Garrard[13], i.e. 6.4 mM Tris, 3.2 mM sodium acetate,
0.32 mM EDTA, pH 8.0. Samples of ionic strength less than 5 mM
were loaded in 10% glycerol, 2 mM EDTA, .025% bromphenol blue.
Sample loads contained on the order of 10-25 µg each of DNA and
protein for one-dimensional slabs and 20-50 µg each of DNA and
protein for one-dimensional tubes. Electrophoresis with recircu-
lating buffer was 15 min at 70 V followed by 4 1/2 hr at 90 V for
slab gels, and 15 min at 50 V followed by 5 1/2 hr at 60 V for
tube gels. Electrophoresis of DNA was carried out according to
Loening[18]. For visualization of DNA, gels were stained with 1
µg/ml of ethidium bromide in water for 1/2 hr and photographed on
a Model C-62 short wave ultraviolet light box (Chromato-Vue trans-
illuminator, UV Products) with Polaroid 107 or 55 film and a red
filter. For visualization of proteins, gels were stained in
Coomassie Blue R-250 following Fairbank's staining procedure[19].

(F) Protamine-release method on Triton/Acid/Urea gels.
Electrophoresis of histone variants was performed at 25°C with
Triton/acid/urea[20] slab gels containing 12% acrylamide (acryl-
amide:bis ratio of 30:0.2), 8 M urea, 6 mM Triton X-100 and 5%
acetic acid. The plates are sealed with 15% acrylamide instead of
agarose since the latter interferes in the staining of the gels.
Slab gels were 13 cm x 13 cm glass plates, and either 0.15 cm
thick if used for one-dimensional gels or 0.30 cm thick if used
for second-dimensional gels. Gels were pre-electrophoresed for 5
hr at 125 V with 5% acetic acid running buffer. The gel was then
scavenged for 1 1/2 hr at 125 V with 0.1 mls of a solution of 8 M
urea, 0.3 M cysteamine and 0.3 mg/ml protamine in each slot for one-
dimensional slabs or a total of 2 ml per gel for second-dimension
slabs. A second scavenge of 15-20 min duration was performed with
a similar volume of 0.6 M cysteamine. The remaining scavenging
solution was removed from the slots and the buffer replaced with
fresh 5% acetic acid.

Protamine release of histones from nuclear digests and nucleosomes into Triton/acid/urea gels was accomplished as follows.
Samples of ionic strength less than 10 mM were made 8 M in urea-5%
in acetic acid-2 1/2% in thioglycolic acid-5% in β-mercaptoethanol
(sample buffer) and 0.1% in protamine sulfate from Salmon-sperm
(histone-free, Sigma) and applied directly to the gel. Electrophoresis was for 8 1/2 hr at 150 V. For two-dimensional electrophoresis to display histones, first-dimension nucleoprotein tube

gels (12 cm x 0.3 cm d.) were soaked for 1/2 hr in 5% acetic acid,
5% β-mercaptoethanol, 2 1/2% thioglycolic acid and 8 M urea. The
tube was laid horizontally across the top of a preformed and
prescavenged Triton slab (13 x 13 x 0.3 cm) which had a 10 cm long
middle slot to hold the tube gel, and a small slot at each end to
serve as sample well for a standard. The tube gel was carefully
overlayered with 200 μℓ of a solution of 8 M urea, 5% acetic acid
acid and 1% protamine sulfate. Electrophoresis was for 13 hr at
100 V.

RESULTS

Separation of nucleosomes. Our first dimensional nucleoprotein gel is a 4% polyacrylamide gel system which utilizes the
low ionic strength buffer system described by Todd and Garrard[13].
We have found that an acrylamide:bis ratio of 25:1 gives excellent
resolution of nucleosomes in the region of monomer and dimer subunits. In Fig. 1 we have electrophoresed samples of monomer (core
particles) and dimer nucleosomes obtained from column and sucrose
gradient fractionation of a micrococcal nuclease digest of H1/H5-
depleted chicken chromatin. On the left the samples were electrophoresed as DNA and the size calibrated with Hae III-PM2 restriction fragments; on the right similar samples were electrophoresed
as nucleoproteins. Note the sharply defined monomer and dimer
nucleoprotein bands in slot B; these bands are as sharp or sharper
than the corresponding DNA bands in slot No. 2 on the left half of
Fig. 1. The dimer nucleosome seen in slots A and B has a DNA size
of 270 ± ∿10 base pairs as measured in slots No. 1 and 2, and
corresponds to the "tight" (spacerless) dimer seen by Tatchell and
Van Holde[24].* Overloading of a core particle sample (slots C and
D) reveals the presence of a small amount of a second monomer
species that migrates slower than the core. The DNA in this minor
band migrates at about 160-170 base pairs (not shown). As can be
seen, when the gel is not overloaded, the nucleoprotein gel is
capable of separating nucleosomes with a resolution nearly comparable to that obtained with the component DNA. The sharpness of
our nucleoprotein bands implies that a broad band or series of
closely spaced bands within the monomer/dimer region would correspond to different forms of nucleosomes.

---

*Klevin and Crothers have reported a dimer subset in H1-
depleted calf-thymus chromatin that has a DNA size of 240 base
pairs[28].

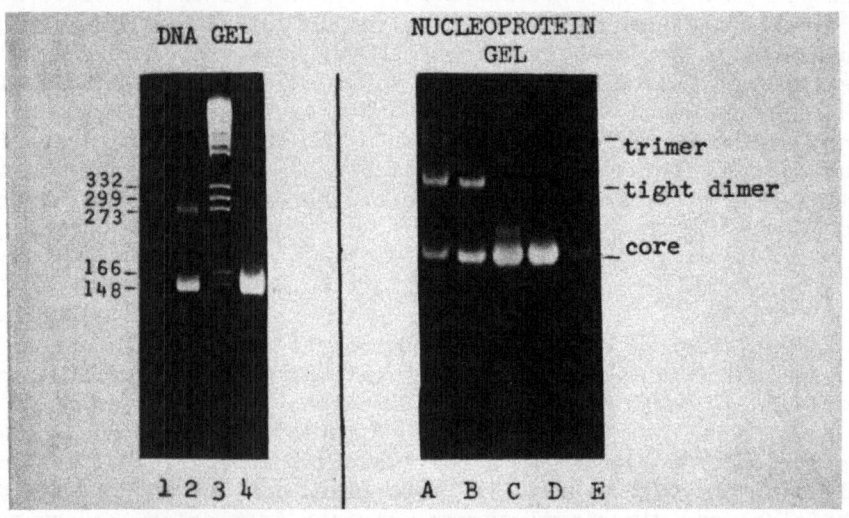

Figure 1. (Right). Nucleoprotein gel showing chicken erythrocyte
(core particle) monomer, dimer and trimer nucleosomes. The
chromatin was H1 and H5 depleted, digested with micrococcal
nuclease, and fractioned on an A-5m column followed by a 10 mM
Tris, 0.7 mM EDTA, pH 7.2 sucrose gradient. Samples were applied
directly to a 4% polyacrylamide (25:1) gel: A. Monomer, dimer
and trimer; B. Monomer and dimer; C. Overloaded monomer containing two species; D. Purer (core particle) monomer (overloaded
on this gel); E. Same sample as D. (Left). DNA extracted from
nucleosome samples of the same sucrose gradient as above and
electrophoresed on 4% polyacrylamide DNA gels.[4,18] 1. Monomer,
dimer and trimer; 2. Monomer and dimer; 3. PM2-Hae III restriction fragments; 4. Overloaded monomer sample.

Protamine Release of Proteins into Gels. In order to analyze
the histone content of nucleosomes, it has been necessary in the
past to either extract the histones from chromatin with acid[21],
or alternatively treat with SDS and electrophorese the proteins on
SDS gels[26]. The method of acid extraction is time consuming and
can result in selective extraction of certain histones[21]. The
SDS electrophoretic method is quite reliable; however, only the
five major histone classes can be resolved. We have developed a
method in which variant and covalently modified histones can be
easily and quanitatively solubilized for analysis on Triton/acid/-
urea gels[15] which resolve at least 15 discrete forms of histones.

In order to release histones from nucleosomes into Triton/-
acid/urea gels we have tried various solubilizing reagents,
including spermine, spermidine, cetyltriethylammonium bromide,
DNAase I (which Bafus et al have recently studied[27]), and protamine
sulfate from salmon sperm. Protamine sulfate is the best releasing agent for our studies. Fig. 2 is a comparison of two of the
most effective protein-releasing reagents: spermine and protamine. The sample applied to each slot was a multimeric subunit
digestion product from H1/H5-depleted chicken chromatin. As seen
on the left side of Fig. 2, histone H4 is poorly released from the
spermine-solubilized sample at all concentrations of spermine
employed. On the right is the same sample using protamine as the
releasing agent. Note that concentrations as low as .05% protamine are effective in displacing histones and other proteins from
the DNA. Perhaps more important for electrophoretic analysis is
that the time needed for protein displacement is quite rapid using
protamine.

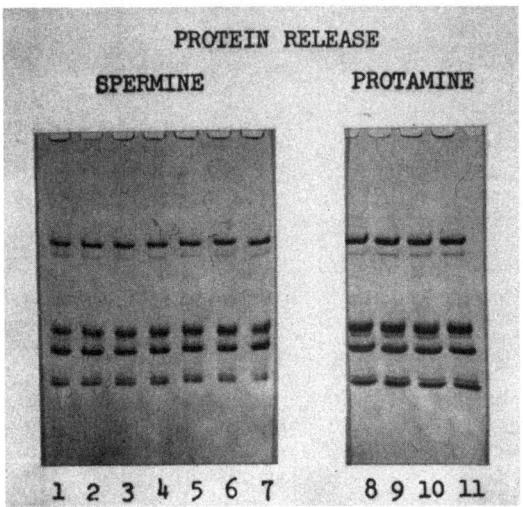

Figure 2. Triton-acid-urea gel illustrating protamine and
spermine release of protein from A-5m multimer fractions of a
micrococcal nuclease digestion (of H1 and H5-depleted chicken
erythrocyte chromatin). Slots No. 1 thru 7 correspond to samples
solubilized for 1 hr in 1, 2, 3, 4, 5, 7.5 and 10% spermine plus
sample buffer. In slots No. 8-11 the samples were solubilized in
sample buffer plus 0.05, 0.1, 0.1 and 1% protamine, respectively.
Whereas samples No. 8 and 9 were incubated at 25°C for 1 hour
prior to electrophoresis, samples No. 10 and 11 were applied to
the gel within 5 minutes after the addition of protamine.

Protamine is equal to or better than acid extraction, as can be seen in Fig. 3. For comparison, four different nucleoprotein samples are electrophoresed, in pairs. Acid extracted samples are the left sample in each pair. In all cases, the protamine treatment equaled or superceded the acid extracted samples in the amount and number of proteins released from the chromatin.

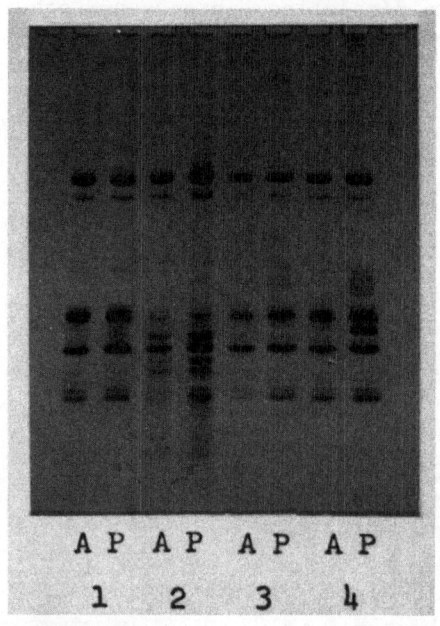

Figure 3. Comparison of acid extracted (A) and protamine (P) solubilized protein from various nucleoprotein samples. The (A) samples were extracted from 1% SDS, 4M urea, 0.4N H_2SO_4 with stirring, precipitated with ethanol[21], and solubilized in sample buffer (no protamine). The (P) samples were used directly. They were either in column buffer (No. 1 and 2) or in 10 mM Tris–3 mM EDTA–0.75 mM $CaCl_2$ – 0.3 M sucrose, pH 7.2 and all were adjusted prior to electrophoresis with two volumes of 1 1/2 x sample buffer plus 0.15% protamine. Pair 1: A–5m multimer fraction of H1/H5–depleted chromatin; Pair 2: Degraded A–5m monomer fraction of a chicken erythrocyte nuclei digestion; Pair 3: Nucleoprotein released from erythrocyte nuclei after a 4 min micrococcal nuclease digestion; Pair 4: Erythrocyte nuclei digested four minutes with nuclease.

Having shown that histones as well as other proteins can be
displaced from chromatin with protamine for analysis on Triton/-
acid/urea gels, we have employed a variation of this method with
second-dimensional gels to screen for nucleosomes that differ in
protein composition. We have used this two-dimensional system to
analyze in Fig. 4 the chicken monomer and dimer sample shown in
Fig. 1, slot B. In the first dimension of Fig. 4 the nucleosomes
were electrophoresed as nucleoprotein. In the second dimension
the proteins were released with protamine from the nucleoprotein
gel and separated on a Triton/acid/urea gel as described in the
Methods. A nucleosome sample with protamine added is placed in

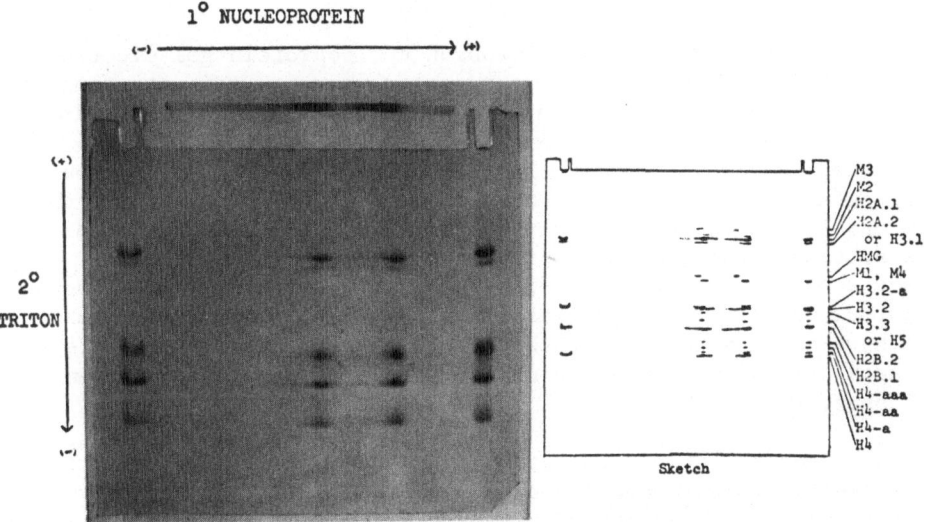

Figure 4. Two dimensional electrophoresis of monomer and dimer
from micrococcal nuclease digested H1/H5-depleted chromatin from
chicken erythrocytes. (See figure 1, slot B) Nucleoprotein was
electrophoresed from left to right in a 0.3 cm diameter tube.
The tube gel was then soaked for 1/2 hr. in 8M urea, 5% acetic
acid, 5% β-mercaptoethanol, and 2 1/2% thioglycolic acid and laid
horizontally across a preformed, prescavenged 0.3 cm thick Triton-
acid-urea slab gel. The tube was overlayered with 0.2 mls of 8M
urea, 5% acetic acid, and 1% protamine and electrophoresed for 13
hrs at 100V. Slots on the left and right correspond to protamine
released proteins from digested H1/H5-depleted chromatin in
solution. (Arrows indicate migration of proteins which are not
visible in the picture.) Nomenclature follows Franklin and
Zweidler[16].

the reference slots at the far left and far right of the gel. The
release of histones from the first dimensional tube is comparable
to the protein in the reference sample. As can be seen, the histone
content of the chicken erythrocyte core monomer and "tight" 270
base pair dimer are practically identical (Fig. 4). This is not
unexpected since the erythrocyte is a quiescent cell. The similarity in histone content of our core particle and "tight" dimer
suggest that the formation of this compact dimer is independent of
the nature of the component histone proteins.

Of special interest in Fig. 4 are the small amounts of nonhistones associated with the monomer and dimer species. These
proteins are difficult to visualize in the picture but are indicated with arrows. Nonhistone proteins have been reported to be
associated with nucleosomes by several groups[27,29,30].

DISCUSSION

In summary, the first dimension nucleoprotein gel electrophoresis system described here has been shown to separate monomer
and dimer nucleosomes with high resolution. Moreover, the histone
as well as non-histone protein content of each class of these
nucleosomes can be readily examined using the second-dimension
Triton/acid/urea gel system described. The effectiveness of the
second dimension gel separation lies in the quantitative release
of proteins from the nucleoprotein complexes with protamine. We
have shown that protamine efficiently releases histones and nonhistone proteins from nuclear digests and from chromatin subunits.
We have taken advantage of this protein solubilization to effect
second-dimension separation of histones from various nucleosome
classes.

Using the method of protamine release to identify the histone
components of nucleosome classes, we have been able to observe
and compare different types of nucleosomes released from chromatin
by micrococcal nuclease. Certain mono- and di-nucleosome classes
contain no H1 or H5; some fractions contain either H1 or H5, and
most nucloesomes contain small amounts of non-histone proteins.[31]
The nucleosomes that carry histones H1 or H5, and nucleosomes that
contain core proteins plus certain non-histone protiens, migrate
slower than core particles on first-dimension nucleoprotein gels.[27,31]

In the chicken erythrocyte chromatin, all classes of nucleosomes (within the limit of sensitivity of our first-dimensional
gels) appear to have similar amounts of variant core histones;
that is, the relative proportion of each core histone variant
found in total chromatin is similar to that found in mononucleosome, dinucleosome and multimer nucleosome classes. The

distinguishing features of chicken erythrocyte nucleosome classes arise primarily from the size of the DNA and the relative amounts of histones H1 and H5.[31] In sea urchin embryo, however, we have observed different types of nucleosome subsets that appear to be derived from different variants of core histones (to be submitted).

In summary, the two-dimensional gel method described here is a very convenient method for the analysis of histone variants and non-histone proteins in nucleosomes. The method has the potential of easily screening numerous samples of nuclease digestion products for the identification of modified and variant nucleosomes. Most important here is the fact that with one gel all the variant and modified histone proteins in nucleosome products produced at one time of digestion can be analyzed, simultaneously, thus eliminating the necessity of collecting and concentrating fractions as one might do with sucrose gradient analysis.

ACKNOWLEDGMENTS

We wish to express our appreciation to Ms. Katherine Marsh for excellent technical assistance. This research was supported by grants from NIH (GM 23681), Duke University Biomedical Research Science Support Grant, and by a Teacher-Scholar Award to B.R.S. from the Camille and Henry Dreyfus Foundation.

REFERENCES

1. Van Holde, K. E., Sahasrabuddhe, C. G. and Shaw, B. Ramsay (1974) Nucl. Acids Res. $\underline{1}$, 1579-1586.
2. Van Holde, K. E. and Isenberg, I., (1975) Accounts of Chemical Research $\underline{8}$, 327-335.
3. Van Holde, K. E., Shaw, B. Ramsay, Lohr, D., Herman T. M. and Kovack, R. T. (1975) in Proceedings of the Tenth FEBS Meeting, Vol. 38 Federation of European Biochemical Societies, pp 57-72.
4. Shaw, B. Ramsay, Herman, T. M., Kovacic, R. T., Beaudreau, G. S. and Van Holde, K. E. (1976) Proc. Nat. Acad. Sci. USA $\underline{73}$, 505-509.
5. Sollner-Webb, B. and Felsenfeld, G. (1975) Biochemistry $\underline{14}$, 2915-2920.
7. Axel, R. (1975) Biochemistry $\underline{14}$, 2921-2925.
8. Kornberg, R. D. (1974) Science $\underline{184}$, 868-871.
9. Oudet, P., Gross-Bellard, M., and Chambon, P. (1975) Cell 4, 281-300.
10. Shaw, B. Ramsay, Herman, T. M., Kovacic, R. T. and Van Holde, K. E. (1976) in Molecular Mechanisms in the Control of Gene Expression, ed. by D. P. Nierlich, W. J. Rutter and C. F. Fox, Academic Press pp. 13-20.
11. Whitloch, J. P. and Simpson, R. T. (1976) Biochemistry $\underline{15}$, 3307-3314.

12. Varshavsky, A. J., Bakayev, V. V. and Georgiev, G. P. (1976) Nucl. Acids Res. 3, 477-492.
13. Todd, R. D. and Garrard, W. T. (1977) J. Biol. Chem. 252, 4729-4738.
14. See Elgin, S. C. and Weintraub, H. (1975) Ann. Rev. of Biochemistry 44, 725-774.
15. Cohen, L. H. , Newrock, K. M. and Zweidler, A. (1975) Science 190, 994-997.
16. Franklin, S. G. and Zweidler, A. (1977) Nature 266, 273-275.
17. Bakayev, V. V., Bakayeva, T. G. and Varshavsky, A. J. (1977) Cell 11, 619-629.
18. Loening, U. E. (1967) Biochem. J. 102, 251-257.
19. Fairbanks, G. Steck, T. L. and Wallach, T., (1971) Biochemistry 10, 2606-2617.
20. Alfageme, C. R., Zweidler, A., Mahowald, A., and Cohen, L. H. (1974) J. Biol. Chem. 249, 3729-3733.
21. Panyim, S. and Chalkley, R. (1969) Biochemistry 8, 3972-3979.
22. Shaw, B. Ramsay, Corden, J. L., Sahasrabuddhe, C. G. and Van Holde, K. E. (1974) Biochem. Biophys. Res. Commun. 61, 1193-1198.
23. Rill, R. L., Shaw, B. Ramsay and Van Holde, K. E. (1978) in Methods in Cell Biology, Vol. 17, ed by G. S. Stein, J. Stein and J. Kleinsmith,, Academic Press, N. Y. pp. 69-101.
24. Tatchell, K. and Van Holde, K. E. (1977) Biochemistry 16, 5295-5303.
25. Noll, M. (1974) Nucl. Acids. Res. 1, 1573-1578.
26. Laemmli, U. K. (1971) Nature 227, 1-4.
27. Bafus, N. L., Albright, S. C., Todd, R. D. and Garrard, W. T. (1978) J. Biol. Chem. 253, 2568-2574.
28. Klevin, L. and Crothers, D. M. (1977) Nucl. Acids Res. 4, 4077-4089.
29. Goldknopf, I. L., French, M. F., Musso, R. and Busch, H. (1977) Proc. Nat. Acad. Sci. USA 74, 5492-5495.
30. Levy W. B. and Dixon, G. H. (1978) Fed. Proceed 37, (6) Abs. No. 2837 p. 1787.
31. Richards, R. G. and Shaw, B. Ramsay (1978) Fed. Proceed 37, (6) Abs. No. 2836 p. 1787.

THE INTERACTION OF HISTONES WITH DNA: EQUILIBRIUM BINDING STUDIES

D.R. Burton, M.J. Butler, J.E. Hyde, D. Phillips,
C.J. Skidmore and I.O. Walker

Department of Biochemistry, University of Oxford
South Parks Road, Oxford OX1 3QU

INTRODUCTION

Chromatin, the interphase form of the genetic material, is
made up of a linear array of repeating substructures called nucleosomes. Each nucleosome is composed of eight core histones, two each
of the histones H2a, H2b, H3 and H4 associated with about 200 base
pairs of DNA and two molecules of histone H1 [1]. The nucleosomes
interact with each other to form a linear fibre about 100 Å in
diameter. Although a wealth of information has accumulated in the
last few years relating to the structure and function of chromatin
[2] there have been few quantitative studies on the binding of
histones to DNA. Such studies are important because they provide a
thermodynamic framework within which the properties of chromatin
can be discussed. Thermodynamic information on the interaction of
histones with DNA can be obtained from equilibrium binding curves
and their variation with temperature, pH, ionic strength and so on.
Histone-DNA interactions in chromatin are, however, highly complex
in the sense that five different proteins interact with each other
and with DNA to form a stoicheiometric complex in the nucleosome.
Thus the construction of binding curves for the binding of individual histones or histone pairs to DNA may not be as revealing as
the binding of all five histones. On the other hand all the histones
bind very tightly to DNA at physiological ionic strengths which
makes the task of constructing equilibrium binding curves very difficult. It is known, however, that the histones may be dissociated
differentially from chromatin by increasing concentrations of
NaCl [3]. In the following the properties of the binding curves obtained in this way are described. In these experiments a sheared
chromatin preparation has been used in which the higher-order

structure has been destroyed since it does not give the characteristic pattern of DNA fragments on digestion with micrococcal
nuclease [4,5]. However, these higher-order interactions appear to
involve only H1 and not the core histones since the same limit
digest pattern is obtained on extensive digestion with micrococcal
nuclease whether the starting material is native chromatin or
sheared chromatin [6,7]. In sheared chromatin, therefore, the core
histone-DNA interactions are preserved whereas the H1-chromatin
interactions have been modified. In what follows the binding of
core histones to DNA *in the presence of dissociated H1* are first
described. Secondly, the binding of H1 to nucleosome core particles
is examined with particular reference to the role that H1 plays in
the higher-order structure characteristic of native chromatin.

EXPERIMENTAL METHODS

Chromatin was prepared from calf thymus tissue by the method
of Zubay and Doty [8] and characterised as described previously
[3,9,10]. Circular dichroic and viscosity measurements were made as
previously described [9,10].

Digestion of chromatin with micrococcal nuclease (Worthington
Biochem. Corp. Ltd.) was carried out at 37°C in Tris-HCl (1 mM,
pH 7.0), $CaCl_2$ (10^{-4} M). The reaction was terminated by adding
EDTA to 10 mM and cooling the reaction tubes in ice. The chromatin
was then incubated with proteinase K (Boehringer Biochem. Corp.),
100 μg/ml, for 1 h at 37°C. The DNA was further deproteinised with
chloroform-isoamyl alcohol (24:1) and then dialysed against Tris-
HCl (40 mM) pH 8.0, sodium acetate (5 mM), EDTA (1 mM) for electrophoresis on agarose or polyacrylamide gels. After electrophoresis,
the DNA was visualised by staining with ethidium bromide. The size
of the DNA fragments was determined by comparison with fragments of
PM2 or *polyoma* DNA of known size digested with restriction nuclease
Endo R Hae III (gift from B. Ponder). *Polyoma* fragments were in
turn calibrated against sequenced $\phi \times 174$ DNA fragments (B. Ponder,
private communication); PM2 fragment sizes were according to Noll
[11].

The dissociation of core histones from DNA was induced by increasing the concentration of NaCl over the range 0.7 to 2.0 Molar.
In all cases the salt solutions contained sodium dihydrogen phosphate, 10 mM and disodium hydrogen phosphate, 10 mM, so that the
final pH of the solutions was 6.2—6.4. The pH meter was calibrated
with sodium hydrogen phthalate, 0.1 M, pH 4.01 and sodium borate
decahydrate, 0.1 M, pH 9.18, all measurements being made at 18—20°C.
The construction of binding curves requires that the concentrations
of the products and reactants in the dissocation reaction be
measured without disturbing the equilibrium. This may be done by
true equilibrium methods which measure the concentrations of the

various species present *in situ* or by kinetic methods which separate
the products and reactants and allow their concentrations to be
measured separately. In the case of the system under study here the
application of equilibrium methods is not possible since there is
no easily measurable property of dissociated histone which distin-
guishes it from bound histone. A kinetic method has therefore been
used which separates dissociated histone from depleted chromatin
using gel filtration with Sepharose 4B. The method relies on the
fact that the rate of dissociation and reassociation is slow compared to the time required to separate the products and reactants
during their passage down the column. The equilibrium is thereby
frozen. A typical elution profile is shown in Fig. 1 which shows
that dissociated histone can be completely separated from chromatin.
In high salt there is a tendency for the histone to dissociate from
the depleted chromatin peak which leads to a variation of histone
to DNA ratio across the depleted chromatin peak. However, the
histone–DNA ratio of the original depleted chromatin may be obtained
by pooling all fractions across the peak and measuring the average
histone–DNA content [12]. Chromatin (0.1–1.0 mg/ml) in 0.7 mM sodium
phosphate, pH 7.0, was dialysed for 3 hours against salt solutions
at either 4°C or 20°C. Aliquots (5 mls) were applied to a column
of Sepharose 4B (50 × 1.0 cm) equilibrated with the appropriate salt

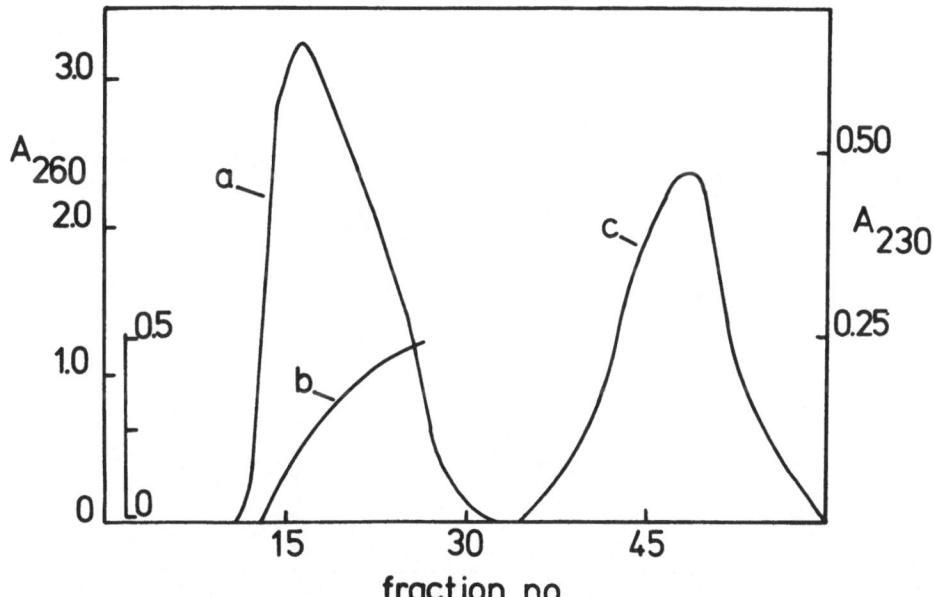

Fig. 1. Elution profile of chromatin in 1.2 M NaCl from Sepharose
4B column: (a) depleted chromatin; (c) dissociated histone; (b) the
variation of the protein:DNA ratio across the first peak (see inset
ordinate).

solution; fractions (2 ml) were collected and analysed.

H1 binding curves were obtained by dialysing chromatin to
equilibrium against solutions of NaCl, 0.7 mM sodium phosphate,
pH 6.8, at either 4° or 21°C. The solutions were centrifuged at 45 K
for 3 h to 6 h. After this time the depleted chromatin was completely removed to the bottom of the tube as a pellet. The amount
of H1 present in the upper two-thirds of the supernatant was determined chemically. The amount in the supernatant did not significantly change with time over a 3 to 14 h period, indicating that
the dissociation had reached equilibrium and did not change during
the time-scale of the separation.

RESULTS

The dissociation curve of histones from DNA as a function of
NaCl concentration is shown in Fig. 2. Dissociation of histone takes
place in three distinct stages. Between 0.7 mM sodium phosphate and
0.7 M NaCl H1 is selectively dissociated as shown previously [13].
Between 0.7 and 1.2 M NaCl H2a and H2b are selectively removed from
the chromatin and between 1.2 and 2.0 M NaCl H3 and H4 are dissociated (Fig. 3 and refs. [3,12]). Dissociated H1 is always present in

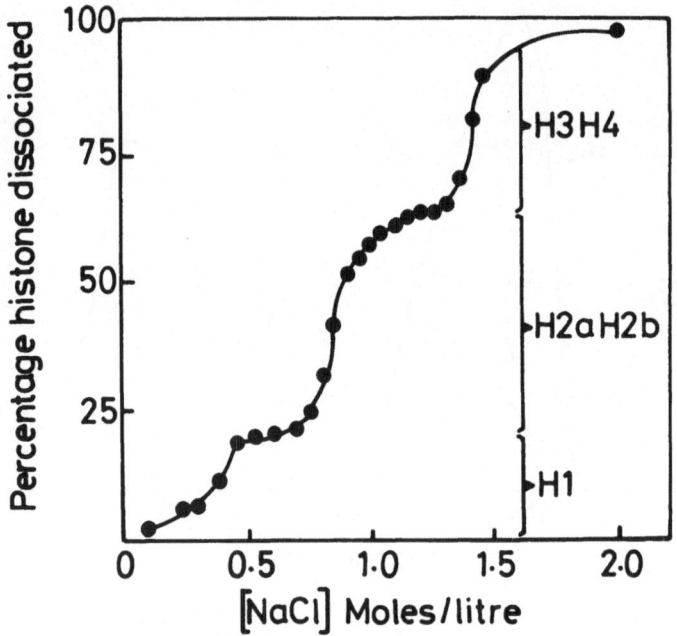

Fig. 2. The dissociation of histone types from chromatin as a function of NaCl concentration.

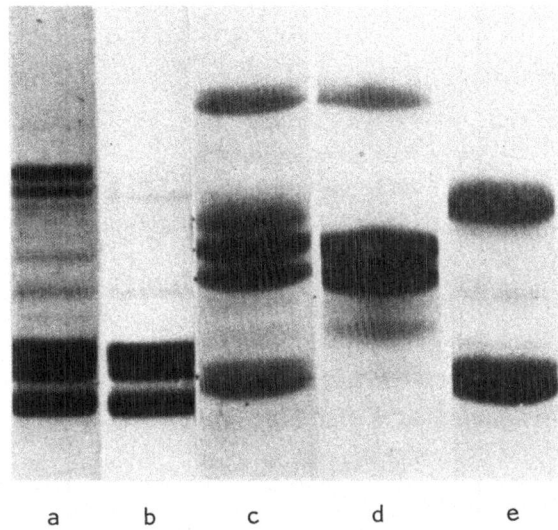

Fig. 3. Polyacrylamide gel electrophoresis patterns of histone types dissociated as a function of NaCl concentration: SDS-polyacrylamide gels of (a) histones in chromatin; (b) histones remaining bound to chromatin in 0.7 M NaCl; polyacrylamide-urea gels of (c) histones in chromatin; (d) histones dissociated in 1.2 M NaCl; (e) histones remaining bound to chromatin in 1.2 M NaCl.

the system during the dissociation and re-association of the other four core histones.

(a) The Dissociation of H2a and H2b

The dissociation curve of H2a and H2b from chromatin is shown in detail in Fig. 4 together with the re-association curve obtained by dialysing samples of chromatin into 1.2 M NaCl, to dissociate completely H2a and H2b, followed by further dialysis into salt solutions of lower concentrations in the range 1.2 to 0.7 M. The two sigmoid curves are equivalent showing that the dissociating species are in reversible thermodynamic equilibrium.

The effect of chromatin concentration on the degree of dissociation was investigated at a constant NaCl concentration of 0.85 M by decreasing the initial chromatin concentration. As would be expected from considerations of mass action effects on a disproportionation reaction, the degree of dissociation increases with

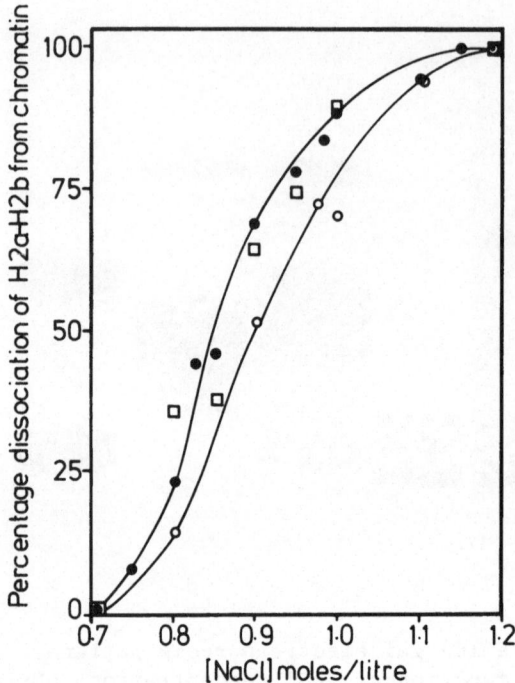

Fig. 4. The dissociation of H2a-H2b as a function of NaCl concentration: — ● — forward dissociation curve at 4°C; — □ — reverse dissociation curve at 4°C; — ○ — forward dissociation curve at 20°C.

increasing dilution as shown in Table 1. The right hand column shows the dissociation constant, K_{app}, predicted by applying the simple

TABLE 1

c_0	α	K_{app}
1010	0.26	9.2
725	0.28	8.2
520	0.29	6.1
250	0.38	5.9

Apparent dissociation constant, $K_{app} = \alpha^2 c_0^2/(1-\alpha)c_0$, as a function of degree of dilution of chromatin at NaCl = 0.85 M. c_0 = initial concentration; α = degree of dissociation of H2a-H2b. K_{app} in arbitrary units; temp. = 4°C.

TABLE 2

[NaCl],M	α	[H2a]/[H2b]
0.8	.25	1.04
0.9	.70	1.02

The ratio of H2a to H2b remaining
bound to chromatin at two different
degrees of dissociation, α. Temp. =
4°C.

mass action law. α, the degree of dissociation, is defined in (d).
The relatively constant value of K_{app} over the four-fold change in
initial concentration leads to the conclusion that H2a and H2b
interact with chromatin non-cooperatively, that is, they bind to
equivalent, independent, non-interacting sites. The effect of temperature on the dissociation is shown in Fig. 4. Increasing the
temperature from 4°C to 20°C decreases the degree of dissociation
over the whole dissociation range although the two dissociation
curves are not parallel. The dissociation reaction is therefore
exothermic but the enthalpy change may vary with the salt concentration possibly due to a contribution to the enthalpy change from
salt-induced conformational changes in the components of the system.

The amounts of H2a and H2b which remained bound to the chromatin at various degrees of dissociation were monitored using polyacrylamide urea gel electrophoresis. The results are shown in Table
2 and suggest that these two histones appear to dissociate conco-
mitantly possibly as an equimolecular complex. This suggestion is
supported by crosslinking studies which show that between 0.7 and
1.2 M NaCl dissociated H2a and H2b may be covalently crosslinked by
formaldehyde to form dimers and higher polymers. No monomers were
detected [3].

The dissociation reaction is also reversible at the level of
the secondary structure of the histones and DNA in the chromatin
complex as judged by the CD spectra of partially dissociated and
reconstructed chromatin. Data were obtained for four different salt
concentrations between 0.7 and 1.2 M NaCl. The spectra of the dissociated and reassociated chromatins were identical in all cases;
that for 0.85 M NaCl is shown in Fig. 5.

(b) The Dissociation of H3 and H4

The dissociation and reassociation curves of H3 and H4 from
DNA is shown in Fig. 6. As in the case of H2a and H2b the curve is

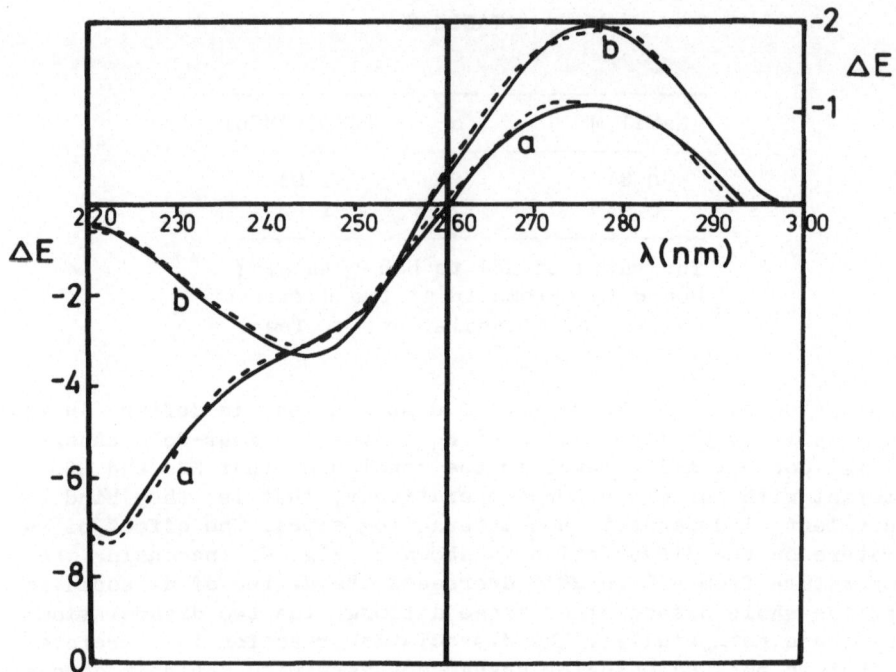

Fig. 5. Circular dichroic spectra of partially dissociated and re-
associated chromatin in (a) 0.85 M NaCl and (b) 1.35 M NaCl:
——— dissociated species, ---- reassociated chromatin; ΔE is the
difference in extinction per mole of nucleotide.

sigmoidal and thermodynamically reversible. It is important to
realise that the dissociation of H3 and H4 from DNA has here been
studied in the presence of unbound H1, H2a and H2b, which have completely dissociated before H3 and H4 begin to dissociate above
1.2 M NaCl (Fig. 2).

The effect of protein concentration on the degree of dissociation was investigated at 1.35 M NaCl, as before, by decreasing the
initial concentration. The equilibrium constants, calculated by
assuming a simple mass action effect on the reaction, are shown in
Table 3. α, the degree of dissociation is defined in (d). Over a
five-fold range of initial chromatin concentration, the value of
K_{app} changes by two orders of magnitude, showing that K_{app} is itself
a function of concentration. This indicates that the binding of H3
and H4 to DNA is accompanied by cooperative interactions between
the reacting species.

Increasing the temperature from 4°C to 20°C provokes a decrease
in the degree of dissociation showing, as with H2a and H2b, that the

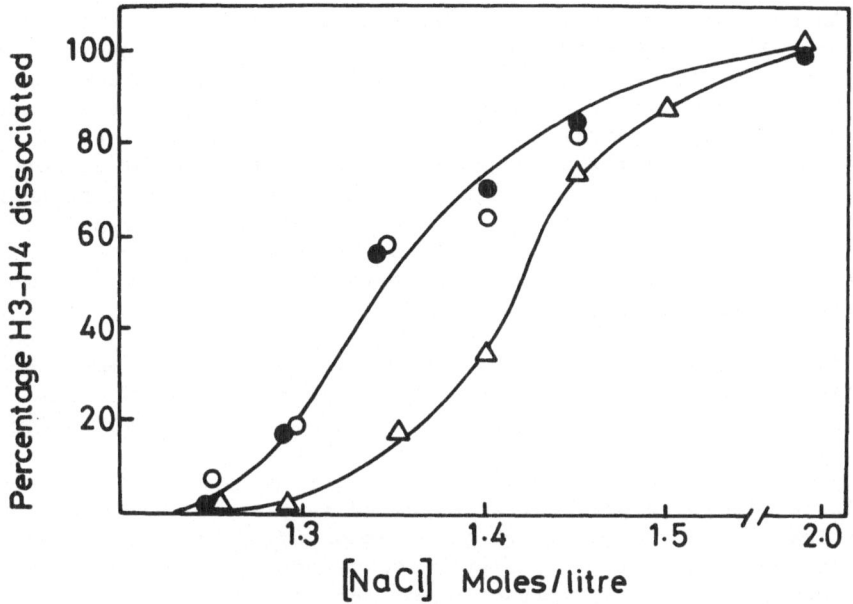

Fig. 6. The dissociation of H3-H4 as a function of NaCl concentration: $-\bullet-$ forward dissociation curve at 4°C; $-\circ-$ reverse dissociation curve at 4°C; $-\triangle-$ forward dissociation curve at 20°C.

TABLE 3

c_0	α	K_{app}
106	0.10	1.17
97.5	0.16	3.02
76.1	0.42	23.0
66.7	0.52	37.5
46.9	0.70	76.4

Apparent dissociation constant,
$K_{app} = \alpha^2 c_0^2/(1-\alpha)c_0$, as a function
of degree of dilution of chromatin at
NaCl = 1.35 M. c_0 = initial concentration; α = degree of dissociation of
H3-H4. K_{app} in arbitrary units; temp.
= 4°C.

reaction is exothermic. The two dissociation curves are not parallel which suggests that there may be a contribution to the overall
enthalpy change from salt-induced conformational changes.

An analysis of the histone types which are dissociated over the range 1.2 to 2.0 M NaCl by gel electrophoresis has shown that H3 and H4 are dissociated and reassociated in equimolar amounts, probably as an equimolar complex, which must therefore be at least a dimer [12]. This result is supported by crosslinking studies of H3-H4 DNA in 2 M NaCl which show that the histones exist as hetero-typic dimers, tetramers and higher polymers [3]. Circular dichroic spectra of partially dissociated complexes and re-associated complexes at the same NaCl concentration show that as regards the secondary structure of the nucleoprotein the interaction is completely reversible (Fig. 5).

The intrinsic viscosity of the native, sheared chromatin preparation was 20 dl/gm, in 0.7 mM sodium phosphate, pH 7.0. The intrinsic viscosity of the DNA in the same solvent was 105 dl/gm. A sample of chromatin was made 2 M in NaCl by the addition of solid and the solution was then dialysed into 0.7 mM sodium phosphate. The intrinsic viscosity of the reconstructed sample was 20 dl/gm. This shows that dissociation and re-association of all five histones takes place reversibly with respect to viscosity which is a measure of the supercoiling and compaction of the molecule [14].

The intrinsic viscosity of H3-H4-DNA in sodium phosphate (0.7 mM, pH 7.0) was 42 dl/gm, indicating that the arginine-rich histone DNA complex has a conformation intermediate between the fully super-coiled native structure and free DNA. H3-H4-DNA which had been dissociated in 2M NaCl and then re-associated by dialysis into 1.1 M NaCl had an intrinsic viscosity of 51 dl/gm (measured in 0.7 mM sodium phosphate). Thus the intermediate, partly supercoiled conformation is recovered after dissociation and re-association of the arginine-rich histone complex.

(c) Interaction of H1 with Chromatin Core Particles

The forward dissociation curve of H1 as a function of NaCl is shown in Fig. 7 together with the re-association curve. This latter curve was obtained by completely dissociating H1 by dialysis into 0.7 M NaCl. The NaCl concentration was then reduced to the required value by further dialysis and the amount of dissociated H1 estimated as before. The coincidence of the two curves shows that the binding process is reversible. The amount of H1 dissociated in 0.7 M NaCl was 21% of the total histone present in the chromatin. Since H1 is completely and discretely dissociated under these conditions, this represents the total amount present. It corresponds to 1.5 moles of H1 bound per mole of nucleosome core particle. The binding curve is bi-phasic, indicating that two distinct modes of binding are involved in the equilibrium process. Furthermore, each phase of the binding curve is sigmoidal which suggests that the binding processes are cooperative with respect to NaCl.

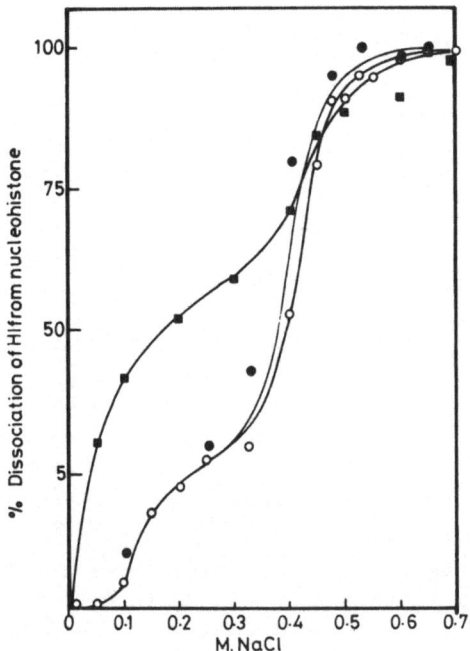

Fig. 7. The dissociation of H1 as a function of NaCl concentration:
— ○ — forward dissociation curve at 4°C; — ● — reverse dissociation
curve at 4°C; — ■ — forward dissociation curve at 21°C.

The CD spectrum of chromatin, after H1 had been completely dissociated and re-associated, was identical to that of native chromatin. Thus the binding of H1 is reversible with respect to secondary
structure.

The effect of increasing the temperature on the binding curve
is shown in Fig. 7. At 21°C the relative proportion of H1 dissociating in the first phase at lower [NaCl] increased from about 25% to
50%, whereas the proportion in the second phase was correspondingly
reduced.

The effect of decreasing the chromatin concentration on the
degree of dissociation of H1 at constant [NaCl] was investigated by
diluting the chromatin with 0.7 mM sodium phosphate, pH 7.0, before
dialysis into the required NaCl solution. At each dilution the
amount of dissociated H1 was estimated as before. By assuming that
the total number of binding sites on the chromatin for H1 is represented by the number of molecules of H1 bound to the native chromatin preparation, i.e. all the sites are occupied in the native state,
the fraction of free and bound sites may be estimated at each

TABLE 4

0.2 M NaCl

c_0	α	K_{app}
870	.19	4.1
500	.29	5.9
250	.32	3.8

0.3 M NaCl

c_0	α	K_{app}
630	.40	1.7
460	.49	2.2
310	.50	1.6
160	.60	1.5
120	.64	1.3

0.4 M NaCl

c_0	α	K_{app}
986	.45	3.6
450	.68	6.5
230	.70	3.7

dilution as described in (d). The degree of dissociation increases with increasing dilution as shown in Table 4. The apparent dissociation constant, K_{app}, remains constant over a 4-5 fold change in concentration at 3 different salt concentrations. This shows that H1 binds non-cooperatively to equivalent, independent binding sites on the core particles. The stoicheiometry of the binding suggests that there are 1.5 H1 binding sites per nucleosome core particle. Symmetry considerations imply that the number should be two [15] but that not all these sites are filled in sheared chromatin probably due to loss of H1 during preparation.

(d) Analysis of Binding Curves

The equation describing the equilibrium between dissociated histone and chromatin as a function of NaCl may be written as:

$$DNH + nNaCl = DNH' + mH$$

where DNH is the molar concentration of undissociated chromatin, H is the molar concentration of dissociated histone, DNH' is the concentration of depleted chromatin, n is the number of moles of NaCl bound to depleted chromatin when m moles of histone are dissociated. The equilibrium constant, K_{eq}, may then be written:

$$K_{eq} = \frac{(DNH')(H)^m}{(DNH)(NaCl)^n} \quad . \tag{1}$$

It follows that

$$\log K_{eq} = \log \left[\frac{(DNH')(H)^m}{(DNH)} \right] - n \log (NaCl).$$

For equivalent, independent, non-interacting sites for histone binding, $m = 1$; cooperativity is indicated if $m > 1$. It is assumed that
the number of binding sites for histone molecules in native chromatin is equal, on a molar basis, to the number of moles of histone
bound. That is, in the native state, for core histones, the binding
sites are fully saturated and binding is stoicheiometric. If c_0 is
the total concentration of a given histone type bound in native
chromatin the concentration of dissociated histone is αc_0 where α
is the degree of dissociation determined from the binding curves;
αc_0 is the concentration of free binding sites on the chromatin and
$(1-\alpha)c_0$ is the concentration of bound sites on the crhomatin. It
follows from (1) that

$$K_{eq} = \frac{(\alpha c_0)^{m+1}}{(1-\alpha)c_0 \; (NaCl)^n} \; . \tag{2}$$

Since it has already been shown that the binding of H2a-H2b to
DNA is non-cooperative (Table 1), $m = 1$ and a graph of log
(DNH')(H)/(DNH) against log (NaCl) will have a slope of n. Such a
graph is shown in Fig. 8, for the data at 4° and 20°C. The relationship is linear in both cases with $n = 30 \pm 1$ per dimer of H2a-H2b.
The intercept at 4°C gives $K_{eq} = .072$ M and $\Delta G_0 = 1.6$ Kcals/mole at
1.0 M NaCl. The temperature-dependence of K_{eq} yields values of ΔH_0
and ΔS_0 equal to -10 Kcals/mole and -31 cals/mole/° respectively at
4°C and 1.0 M NaCl.

Thus the dissociation of one mole of H2a-H2b dimer results in
the net binding of 30 moles of NaCl. The standard free energy of
dissociation is positive which is to be expected since molar concentrations of NaCl are required to dissociate histones from chromatin in micromolar concentrations. Since the enthalpy change is
negative, the positive free energy term is dominated by the large
negative entropy change on dissociation. This presumably arises
primarily as a result of the ordering of solvent water molecules on
breaking electrostatic bonds and hydrophobic interactions in the
dissociation reaction. The number of sodium chloride molecules bound
as one molecule of histone dimer dissociates is very close to the
number expected (27) if all the lysine and arginine side chains in
the amino terminal domains of the histones defined as in ref. [15]
form salt links with phosphate groups on the DNA.

Since the dissociation of H2a-H2b is non-cooperative with
respect to histone the sigmoidal nature of the dissociation curve
(Fig. 4) shows that the dissociation with respect to NaCl is

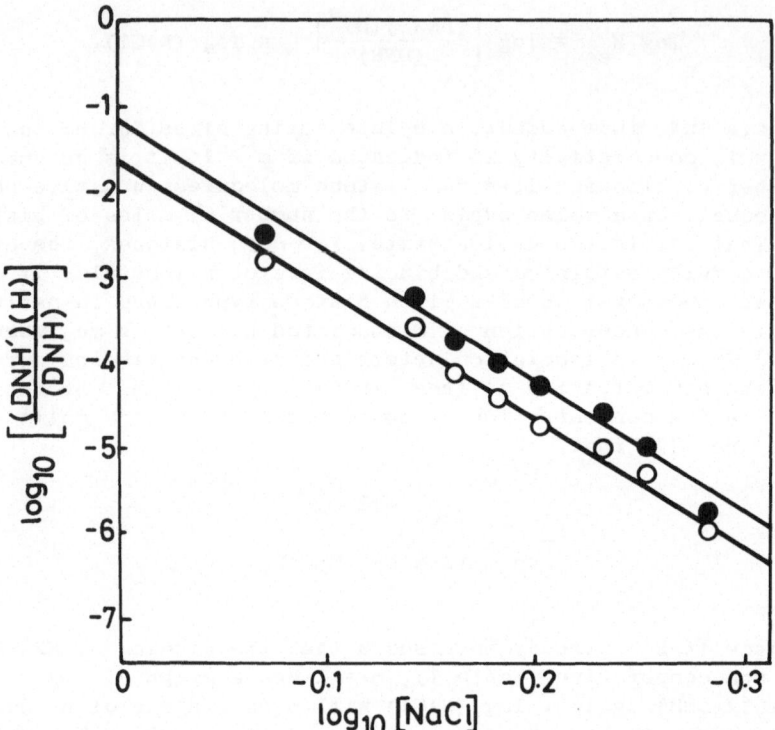

Fig. 8. The variation of $\log_{10}[(DNH')(H)/(DNH)]$ with $\log_{10}$ (NaCl): — • —, 4°C; — ○ — 20°C.

cooperative. It is important to realise that n in Equation (1) is not a Hill coefficient and is therefore not a measure of the degree of cooperativity. It is not possible to construct Hill plots directly from the data presented here because there is no direct measure of the amount of sodium chloride bound as a function of the sodium chloride concentration. However, by making the assumption that the amount of sodium chloride bound (to either DNA or histone or both) is directly measured by the degree of dissociation of the histone it is possible to construct a Hill plot with a slope of 11.5, which indicates a high degree of cooperativity for sodium chloride binding.

The data of Table 3 show that the binding of H3-H4 to DNA is cooperative and that m > 1 in Equation (1). It follows from (2) that

$$K_{eq} = \frac{\alpha}{1-\alpha} \cdot (\alpha c_0)^m \cdot \frac{1}{(NaCl)^n} \quad ,$$

and $\log K_{eq} = \log \alpha/1-\alpha + m \log (\alpha c_0) - n \log (NaCl)$. At constant (NaCl)

$$\log K_{eq} = \log \frac{\alpha}{1-\alpha} + m \log (\alpha c_0) + \text{Constant}.$$

Thus m may be obtained from the slope of a graph of $\log \alpha/1-\alpha$ versus $\log \alpha c_0$. This is shown on Fig. 9 and gives slope of m = 2.

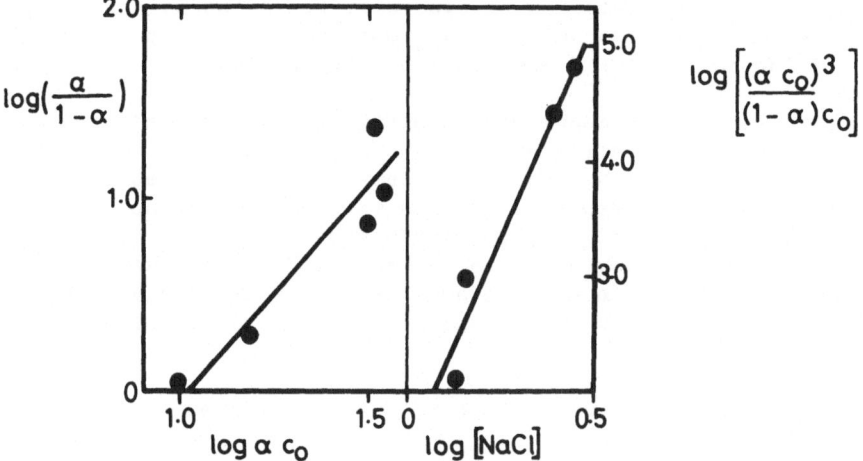

Fig. 9. Left: the variation of $\log (\alpha/1-\alpha)$ with $\log \alpha c_0$; right: the variation of $\log [(\alpha c_0)^3/(1-\alpha)c_0]$ with $\log (NaCl)$.

From equation (2),

$$\log K_{eq} = \log \left[\frac{(\alpha c_0)^{m+1}}{(1-\alpha)c_0} \right] - n \log (NaCl) .$$

Thus a graph of $\log [(\alpha c_0)^{m+1}/(1-\alpha)c_0]$ with m = 2, versus $\log (NaCl)$ gives a slope of n. This is shown in Fig. 8 and gives n = 14.0 per mole of H3-H4 dimer. As before this number represents the *net* number of sodium chloride molecules bound when one mole of H3-H4 dissociates. It may be compared to the number of basic amino acid side chains (29) in the N-terminal domains of H3-H4 (see ref. [15]) and indicates that, in contrast to H2a-H2b, displacement of the histone-DNA salt linkages by bound counter ion is not stoicheiometric. The apparent equilibrium constant at 1 M NaCl is 4.3×10^{-6} M, corresponding to $\Delta G_0 = 7.4$ Kcals/mole; $\Delta H_0 = -15$ Kcals/mole at 1.4 M NaCl.

The data of Table 4 show that the binding of H1 is non-

cooperative and that m = 1 in equation (1). The variation of log
(DNH')(H)/(DNH) with log (NaCl) is shown in Fig. 10 at two different temperatures; the data refer to that shown in Fig. 7. It is
apparent that n and K are not constant across the dissociation
range. At 4°C n = 4 at low salt concentrations but from a point corresponding to a concentration of about 0.4 Molar the value of n
increases abruptly to 10. At 21°C, n = 1 at low salt and again increases sharply to 10 at high salt with a transition point at about
0.4 Molar NaCl. Above this point the two curves are identical. This
indicates that the enthalpy of dissociation in the high salt range
($>$ 0.4 M) is zero. The value of K_{eq} obtained by extrapolating this
part of the curve to 1 M NaCl is 1.0, giving ΔG_0 = 0.

In deriving the above equations the assumption has been made
that the activity coefficients of the macromolecular components
are unity. However, the ionic environment of these components and
in particular the dissociated H1, is varying across the dissociation range from one of very low ionic strength to one of relatively
very high ionic strength. If this change of environment induces conformational changes in H1 or otherwise alters the activity coefficient, there will be a marked change to the activity of the dissociated histone.

At low ionic strength the conformation of free histones is

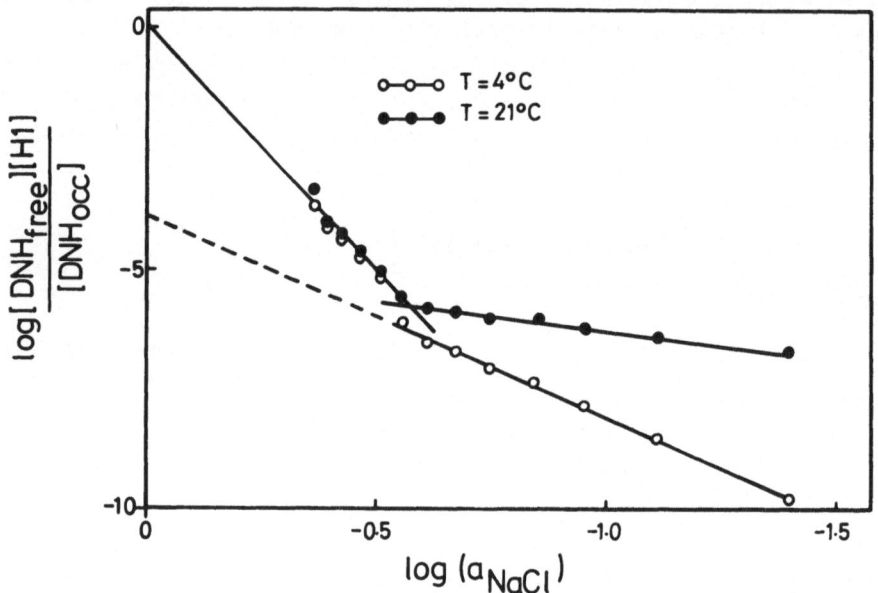

Fig. 10. The variation of $\log_{10}$[(DNH)(H1)/(DNH)] with $\log_{10}$ (NaCl
activity).

largely random coil. Secondary structure, especially α-helical
structure, similar to that found in histone bound in the nucleohistone complex, is only apparent at high ionic strength. In the
case of H1 there is a large decrease in ellipticity at 222 nm as
the ionic strength is reduced (data not shown), which indicates
that the molecule undergoes a conformational change leading to loss
of α-helical secondary structure. The mid-point of the change is
centred on 0.3 M NaCl. This evidence suggests that histone H1, dissociated from chromatin below 0.4 NaCl, is in an unfolded, largely
unstructured conformation, whereas H1 dissociated above 0.4 M NaCl
is in a more structured conformation similar to the conformation
of bound H1. If the structure of bound H1 is considered the native
structure and the low-ionic-strength form a denatured structure,
then the results shown in Fig. 7 allow the equilibrium below 0.4 M
NaCl at 4°C to be represented by:

$$(\text{Native H1})_{\text{bound}} + 4 \text{ NaCl} \rightleftharpoons (\text{Denatured H1})_{\text{free}} \ ,$$

and above 0.4 M NaCl by:

$$(\text{Native H1})_{\text{bound}} + 10 \text{ NaCl} \rightleftharpoons (\text{Native H1})_{\text{free}} \ .$$

From this it follows that

$$(\text{Denatured H1})_{\text{free}} + 6 \text{ NaCl} \rightleftharpoons (\text{Native H1})_{\text{free}} .$$

Thus the refolding of the denatured to the native form is accompanied by the net binding of six sodium chloride molecules. These
changes are shown schematically in Fig. 11.

In order to investigate the binding of H1 to DNA H1 was

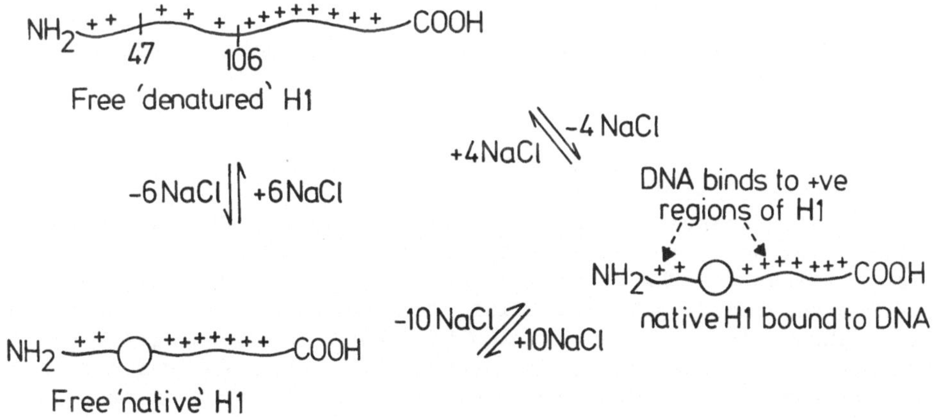

Fig. 11. Salt-dependent equilibria between native and denatured,
free and bound H1.

dissociated and separated from the chromatin by dialysis into 0.7 M
NaCl followed by gel filtration chromatography [13]. An amount of
DNA was then added to the solution of H1 equivalent to the amount
of DNA present in the original chromatin. The mixture was then dialysed into solutions of lower NaCl concentration in the range 10^{-3}
to 0.6 Molar. Samples were removed and applied to a Sepharose 4B
column. The DNA-H1 complex eluted at the exclusion volume of the
column whereas free H1 eluted as a separate, later fraction. The
amount of H1 bound to DNA was determined, the amount of H1 free
estimated by difference, and the association curve obtained. The
binding curves of H1 to DNA and chromatin as a function of NaCl
were very similar. This leads to the conclusion that the binding
sites for H1 on chromatin are primarily formed by the pattern of
free phosphate groups on the DNA created by the binding of the other
four histones [15] and that there is little contribution to the
binding free energy from interactions of H1 with the other four
histones.

(e) Circular Dichroism Studies

It is well known that the circular dichroic spectrum of nucleohistone in the region 300-260 nm is very much reduced in intensity
compared to that of DNA [10,16]. This difference has been attributed to short range effects of the bound histones on the DNA conformation [10]. The separate contributions of the lysine rich and
the arginine rich pair to the decrease in dichroism at 280 nm is
shown as a function of the amount of each histone type bound in
Fig. 12. The removal of H2a-H2b caused a proportional increase in
the dichroism which is consistent with the non-cooperative mode of
binding revealed by the binding curves. By contrast the dissociation of H3-H4 caused a disproportionate change in the dichroism of
DNA showing that the interaction of H3 and H4 with DNA produces long
range cooperative changes in the conformation of the DNA. These conformational changes may be related to the cooperative binding curves
described earlier.

(f) Nuclease Digestion Studies

Nuclei, sheared chromatin, H1-depleted and H1-H2a- and H2b-
depleted chromatin were digested with micrococcal nuclease for
various times, the DNA extracted and the fragments separated and
sized on acrylamide gels (Fig. 13). The nucleosome repeat length for
native chromatin in nuclei was 200 base pairs, in good agreement
with previous estimates [1]. As others have found, sheared chromatin
and depleted chromatin did not give the regular ladder of base pair
repeats observed for the native material [5,6]. H1-depleted chromatin showed discrete bands at 570, 420, 290 and 160 base pairs. The
larger fragments decreased in amount with time to produce a limit

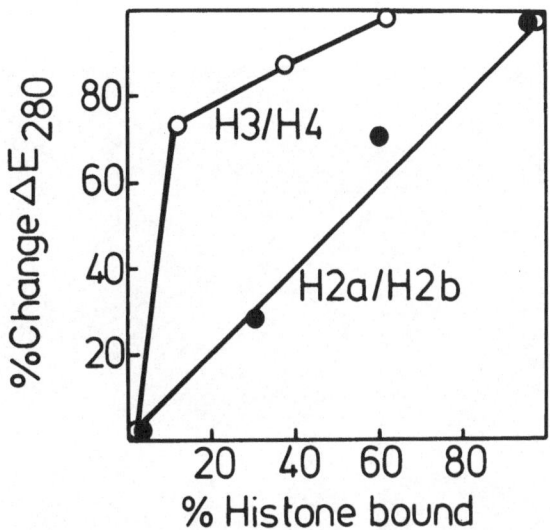

Fig. 12. The percentage change in ΔE at 280 nm for the dissociation
of H2a-H2b over the range 0.7 to 1.2 M NaCl and for the dissociation
of H3-H4 over the range 1.2 to 2.0 M NaCl. The latter spectra were
all measured at 1.30 M NaCl. ΔE is the difference in extinction for
left and right circularly polarised light per mole nucleotide.

digest with bands at 145, 130, 108, 99, 77, 65, 55 and 45 base pairs.
At early stages in the digestion the oligonucleosome patterns indicate an internucleosome spacing of 140-150 base pairs which is considerably less than the repeat length in the native chromatin (200
base pairs). Thus, removing H1 appears to have caused the nucleosome core particles to slide closer together. An alternative explanation might be that the process of digestion and degradation of
the DNA itself promotes a rearrangement of the core particles so
that they move closer together. The oligomer bands are very broad
with a sharper leading edge than trailing edge. For example the
width of the trimer band corresponds to 50 base pairs and the material in the band is clearly very heterogeneous arising from fragments
as large as 450 base pairs and as small as 390 base pairs. It is
possible that at very early times of digestion the repeat length
could approach that of the native chromatin. After longer digestion
times core particles are produced with maximum DNA lengths of 145
base pairs, as observed by others [6,7]. Extensive cutting of intra-
nucleosomal DNA has occurred to give the well-characterised subnucleosomal lengths.

The gels of H3-H4 DNA showed a smear at early times of digestion but later on discrete bands were observed at 260, 194, 145,
135, 129, 105, 73 and 50 base pairs. A metastable limit digest was

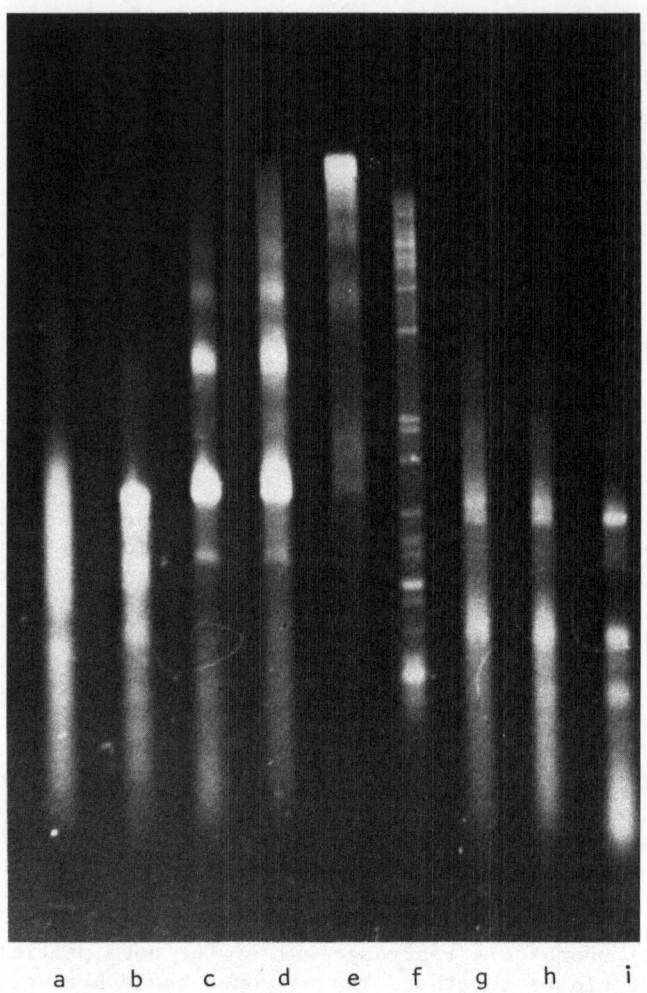

Fig. 13. Digestion of H1-depleted chromatin and H3-H4 DNA with micrococcal nuclease followed by electrophoresis on 5% polyacrylamide gels. a-c, time course of digestion of H1-depleted chromatin: a, 60 mins (36% digestion); b, 15 mins (24% digestion); c, 10 mins (14% digestion). d, partial digest on native chromatin; e, Polyoma DNA digested with Hae III; f-g, time course of digestion of H3-H4 DNA: f, 7 mins (36% digestion); g, 10 mins (38% digestion); h, 40 mins (56% digestion). The extents of digestion were determined in all cases from the absorbance of the sample at 260 nm, after digestion and dialysis, relative to a zero-time control.

reached at about 60% digestion with major bands at 145, 135, 73 and 50 base pairs. It was not possible to estimate a repeat length for the H3-H4 core particle from this data but it is clear that the arginine-rich complex is capable of transiently protecting DNA fragments which are longer than the DNA lengths associated with complete nucleosome core particles (140 base pairs) and longer even than the DNA repeat in native chromatin. From the size of the limit-digest bands it is clear that H3-H4 alone is capable of protecting DNA lengths the size of the complete nucleosome core particle.

DNA, reconstituted by salt dialysis with H2a-H2b or H1, H2a-H2b, and then digested with micrococcal nuclease gave rise to smears on gels; no discrete bands were seen in keeping with other reports [17].

(g) Separation of Limit-digest Products

H3-H4 DNA in 0.7 mM sodium phosphate, pH 7.0, was digested for 30 mins at 37°C with micrococcal nuclease, the reaction terminated by the addition of 10 mM EDTA, pH 7.0, and the products applied to a column of Sephadex G-200 equilibrated with 10 mM EDTA, pH 7.0 at 4°C. The elution profile showed that the column had resolved three fractions of material absorbing at 260 nm. The protein and DNA content of the three fractions were determined and the size of the DNA fragments in each fraction were measured by electrophoresis on polyacrylamide gels (Fig. 14). These experiments showed that the first fraction which eluted from the column had a DNA protein ratio of 1.9. The DNA fragments in this fraction were resolved into three components of length 145, 125 and 104 base pairs present in approximately equal amounts. The amount of protein associated with these fragments suggests that a tetramer of H3-H4 is associated with 104-145 base pairs of DNA. The sedimentation coefficient of fraction I was 7.18 in 10 mM Tris, pH 7.0.

DISCUSSION

The results presented here show that the interaction of core histones with DNA is a thermodynamically reversible process with the respect to the amount and type of histone bound and with respect to the secondary and tertiary structure of the nucleo-protein complex. Other studies have shown that it is also reversible with respect to the intimate association of the histones with DNA as revealed by the protection to digestion of the DNA with micrococcal nuclease [17]. Dissociation and re-association of core histones takes place in two discrete stages. When the NaCl concentration is increased above 0.7 M, the histones H2a and H2b dissociate as an equimolar complex, probably a dimer, in a non-cooperative manner. When the dissociation of this histone pair is complete, H3 and H4

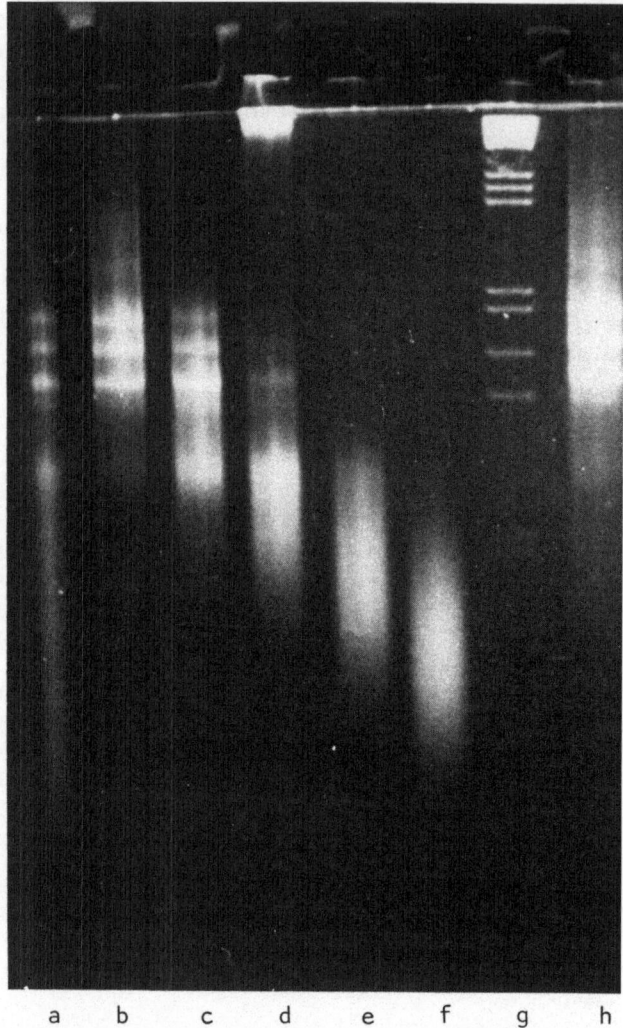

a   b   c   d   e   f   g   h

Fig. 14. Gel electrophoresis patterns of H3-H4 DNA digested with micrococcal nuclease and then fractionated by gel filtration with Sephadex G-200: a, unfractionated digestion mixture; b, c, d, fractions across Fraction I (1st peak from column); e, f, fractions across Fraction II (2nd peak from column); g, PM2 DNA digested with Hae III; h, same as b.

dissociate cooperatively at higher NaCl concentrations as an equimolar complex. The cooperativity is manifested not only in the free energy of interaction between the histones and DNA but also in the conformational change produced in the DNA of the nucleoprotein complex, which shows that the H3-H4 complex is capable of inducing

long range changes in the DNA to produce a conformation which is
intermediate between free DNA and the conformation it adopts in
tightly supercoiled chromatin. It is possible that DNA in 2 M NaCl
exists in a variety of conformational states some of which may
approximate to the supercoiled conformation found in chromatin. If
H3-H4 binds preferentially to supercoiled DNA, the arginine-rich
complex could stabilise such a conformation thus accounting for the
changes in circular dichroism. The binding would be further stabilised by histone-histone interactions between bound tetramers. Such
a mechanism would also explain the decreases in viscosity of the
H3-H4 DNA which, on this view, result from the folding of the DNA
into a supercoiled conformation. Evidence in support of direct
histone-histone interactions has been obtained from crosslinking
studies on H3-H4 DNA at various ionic strengths [3]. It is interesting to note in this context that Bina-Stein and Simpson [26]
have recently reported that H3-H4 histones can compact SV40 DNA by
2.6 fold into nucleosome-like particles. On extensive digestion with
micrococcal nuclease a tetramer of H3-H4 complexed to 104, 125 and
145 base pair lengths of DNA may be isolated. The larger of these
protected fragments is identical in size to the DNA protected in
nucleosome core particles (140 base pairs). This suggests that the
H3-H4 tetramer spans the nucleosome core particles, thus anchoring
the ends of the DNA. The time course of digestion with nuclease
suggests that lengths of DNA greater than the H3-H4 core particle,
140 base pairs, or even the complete nucleosome, 200 base pairs,
can be transiently protected from digestion. This implies that there
are long range interactions between H3-H4 tetramers and higher
polymers which may generate ordering or phasing of the H3-H4 core
along the DNA. The phasing interval cannot be less than 140 base
pairs and, on average, cannot be greater than 200 base pairs, the
repeat length in native calf thymus chromatin. It is tempting to
suggest that the cooperative binding of H3-H4 provides a mechanism
for phasing the completed nucleosomes at 200 base pair intervals
[12]. This argument is, however, difficult to sustain in the light
of the highly conserved amino acid sequences of H3 and H4, presumably reflecting a highly conserved structure and function, and the
variability in the repeat length of nucleosome spacings in chromatin from different sources [1]. Alternatively, it is possible that
H2a and H2b space out the final nucleosome repeat lengths when they
bind to the H3-H4-DNA complex. They probably have sufficient variability in structure to account for the variation in spacer lengths.
In correctly spacing the nucleosomes they would have to interact
with regularly spaced H3-H4-DNA tetramers, already bound to DNA,
and slide them along the DNA. In principle the change in free
energy for the sliding reaction is zero so that the binding of H2a-
H2b dimers would in itself provide sufficient free energy to drive
the reaction until H2a-H2b dimers made contact along the length of
the fibre to produce the correct spacing. Thus each nucleosome core
particle would be associated with 200 base pairs of DNA even though
the core histones in the limit only protected 140 base pairs from

nuclease digestion. However, to achieve this regular spacing requires the H2a-H2b molecules to interact between nucleosomes and
the binding studies reveal no evidence for such interactions either
at the histone-histone level or between histone and DNA.

The proper binding of the H2a-H2b complex to DNA appears to
require the presence of H3-H4. The binding of H2a-H2b and/or H1 to
DNA by itself does not lead to the discrete protection of DNA
fragments from nuclease digestion, unlike H3-H4, as observed by
others [17]. This difference appears to be due to the differential
stability of the secondary structure in the two histone pairs.
Whereas, at low ionic strength, the secondary structure of the histones in the H3-H4-DNA complex is retained, the H2a-H2b-DNA complex
loses structure at low ionic strength. Thus in the former case DNA
maintains the stability of the bound histones, whereas in the latter
it does not. On the other hand, when H2a-H2b binds to the H3-H4-DNA,
the secondary structure is stabilised at low ionic strength, presumably due to histone-histone interactions between the arginine-rich
and the lysine-rich pair. This suggests that the binding site for
the H2a-H2b dimer is created by and formed from the H3-H4-DNA complex, such that binding occurs not only to the histone but also to
the DNA. These conclusions confirm and support the observations of
Sollner-Webb *et al.* [17], Camerini-Otero *et al.* [18] and Oudet *et
al.* [19]on the central importance of the H3-H4 arginine-rich complex in the structure of the nucleosome core particle. The thermodynamic analysis shows that the binding of H2a-H2b is non-
cooperative with respect to histone-histone interactions but highly
cooperative with respect to NaCl. A further indication of cooperativity comes from considerations of the rates of dissociation and
association. The rate of dissociation of core histones from DNA is
a relatively slow process with a half time of at least one hour.
This property has been utilised in separating the products of the
reaction by gel exclusion chromatography in order to construct the
binding curves. Thus the dissociation rate constant is of the order
of 10^4 sec^{-1}. From this and the equilibrium constant (in 1.0 M NaCl
equal to 0.072 M), we estimate that the association rate constant
must be in the range 10^5-10^6 M sec^{-1}. Both rate constants are small
compared to the values expected for simple ionic reactions involving unlike charges. These have very high frequency factors, low
activation energies and large negative activation free energies. By
contrast, the histone-DNA binding, which is dominated by ionic
interactions, appears to involve large free energies of activation.
This may be interpreted to mean that a large number of weak bonds
have to be made or broken simultaneously in the activated complex
for interaction to occur. In other words, the reaction is highly
cooperative.

It is generally assumed that the interaction between histone
and DNA involves electrostatic bonds between the positively charged
lysine and arginine side chains and the phosphate groups in the DNA.

The positive charges on the core histones are clustered mainly in
the N-terminal and C-terminal parts of the molecules [20]. It has
been suggested on the basis of nmr studies that these regions are
structureless and exist, in the uncombined histones, as random
coils which are thought to wrap around the grooves of the DNA when
the histones interact [21,22]. If the extent of the tails is as
defined by Hyde and Walker [18] then the number of basic side chains
varies between 12 in H2a and 15 in H2b and H4. If these basic side
chains neutralise adjacent phosphates on both chains of DNA, they
will interact with at least one half to three quarters of a turn of
DNA helix. Thus, on the assumption that the histone tails wrap into
the DNA grooves, the tails must be flexible enough to wrap around
at least half a turn of DNA and the binding mechanism must involve
a 'zipper' interaction in which successive segments of the molecule
bind consecutively. Such binding would involve fast 'on' and 'off'
rate constants and show a low degree of cooperativity [23]. Alternatively, binding may occur in a single step with the ligand (the
histone basic tale) rigidly and correctly positioned with respect
to the phosphate binding site. This mode of binding could occur if
binding takes place *to one face of the DNA helix only* so that the
tails do not wrap into and round the DNA grooves. Thus, interaction
takes place on that surface of the DNA which faces into the histone
core leaving the other, outer-facing half freely accessible to
enzymes and other small molecules. Such a binding process would be
slower than the 'zipper' mechanism and highly cooperative since all
the corresponding subsites on the macromolecules interact simultaneously. Both 'on' and 'off' rate constants would in this case be
small. The experimental data, such as it is, is more consistent
with the latter kinetic model.

THE MECHANISM OF CORE PARTICLE ASSEMBLY

It is evident from the data presented here that the formation
of core particles from free histone and DNA by dialysis from high
salt is an example of a self-assembling process. The key step in
this assembly appears to be the association of an equimolar H3-H4
complex with DNA; this is followed by the binding of H2a-H2b. Evidence has been presented to show that the core histones in 2 M NaCl
exist in the form of a heterotypic tetramer or octamer [24,25]. If
such species do exist in high salt, it is clear that they are not
on the main thermodynamic pathway of assembly of core particles and
are not a necessary requirement for the *in vitro* assembly. The
situation *in vivo* will presumably depend very strongly on local concentrations of histones, DNA and counter ion at or close to the DNA
replication points.

Several models for chromatin structure have been described
which have as their basis the symmetry elements which arise as a
result of specific interactions between the eight histones in the

nucleosome core particle [15,27,28]. The model described by Hyde
and Walker [15] is based on a central core of histones composed of
two helical polymers of H3-H4 and H2a-H2b. The DNA is supercoiled
around the central core with a pitch determined by the pitch of
the helical polymers. The models described by Weintraub et al. [27]
and Worcel et al. [28] for the 100 Å fibre are generally rather
similar to that described in [15] and will not be discussed further.
It is, however, instructive to see how the assembly of core particles
can be explained in terms of the structure described in [15]. In
high ionic strength solutions DNA may be visualised as an equilibrium mixture of different conformations one of which may approximate to the supercoiled conformation found in chromatin. The core
histones also exist as an equilibrium mixture of heterotypic dimers,
tetramers and higher polymers in equilibrium with homotypic species.
As the salt concentration is decreased these equilibria are disturbed by the cooperative formation of the H3-H4-DNA complex. In
1.2 M NaCl the H3-H4 dimers bound to DNA are in contact and may be
crosslinked with HCHO to form high polymers [3]. The core particles
are completed, on further lowering the salt concentration, by the
association of H2a-H2b oligomers (probably dimers) to specific non-
interacting sites formed by the H3-H4 DNA. This results in a tightening of the structure, shown in a decrease in viscosity, a further
tightening of the supercoil which leads to changes in the circular
dichroic spectrum and the formation of an interacting, linear array
of nucleosome cores to form the 100 Å fibre (Fig. 15).

Native H1 binds to nucleosome core particles reversibly and
non-cooperatively. The binding sites (probably two per core particle) are identical and are formed primarily by a specific pattern
of free DNA phosphate groups. Since the lengths of DNA protected
from micrococcal nuclease digestion increase from 140 to about 180
base pairs when H1 is bound [30], the binding sites are probably
situated at the extreme ends of the DNA in the core. The enthalpy
of H1 binding is zero, in contrast to the negative enthalpy changes
observed on binding the lysine-rich and arginine-rich pairs. This
difference may be due to the possibility that H1 is involved in
purely ionic interactions whereas the core histones bind via ionic
bonds with DNA but also via protein-protein bonds with each other.
The dissociation of H1 involves the net binding of 10 moles of
NaCl per mole of H1 dissociated. Clearly, salt binding is not
stoicheiometric since there are a total of about 60 positive charges
per molecule of H1.

The denatured form of H1 present at NaCl concentrations less
than 0.4 M does not form a complex with chromatin directly, although
it will bind to form a fully renatured complex. The greater extent
of dissociation at low ionic strengths is therefore caused by a
displacement of the equilibrium between bound and free native H1 by
the equilibrium between free native and free denatured H1. The salt-
induced renaturation of H1 involves the net binding of 6 moles of

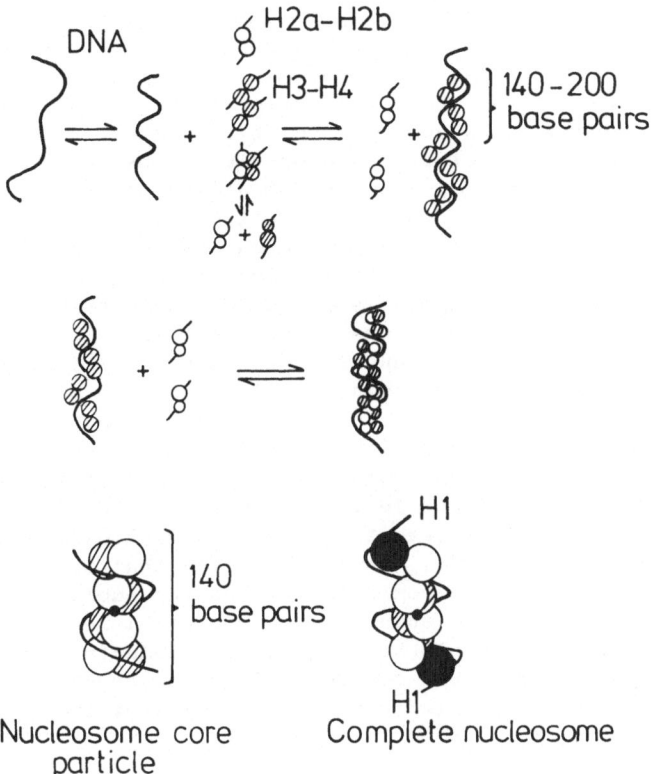

Fig. 15. The association of the arginine-rich and lysine-rich histone pairs to form the 100 Å fibre in terms of the chromatin model described in ref. [15]. Bottom left: details of a nucleosome core particle; bottom right: a complete nucleosome with molecules of bound H1.

NaCl per mole of H1. This shows that electrostatic interactions are important in maintaining the native tertiary structure of the H1. Since the low-ionic-strength form of denatured H1 does not bind to core particles whereas the native, structured form does, the native conformation which includes α-helical structure, is essential for proper binding to occur. It has been suggested that the primary site for H1-chromatin interaction is the c-terminal half of the molecule or the N-terminal, since these are the most basic regions, whereas structure formation takes place in the central region [29]. These results do not support this conclusion. They show either that there must be structured regions in the COOH and NH$_2$ terminal ends, if these are the only sites of interaction, or that the central structured region plays an important role in the binding of H1.

Although H1 binds reversibly to sheared chromatin it is clear
that the removal and rebinding of this histone to core particles
does not restore the 'native' properties of chromatin. Conditions
have not yet been found which will reverse the change which occurs
when native chromatin is rendered soluble by shearing. Until such
time as the native structure of chromatin can be fully restored
the role of H1 in supercoiling the 100 Å fibre must remain unclear.
Despite this drawback one possible function for H1 becomes apparent
from a consideration of the way in which the 100 Å fibre must fold
or coil to form the 300 Å supercoil. Crosslinking studies using
very short crosslinking agents such as HCHO have shown that nucleosome core particles in solutions of moderate ionic strength make
contact through histone-histone interactions [3]. If it is assumed
that the core particles are identical in structure and that they
make contact through specific interactions then each core particle
is related to its neighbour along the fibre by symmetry rules
which give rise to a linear structure [15]. It is not possible to
coil the linear 100 Å fibre so generated into a 300 Å fibre without
breaking the symmetric interactions between nucleosomes despite
claims to the contrary [28]. It is suggested that the role of H1
is to break the symmetric contacts between the nucleosome cores by
binding to the DNA at the ends of the core particles and tightening
or loosening the DNA double helix (that is, inducing superturns).
This process would have the combined effect of breaking the internucleosome contacts and also providing sufficient free energy to
further coil the 100 Å fibre into the 300 Å fibre.

REFERENCES

1. Kornberg, R.D., Ann. Rev. *Biochem.* (1977), 931-954.
2. Felsenfeld, G., *Nature* (1978), **271**, 115-122.
3. Hyde, J.E. and Walker, I.O., *FEBS Letts.* (1975), **50**, 150-154.
4. Noll, M., *Nature* (1974), **251**, 249-
5. Noll, M., Thomas, J.O. and Kornberg, R.D., *Science* (1975), **187**, 1203-1206.
6. Sollner-Webb, B. and Felsenfeld, G., *Biochemistry* (1975), **14**, 2915-2920.
7. Axel, R., *Biochemistry* (1975), **14**, 2921-2925.
8. Zubay, G. and Doty, P., *J. Mol. Biol.* (1959), **1**, 1-21.
9. Henson, P. and Walker, I.O., *Eur. J. Biochem.* (1970), **14**, 345-350.
10. Henson, P. and Walker, I.O., *Eur. J. Biochem.* (1970), **16**, 524.
11. Noll, M., *Cell* (1976), **8**, 349-355.
12. Burton, D.R., Hyde, J.E. and Walker, I.O., *FEBS Letts.* (1975), **55**, 77-80.
13. Skidmore, C.J., Walker, I.O., Pardon, J.F. and Richards, B.M., *FEBS Letts.* (1973), **32**, 175-178.
14. Henson, P. and Walker, I.O., *Eur. J. Biochem.* (1971), **22**, 1-4.
15. Hyde, J.E. and Walker, I.O., *Nucleic Acids Res.* (1975), **2**, 405-421.

16. Wilhelm, X. and Champagne, M., *Eur. J. Biochem.* (1969), **10**, 102.

17. Sollner-Webb, B., Camerini-Otero, R.D. and Felsenfeld, G., *Cell* (1976), **9**, 179-193.

18. Camerini-Otero, R.D., Sollner-Webb, B. and Felsenfeld, G., *Cell* (1976), **8**, 333-347.

19. Oudet, P., Germond, J.E., Sures, M., Gallwitz, D., Bellard, M. and Chambon, P., *Cold Spring Harbour Symp. Quant. Biol.* (1977) in press.

20. Elgin, S.R. and Weintraub, H., *Ann. Rev. Biochem.* (1975), **44**, 725-794.

21. Bradbury, E.M. and Rattle, H.W.E., *Eur. J. Biochem.* (1972), **27**, 270-281.

22. Lilley, D.M.J., Howarth, W., Clark, V.M., Pardon, J.F. and Richards, B.M., *Biochemistry* (1975), **14**, 4590-4600.

23. Burgen, A.S.V., Roberts, G.C.K. and Feeney, J., *Nature* (1975), **253**, 753-755.

24. Weintraub, H., Palter, K. and Van Lente, F., *Cell* (1975), **6**, 85-110.

25. Thomas, J.O. and Butler, P.J.G., *J. Mol. Biol.* (1977), **116**, 769-781.

26. Bina-Stein, M. and Simpson, R.T., *Cell* (1977), **11**, 609-618.

27. Weintraub, H., Worcel, A. and Alberts, B., *Cell* (1976), **9**, 409-417.

28. Worcel, A. and Benyajati, C., *Cell* (1977), **12**, 83-100.

29. Bradbury, E.M., Cary, P.D., Chapman, G.E., Crane-Robinson, C., Danby, S.E., Rattle, H.W.E., Bublik, M., Palau, J. and Avites, F.J., *Eur. J. Biochem.* (1975), **52**, 605-613.

30. Kumar, H. and Walker, I.O., unpublished observations.

NUCLEOSOME SHAPE AND STRUCTURE IN SOLUTION FROM FLOW BIREFRINGENCE

R. E. Harrington

University of Nevada, Reno

Reno, Nevada 89557

As our knowledge of the general features of chromatin structure at all levels improves, we will necessarily wish to shift our
interest more and more toward an understanding of those dynamical
characteristics of chromatin structure that relate directly to
biological function. There is an increasing body of evidence, both
theoretical and experimental, that structural dynamics in chromatin
are to a degree controlled, and certainly modulated, by solvent
factors, particularly ionic strength and specific ion polyelectro-
lyte effects.[1] Thus, an increasingly heavy experimental burden
will fall upon those biophysical methods capable of giving structural information in solution, and especially upon those capable of
interpretation under more-or-less arbitrary solvent conditions.

The number of biophysical methods which can provide this kind
of structural information on complex systems such as nucleosomes
or higher order chromatin chains in solution is not large. Spectroscopic and low angle scattering experiments involving X-rays and
neutrons have already made significant contributions to the structural problem. Hydrodynamic property measurements, particularly
sedimentation and translational diffusion coefficients from inelas-
tic light scattering, have provided good molecular weight data but
have been less successful, so far at least, in characterizing
particle shape and structure. Much of the latter difficulty comes
from the fact that the frictional coefficient must be related to an
assumed hydrodynamic model, and especially for more complex structures, the experimental data are only rarely consistent with a
single, unique structure. However, these methods can be exceedingly
useful for detecting structural changes in nucleosomes and chromatin chains in solution.[2-4]

Most of the hydrodynamic property work to date on chromatin and its subunits has involved the complementary methods of sedimentation and translational diffusion. For characterizing asymmetric rigid particles and stiff chains, however, hydrodynamic properties based upon rotational diffusion offer certain advantages both in sensitivity and in clarity of interpretation. Two such measurements are flow birefringence (or dichroism) and intrinsic viscosity. These methods are complementary in the sense that they can be combined to remove much of the ambiguity associated with the rotational frictional coefficient analogously to the more usual combination of sedimentation and translational diffusion.

Flow birefringence is an ancient and venerable method which has received periodic attention by polymer physicists and fluid dynamicists since its simultaneous discovery in 1873 by Mach[4] and by Maxwell.[5] However, it was not until well into the present century that the correct hydrodynamic basis of the so-called "Maxwell effect" was elucidated, and a complete dynamical theory for a suspension of rigid ellipsoids of revolution in 3-dimensional velocity gradient flow based upon rotational diffusion in continuum fluids did not appear until 1939.[7] This theory, by Peterlin and Stuart, remains the accepted treatment for rigid particle systems at the present time. More recent studies of the flow birefringence phenomenon in 3-dimensions appears to have established beyond question both the correctness and the completeness of the Peterlin and Stuart hydrodynamic treatment.[8]

The associated optical treatment, based upon the theory of macroscopic continuum dielectrics, has been criticized in its application to molecular systems, particularly in its ability to apportion correctly the observed birefringence between the intrinsic and form or shape components. This question is of importance because the form birefringence disappears only when the absolute refractive indices of solvent and particles become equal, an experimental condition not usually attainable with most biological systems in physiological or quasi-physiological solvents. Experimental studies have demonstrated that the Peterlin and Stuart optical treatment evidently is satisfactory in its application to DNA, however.[9,10] In any event, the interpretation of the extinction angle is based entirely upon hydrodynamic theory. Hence, this quantity permits the rotational diffusion coefficient and the size and shape parameters of the particle to be obtained unambiguously from accurate experimental measurements, and thereby provides, in favorable cases, an internal check upon the optical theory of the birefringence.

The first really productive applications of flow birefringence to biological systems were the pioneering studies on fibrous and globular proteins, first by Edsall and subsequently by Scheraga and others. These studies, along with others involving viruses and other rigid systems, are documented along with associated bibliography in several excellent reviews.[11-16] The Peterlin and Stuart rigid particle theory seems to describe remarkably well the flow birefringence of these systems when formal consideration of polydispersity is included in the treatment.[17] Thus, a fairly extensive precedent exists for using flow birefringence to characterize the size and shape of rigid particle systems in solution through the analysis of the rotational diffusion coefficient and the intrinsic and shape birefringence. However, the earlier studies were limited, sometimes seriously, by available sensitivity in the measurement of the flow birefringence and extinction angles, and the earlier results may also have been complicated by the nonlinearity inherent in the optical instruments used. Further application of the method to small rigid macromolecules and structures of low asymmetry in solution had to await the development of more sensitive photoelectric instrumentation.

More recently, flow birefringence has provided considerable information on DNA structure and conformation in aqueous solution. DNA is a stiff-chain polymer under these conditions which assumes an expanded coil conformation at high molecular weights. Its rotational dynamics are therefore complicated and involve multiple relaxation processes. The application of flow birefringence to chain macromolecules has paralleled the development of chain dynamical theory following the demonstration by Kuhn and Grun[18] that the optical anisotropy of a chain element can be taken as proportional to its mean square length. Current dynamical theory is based upon the Gaussian subchain model as developed by Bueche, Rouse, Zimm, Tschoegl and Peterlin. This model is equivalent in the high molecular weight limit to the Kratky-Porod wormlike coil model which has been used quite successfully to characterize many solution properties of DNA. These models and their applicability to biological systems including DNA have been reviewed by Bloomfield et al.[19] and by Harrington.[20]

Because of limitations inherent in the Gaussian subchain model, real progress in the application of available theory to experimental flow birefringence data on DNA could not occur until the development of highly sensitive instrumentation capable of accurate measurement near the limits of zero concentration and shear. In those laboratories where satisfactory instrumentation was developed, however, progress was relatively rapid. Tsvetkov, Frisman and others in the Soviet Union used flow birefringence to characterize a variety of solution properties of DNA.[14] Harrington in the United States has determined the flow birefringence and extinction angles of DNA both at very low[9,10] and at very high[20] molecular weights. These investigations have provided considerable experimental information on the structure of DNA in neutral aqueous buffers of near physiological ionic strength including evidence in support of the Watson-

Crick B-form structure. Perhaps the most useful information to
come from flow birefringence studies upon high molecular weight
DNA was the characterization of chain stiffness. The Gaussian subchain dynamical theory predicts that both the intrinsic birefringence[21] and the form birefringence[22] are related in a similar way
to the persistence length of the chain. These factors provide the
basis for Tsvetkov's theoretical prediction that the ratio of the
Maxwell constant to the intrinsic viscosity should be directly proportional to persistence length.[14] This prediction has been thoroughly substantiated for DNA.[14,20]

     It is clear that flow birefringence has developed experimen-
tally and theoretically to the point where it offers real potential
to investigate chain stiffness and its relation to higher order
structure in chromatin fragments in dilute solution. Moreover,
since flow birefringence is a relatively simple yet highly sensitive experiment using contemporary instrumentation and techniques,
it offers exciting possibilities for the extensive study of structural dynamics in chromatin as induced by solvent effects and the
binding of cationic proteins. Such investigations could provide
the conceptual precursors for understanding the true structure-
function relationships in this most labile and dynamic material.

THE NUCLEOSOME SHAPE IN SOLUTION: FLOW BIREFRINGENCE STUDIES

     We have utilized the high available sensitivity of our photo-
electric flow birefringence apparatus to study the extinction angle
and flow birefringence of isolated, intact nucleosomes from chicken
erythrocytes. Because of the relatively small size and high symmetry of these particles, flow orientation is small and extremely
stable and sensitive equipment is required.

                              Materials

     Isolated, fractionated nucleosome monomers from chicken eryth-
rocytes were prepared using standard methods[23,24] at the Division
of Biology, Oak Ridge National Laboratory, using facilities generously provided by Dr. Donald E. Olins. The whole nucleosome
preparations so obtained were dialyzed exhaustively against 0.1 M
KCl 0.2 mM EDTA solvent. The soluble nucleosome fraction obtained
from this treatment has been shown previously to contain negligible
H1 histone.[25]

     Samples were dialyzed against the KCl-EDTA solvent immediately
prior to flow birefringence runs. Final dialyzate was used as
reference solvent in all cases. In order to conserve material, a
total of 5 replicate runs were made at each sample concentration.
This latter procedure was required because of the relatively large

volume of the optical flow cell (~5ml). As an added precaution, samples were dialyzed overnight against pure solvent between runs. No changes in the flow birefringence, intrinsic viscosity, absorbance ratios at 230, 260 and 280 nm or circular dichroism were noted with successive flow birefringence determinations. Hence, the method as we employed it seems to be completely non-destructive.

General Methods

Concentrations of nucleosome solutions were determined on a Zeiss PMQ-II UV-visible spectrophotometer using an extinction coefficient for a 1% solution $E_{260}^{1\%}$ = 93.1.[25] The specific refractive index increment of KCl soluble nucleosomes in the 0.1 M KCl 0.2 mM EDTA solvent was determined as dn/dc = 0.184 ± 0.006 cm^3/gm using a Phoenix differential refractometer with sample concentrations determined spectrophotometrically as described above. The solvent refractive index n_0 = 1.33387 was determined on the same differential refractometer using Phoenix standard KCl calibration data.

The experimental intrinsic viscosity [η] = 11.2 cm^3/gm was measured in our laboratory at low shear rates using a cartesian diver Couette-type viscometer.[20] The partial specific volume of nucleosomes $\bar{v}$ = 0.661 cm^3/gm was measured in the laboratory of Professor Walter E. Hill of the University of Montana on a Mettler-Parr densitometer. Both of the above values have been reported previously.[23]

Measurement of Flow Birefringence

In the classical method of measuring birefringence, the birefringent medium is placed between crossed polarizers. The transmittance of this system is related to the phase retardation along the slow axis of the medium and hence to the birefringence, Δn, given the optical path length, L, of the medium.

$$T = \tfrac{1}{2} \sin^2 \frac{\pi L \Delta n}{\lambda} (1 - \cos 4\beta) \tag{1}$$

In this expression, β is the angle between the slow axis of the birefringent medium and the plane of polarization of the polarizer, and λ is the wavelength of the light. This equation shows that elliptically polarized light is transmitted except when the slow axis of the medium is along either polarization axis. Under these conditions, a four-fold symmetric region of extinction is observed corresponding to the symmetry in β. If the birefringence is due to the hydrodynamic orientation of suspended anisotropic particles, as it is in flow birefringence, it is clear that if the principal optical and geometrical axes are coincident, the extinction angle, χ, is also the mean orientation angle of the particles with respect to the streamlines of flow. These relations are illustrated for the case of Couette (concentric cylinder) hydrodynamic flow in Figure 1.

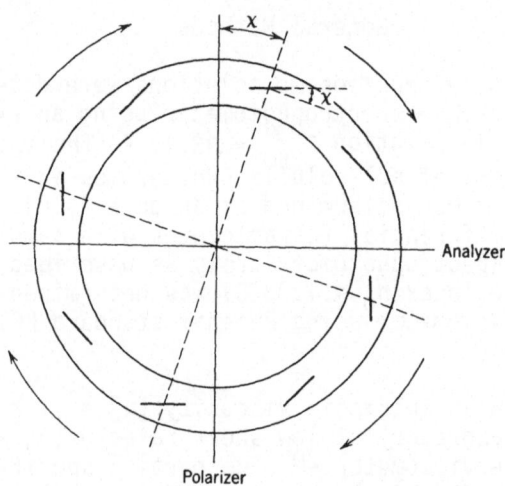

Figure 1. Mean molecular orientation in the velocity gradient of a Couette device (outer cylinder rotating clockwise as shown) for rodlike particles. Relationships are indicated for the extinction angle, χ, with respect to the flow streamlines and for the cross of isocline with respect to the polarizer plane.

In the optical system just described, the birefringence has a quadratic dependence upon the transmittance (for small birefringence) so that different sources of birefringence such as that due to the solute in solution, the solvent alone and various contributions from the optical components of the instrument itself mix nonlinearly and cannot in general be separated. Such an optical system cannot be used, therefore, to determine accurately very small amounts of solute flow birefringence due to inherent noise levels in any flow birefringence instrument. This problem affects all optical flow birefringence equipment presently in use and severely limits their usefulness in the study of small biological macromolecules and aggregate structures.

The transmittance of the optical system just described can be linearized if the polarizer and analyzer are at an angle somewhat different from 90° and if a large amount of fixed birefringence is deliberately introduced into the system. Under these conditions, the transmittance is given by three terms: a large term of constant transmittance independent of the birefringence; a term linear in the birefringence; and a term quadratic in the birefringence with a four-fold symmetry in β as in the crossed polaroid case above.

We have developed a highly sensitive photoelectric flow birefringence instrument based upon the general principles just described. In our instrument, the entire annulus of a Couette-type flow cell is scanned by two spots of constant intensity light located 180° opposite to one another. The transmitted intensity is detected photoelectrically, and the linear term is isolated electronically. A schematic representation of the instrument is given in Figure 2.

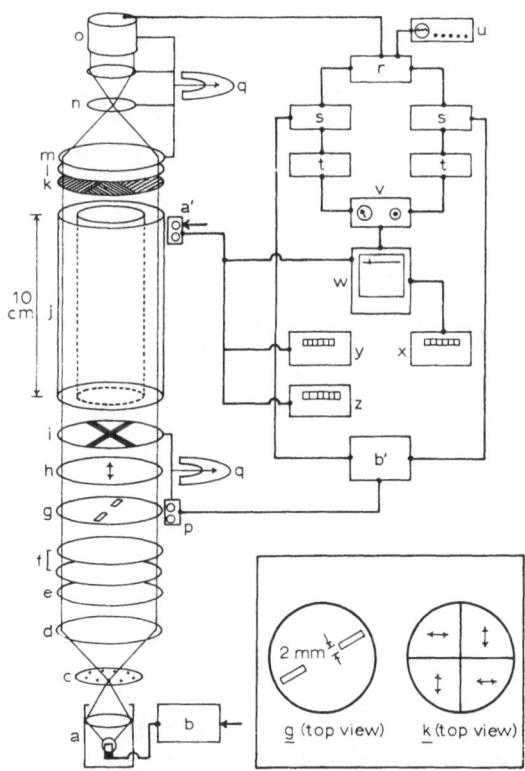

Figure 2. Schematic representation of the flow birefringence instrument: (a,b,c) light source; (d) projection lens; (e) 546 nm sharp cut-off filter; (f) light field compensation plates; (g) scanning plate; (h) polarizer; (i) cross bars; (j) concentric cylinder flow cell (rotating outer cylinder); (k) calibration plate; (l) quarter wave plate; (m) analyzer; (n,o) photocell and optical system; (q) angular scales; (p,r-b') electronic Fourier analyzer

Figure 2. (continued)
and time averaging system. Items (m,n,o) are rotated together to
set the angular variable γ. Items (i,p) are rotated together to
set θ.

The intensity of light due to the linear term in the transmittance is obtained by integrating the transmittance over the
dimensions of the scanning light beams.

$$I_{linear} = \int_{\theta-\phi}^{\theta+\phi} T_{linear} d\theta = \pm D \sin \frac{2\pi L \Delta n}{\lambda} \sin 2\gamma \sin 2\phi \sin 2(\theta-\chi) \quad (2)$$

In this expression, D. is the annular gap width, 2ϕ is the annular
region defined by the width of the scanning slits (g in Figure 2),
χ is the extinction angle, γ is the angle between the planes of the
polarizers, θ is the annular position relative to the plane of the
first polarizer (h in Figure 2) and the positive and negative signs
are associated with clockwise and anticlockwise rotation respectively. Thus, scanning the annulus produces a photomultiplier
signal whose intensity as a function of angle is proportional to
the flow birefringence (for small birefringence such that $\sin\Delta n \approx \Delta n$)
and whose phase is proportional to the extinction angle.

It is clear from equation (2) that both extinction angle and
flow birefringence can be determined from independent measurements
at two values of θ. However, data reduction and time averaging
procedures are simplified if measurements are made at exactly
$\theta = 0°$ and $\theta = 45°$. Under these conditions, equation (2) gives the
following for the photomultiplier output.

$$E_c(45) = k(45)\Delta n_c \cos 2\chi_c$$
$$E_c(0) = -k(0)\Delta n_c \sin 2\chi_c$$
$$E_a(45) = -k(45)\Delta n_a \cos 2\chi_a \quad (3)$$
$$E_a(0) = k(0)\Delta n_a \sin 2\chi_a$$

In these equations, the subscript c refers to clockwise rotation of
the flow cell and a to anticlockwise rotation, and k is an instrument constant determined by calibration against an optical standard
of known birefringence and phase (k in Figure 2).

Equations (3) will in general contain both the flow birefringence of the solution and extraneous birefringence Δn_2 with phase
$\cos 2\chi_2$ due to the instrument itself. Since the instrument responds
only to the linear term in the birefringence, these effects are

additive. However, the flow birefringence changes sign with rotational sense of the flow cell while the extraneous signal is constant. Thus, the photomultiplier outputs of equation (3) can be
combined to isolate the extraneous term.

$$2A = E_c(45) + E_a(45) = k(45)[\Delta n_c \cos 2\chi_c - \Delta n_a \cos 2\chi_a]$$

$$+ 2k'\Delta n_2 \cos 2\chi_2$$

$$= -2k(45)\Delta n \cos 2\chi + 2k'\Delta n_2 \cos 2\chi_2 \qquad (4)$$

$$2B = E_c(0) - E_a(0) = k(0)[-\Delta n_c \sin 2\chi_c - \Delta n_a \sin 2\chi_a]$$

$$= -2k(0)\Delta n \sin 2\chi$$

Δn and χ are now the true birefringence and extinction angle
obtained, in effect, as the average of clockwise and anticlockwise
flow cell rotation.

Since B in equation (4) contains no extraneous birefringence,
its intercept at zero velocity gradient, G, will be due to minor
errors in setting θ or in measurement of flow cell angular velocity.
Such intercepts can always be removed adequately by fitting B to a
quadratic polynomial in G and subtracting the zero shear intercept.
However, particularly at extinction angles near 45°, the A function
in equation (4) may be quite nonlinear and determination of the
zero shear intercept may be difficult.

A satisfactory procedure for obtaining $2k' \Delta n_2 \cos 2\chi_2$ in
equation (4) has recently been described by Barrett and Harrington.[26]
Since the extinction angle is particularly sensitive to the extraneous term in A, the procedure, in effect, maximizes the fit of
extinction angle versus shear rate data to the best smooth curve.
Although the method was developed for large chain macromolecules,
it is equally applicable to rigid particle systems such as
nucleosomes.

Most theories for the extinction angle predict a dependence
upon shear of the form

$$\cot 2\chi = \frac{A_{corr}}{B_{corr}} = \frac{A + c}{B_{corr}} = k_1 G + k_2 G^2 + \cdots \qquad (5)$$

where the subscript corr refers to an intercept-corrected value
and c is the residual correction to A. The constants k_1 and k_2
are functions of shear dependent reduced viscosity and particle
shape parameters only, and are very nearly independent of shear.

To a first approximation, therefore,

$$\frac{A}{B_{corr}} = k'_0 + k'_1 G + k'_2 G^2 + \cdots \tag{6}$$

Subtracting equation (6) from equation (5) and rearranging with neglect of terms in G higher than the first power, one obtains

$$B_{corr} G = \frac{k'_0}{k_1 - k'_1} B_{corr} + \frac{c}{k_1 - k'_1} \tag{7}$$

Thus, a plot of A/B_{corr} versus G gives k'_0 as the intercept. This value is then used along with a plot of $B_{corr} G$ versus B_{corr} in a quadratic or higher order regression analysis to obtain the residual correction.

$$c = \left(\frac{\text{constant coefficient}}{\text{linear term coefficient}} \right) k'_0 \tag{8}$$

This procedure is performed by computer and reiterated until c is a constant to the desired degree of accuracy. When a satisfactory value of A_{corr} is obtained, the true extinction angle is calculated from equation (5) and the flow birefringence from the relation

$$\Delta n = (A^2_{corr} + B^2_{corr})^{\frac{1}{2}} \tag{9}$$

This data reduction procedure results in an exceedingly high level of experimental precision when the instrument functions A and B are obtained from effectively time averaged values of E in equation (3).

Because of the photoelectric detection feature and the fact that the entire annulus is utilized in the detection of flow birefringence, the instrument described is capable of extremely high theoretical sensitivity. The principal limitation in this respect is the available signal-to-noise ratio. Instrumental noise ranges from the steady extraneous instrumental contribution and very slow drift due to thermal changes in the optical bench components to rapid optical noise in the flow cell and electrical noise in the photomultiplier housing. Most of the noise is optical, however, and comes from light reflections in the flow cell, scattering due to particulate material in the flowing solution, and most importantly, refractive gradients in the flowing solution due to heating at high shear rates. The signal-to-noise ratio is significantly improved by time averaging the instrument outputs in equation (3) over the same integral number of flow cell revolutions. In our instrument, this is done automatically with an integrating recorder as shown in Figure 2.

The instrumental factors just described become particularly acute in studies on isolated nucleosomes in solution at low concentrations. Because of the small size and relatively low asymmetry of these particles, both the flow orientation and the flow birefringence are very small even at high velocity gradients. Thus, the extraneous instrumental birefringence and the optical noise peaks at the higher shear rates may be many times larger than the true solution flow birefringence, and effective noise averaging and correction to the A function becomes exceedingly critical. This is especially true for the extinction angles; equation (5) shows that if the extinction angle is very near 45°, the theoretical limit for dominant rotation diffusion, the B function becomes vanishingly small and determining its magnitude in the presence of high levels of optical noise becomes quite difficult.

Flow Birefringence Run Protocol

We have determined the flow birefringence and extinction angles of nucleosomes in the velocity gradient range 1,000 to 16,000 sec^{-1} (the present non-turbulent upper limit of our apparatus with these solutions). Although our flow cell components including the support bearings are effectively thermostatted, appreciable sample heating always occurred during runs due to shearing at the higher shear rates. After a minute or more of such shearing, thermal gradients were produced leading to severe optical noise and drift. To minimize such effects, a standard run protocol was developed: (1) the flow cell was brought to speed and equilibrated for 10 sec; (2) the output signal (equation (3)) was then time averaged for 25 sec; (3) the system was allowed to re-equilibrate thermally at a low flow cell speed for 5 min or longer. Under these conditions, sample heating was kept to less than $\frac{1}{2}$°C and thermal noise and drift were held to tolerable levels.

Extinction Angle Data

Extinction angles extrapolated to zero concentration are shown in Figure 3 as a function of shear rate. These data are remarkable for their precision at very low particle orientation. The data are compared to calculated lines for ellipsoids of axial ratios $\frac{1}{2}$ (oblate) and 2 (prolate) using rigid particle theory.

The rotational diffusion coefficient for a rigid ellipsoid of revolution is given as [10,27]

$$D_r = \frac{\delta RT}{6\eta_0 [\eta] M} \tag{10}$$

in which $\delta = J\nu$ is a relatively insensitive function of axial ratio, p, and J and ν are as defined by Perrin[28] and Simha[29] respectively. The extinction angle is calculated from the rotational diffusion coefficient using the theory of Peterlin and Stuart.[7]

$$(G/D_r \leq 1.5) \qquad X = \frac{\pi}{4} - \frac{G}{12D_r}\left[1 - \frac{G^2}{180D_r^2}\left(1 + \frac{24d^2}{35}\right) + \cdots\right] \qquad (11)$$

$$d = (p^2 - 1)/(p^2 + 1)$$

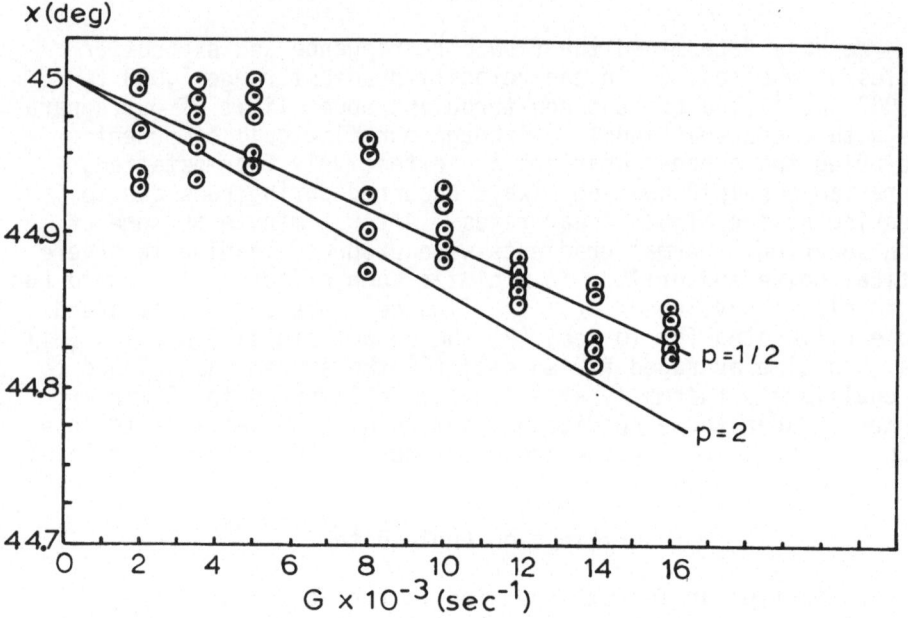

Figure 3. Experimental extinction angle versus velocity gradient data. Solid lines: theoretical trends for oblate ($p = \frac{1}{2}$) and prolate ($p = 2$) ellipsoids of revolution.

The solid lines in Figure 3 are obtained from equation (11) and rotational diffusion coefficients as given by equation (10).

Table I

oblate ellipsoid	prolate ellipsoid
$p = \frac{1}{2}$	$p = 2$
$\delta = 2.52$	$\delta = 1.93$
$D_r = 4.49 \times 10^5$	$D_r = 3.44 \times 10^5$

It is evident that the extinction angle data are most consistent with the oblate model; the disagreement with the prolate
model is outside experimental error except at the lowest shear
rates. However, the scatter in the data are still large on the
scale required, and although the extinction angles definitely
imply an oblate model with axial ratio approximately $\frac{1}{2}$, they are
not sufficiently precise to establish the axial ratio at a useful
level of precision.

Flow Birefringence Data

Flow birefringence data are shown in Figure 4 as functions of
shear rate. Each data point in this case represents the average
of 5 replicate determinations. The precision of the data is excellent in comparison to previous flow birefringence work at high
shear rates, and is particularly so in view of the relatively small
birefringences measured.

The shear dependence of the flow birefringence was fitted by
least squares to a quadratic polynomial: $\Delta n = a + bG + cG^2$. The
linear term coefficients are given in Table II.

Table II

c (ugm/ml)	$b = (\Delta n/G)_{G \to 0}$
100	-9.028×10^{-14}
300	-2.709×10^{-13}
500	-4.514×10^{-13}
700	-6.320×10^{-13}

The data in Table II are linear with zero intercept within experimental error. A linear least squares analysis gives the Maxwell
constant from the slope.

$$[n] = \left(\frac{\Delta n}{cG \, \eta_o}\right)_{c,G \to 0} = -1.010 \times 10^{-7} \text{ (cgs units)}$$

$$\frac{[n]}{[\eta]} = -8.418 \times 10^{-9} \text{ (cgs units)}$$

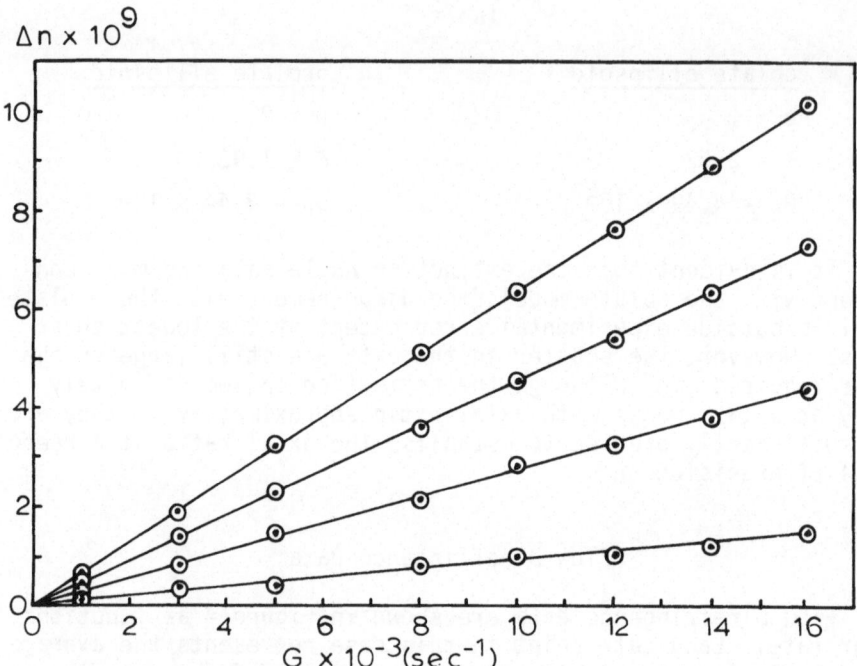

Figure 4. Experimental flow birefringence versus velocity gradient data. Sample concentrations are upper line to lower line respectively: 700 ugm/ml; 500 ugm/ml; 300 ugm/ml; 100 ugm/ml.

The experimental data can be interpreted using the theory of rigid particles.[10] The Maxwell constant is related to the anisotropy in polarizability per unit volume

$$[n] = \left(\frac{\Delta n}{cG \; n_0}\right)_{c,G \to 0} = \frac{2\pi \bar{v} d}{15 n n_0 D_r} (g_1 - g_2) \tag{12}$$

where $\bar{v}$ is the partial specific volume of the particles, and n and n_0 are respectively the refractive index and absolute viscosity of the solvent. Certain specific details of the theoretical model are removed by replacing the rotational diffusion coefficient by the experimental intrinsic viscosity.

$$\frac{[n]}{[\eta]} = \frac{4\pi}{5nkT} \frac{\bar{v}M}{\delta N_A} (g_1 - g_2) \tag{13}$$

In equation (13), M is the particle molecular weight, N_A is Avogadro's number, k is the Boltzmann constant and T is absolute temperature.

The individual polarizability components can also be determined from the experimental refractive index increment.

$$\frac{dn}{dc} = \frac{2\pi\bar{v}}{3n} (g_1 + 2g_2) \tag{14}$$

Thus, the principal components of the absolute real refractive index ellipsoid can be calculated from the relations

$$(\gamma_1 - \gamma_2) = \frac{M\bar{v}}{N_A} (g_1 - g_2)_{int} \tag{15}$$

$$\gamma_i = \frac{v_p}{4\pi} \frac{(n_i^2 - n^2)}{1 + [(n_i^2 - n^2)/n^2]L_i} \tag{16}$$

$$L_1 = \frac{(1 - e^2)}{e^2} \left[\frac{1}{2e} \ln \frac{1 + e}{1 - e} - 1 \right]$$

$$e^2 = \frac{(a^2 - b^2)^{\frac{1}{2}}}{a}$$

$$L_1 + 2L_2 = 1$$

Using equations (13)-(16), the absolute optical parameters for nucleosomes are computed from the experimental data and specific refractive index increment dn/dc = 0.184. Such calculations are shown for oblate and prolate models in Table III. In both cases, axis 1 is the "symmetry" axis.

Table III

	oblate	prolate
$(g_1 - g_2)$	3.26×10^{-3}	-2.49×10^{-3}
g_1	6.12×10^{-2}	5.73×10^{-2}
g_2	5.79×10^{-2}	5.98×10^{-2}
n_1	1.665	1.598
n_2	1.608	1.640
$(n_1 - n_2)$	5.70×10^{-2}	-4.20×10^{-2}

Comparison of the Data with Various Nucleosome Models

Several models for nucleosome shape and structure have recently emerged based upon low angle neutron[30,31] and X-ray[25] scattering data and upon X-ray crystallographic results.[32] All these models are consistent with an oblate ellipsoidal shape having an axial ratio of roughly $\frac{1}{2}$ and with the DNA wound in an overall equivalent circular fashion around the minor (symmetry) axis. However, none of these studies have yet been performed at high enough resolution to determine the nucleosome structure unambiguously. Low angle neutron scattering with H_2O/D_2O contrast matching definitely indicates a concentric structure with DNA on the outside of an inner histone core. Furthermore, this core consists of an octamer containing 2 each of the inner histones H2a, H2b, H3 and H4.[33] Some evidence is available that this octamer may consist of two heterotypic tetramers as basic subunits.[34] This concept, combined with the DNA structural model of Pardon et al. and Klug et al. in which 140 nucleotide pairs of DNA wind in a superhelical trajectory of 1-3/4 turns around the histone core, offers an attractive hypothesis for the first level of structural transformation in nucleosomes during the dynamic phases of transcription and replication: a transition from a closed nucleosome to a more open structure of some kind. Some experimental evidence for such a transition already has been reported.[2,35]

In view of the uncertainty still surrounding the nucleosome structure, it will be of interest to compare the flow birefringence results to current nucleosome models. If we model the nucleosome after Pardon et al. and Klug et al., we may define principal refractive indices as shown in Figure 5.

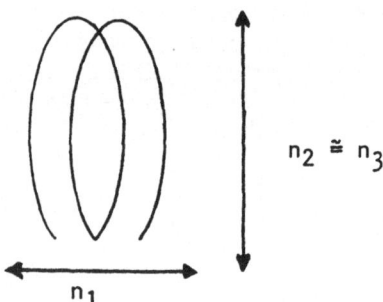

$$n_2 \cong n_3$$

$$n_1$$

Figure 5. DNA trajectory in the nucleosome based upon the oblate ellipsoid model of Pardon et al.[30] and Klug et al.[32] The core histone octamer is not shown, but is presumed to be optically isotropic.

The absolute refractive indices of DNA have been determined previously from flow birefringence data as $n_1 = n_{\shortparallel} = 1.622$ and $n_2 = n_{\perp} = 1.749$, where the subscripts are with respect to the helical axis.[10] If a constant 0.084 is added to the oblate refractive indices in Table III, the following relationships are obtained.

$$n_1 + 0.084 = 1.749 = n_2 \text{ (DNA)} \tag{17}$$

$$n_2 + 0.084 = 1.692 \cong \frac{n_1 \text{ (DNA)} + n_2 \text{ (DNA)}}{2} \quad (=1.686)$$

This degree of agreement with the oblate model may be in part fortuitous, although it must be remembered that the same methods and same theory were used to obtain the refractive indices of both nucleosomes and DNA. The results are certainly highly suggestive of an oblate model with the DNA wound superhelically on the outside of an optically isotropic inner histone core. The equivalent prolate model would show birefringence and anistropy of opposite sign to that actually observed.

An alternative model consistent with the algebraic sign of the observed flow birefringence would be a prolate model with the DNA wound along the symmetry axis in some fashion. A possible schematic representation for such a model is given in Figure 6.

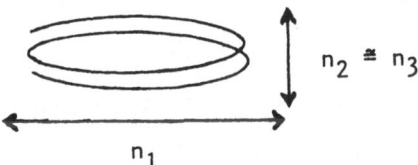

Figure 6. Possible DNA trajectory in a nucleosome model based upon a prolate ellipsoid with the DNA wound along the major (symmetry) axis. The core histone octamer is not shown.

For this

$$n_1 + 0.109 = 1.707 \tag{18}$$

$$n_2 + 0.109 = 1.749 = n_2 \text{ (DNA)}$$

In this case, the sum $n_1 + 0.109 = 1.707$ is quite different from n_1 (DNA) $= 1.622$ and this prolate model would therefore be difficult to justify unless the histone core were highly optically anisotropic.

The above results are suggestive of an oblate model for the nucleosome in 0.1 M KCl along the lines described by Pardon, Klug and others. In spite of the fact that the optical theory employed is based upon macroscopic continuum dielectrics, we believe the results have significance since the optical theory seems to provide a high degree of consistency between nucleosomes and DNA, and the experimental and theoretical methods applied to both systems are similar. Although the conclusions are highly preliminary, we are encouraged by the degree of internal consistency between the extinction angle and flow birefringence results and by the agreement with other experimental methods, especially neutron scattering from solutions and X-ray crystallography. In spite of the particular experimental difficulties in flow birefringence studies upon such small, relatively symmetric particles such as nucleosomes, the method is still a rapid and simple one compared to many other structural techniques. It shows considerable promise, we believe, in making a significant contribution to the vast array of structural problems in chromatin which must be solved if we are ultimately to understand in depth the dynamic character of chromatin structure.

REFERENCES

1. G. S. Manning (1978), Quart. Rev. Biophys. (in press)

2. V. C. Gordon, C. M. Knobler, D. E. Olins and V. N. Schumaker (1978), Proc. Natl. Acad. Sci., U.S.A., 75, 660

3. K. S. Schmitz and B. Ramsay-Shaw (1977), Biopolymers, 16, 2619

4. K. S. Schmitz and B. Ramsay-Shaw (1978), Biophys. J., 21, 96a

5. M. E. Mach (1873), Optische-Akutische Versuche, Calve, Prague

6. J. C. Maxwell (1873), Proc. Roy. Soc. (London), A-22, 46

7. A. Peterlin and H. A. Stuart (1939), Z. Physik, 112, 1, 129

8. P. J. Oriel and J. A. Schellman (1966), Biopolymers, 4, 469

9. J. L. Sarquis and R. E. Harrington (1969), J. Phys. Chem., 73 1685

10. R. E. Harrington (1970), J. Amer. Chem. Soc., 92, 6957

11. J. T. Edsall (1942), Advan. Colloid Sci., 1, 269

12. R. Cerf and H. A. Scheraga (1952), Chem. Rev., 51, 185

13. H. A. Scheraga and R. Signer (1960) in Physical Methods of Organic Chemistry, Vol. I, Part III (A. Weissberger, Ed.), Interscience, New York, Chapter 35

14. V. N. Tsvetkov (1963) in Newer Methods of Polymer Character-
    ization (B. Ke, Ed.), Interscience, New York, Chapter 14

15. R. E. Harrington (1967) in Encyclopedia of Polymer Science and
    Technology (N. M. Bikales, H. F. Mark and N. G. Gaylord, Eds.)
    Interscience, New York, Vol. 7, p. 100

16. A. Peterlin and P. Munk (1972) in Physical Methods of Chemistry,
    Part IIIC (A. Weissberger and B. W. Rossiter, Eds.), Inter-
    science, New York

17. E. D. Alcock and C. Sadron (1935), Physics, 6, 92

18. W. Kuhn and F. Grun (1942), Kolloid-Z., 101, 248

19. V. A. Bloomfield, D. M. Crothers and I. Tinoco, Jr. (1974) in
    Physical Chemistry of Nucleic Acids, Harper and Row, New York,
    Chapter 5

20. R. E. Harrington (1970), Biopolymers, 9, 159

21. B. H. Zimm (1956), J. Chem. Phys., 24, 269

22. R. Koyama (1961), J. Phys. Soc. Japan, 16, 1366

23. D. E. Olins, P. N. Bryan, R. E. Harrington, W. E. Hill and
    A. L. Olins (1977), Nucleic Acids Res., 4, 1911

24. A. L. Olins, J. P. Breillatt, R. D. Carlson, M. B. Senior,
    E. B. Wright and D. E. Olins (1977) in The Molecular Biology
    of the Mammalian Genetic Apparatus (P.O.P T'so, Ed.),
    Elsevier/North Holland, Amsterdam

25. A. L. Olins, R. D. Carlson, E. B. Wright and D. E. Olins (1976),
    Nucleic Acids Res., 3, 3271

26. T. W. Barrett and R. E. Harrington (1977), Biopolymers, 16, 2167

27. H. A. Scheraga and L. Mandelkern (1953), J. Amer. Chem. Soc.,
    75, 129

28. F. Perrin (1936), J. Phys. Radium, 7, 1

29. R. Simha (1940), J. Phys. Chem., 44, 25

30. J. F. Pardon, D. L. Worcester, J. C. Wooley, R. I. Cotter,
    D.M.J. Lilley and B. M. Richards (1977), Nucleic Acids Res., 4,
    3199

31. R. P. Hjelm, G. G. Kneale, P. Suau, J. P. Baldwin and
    E. M. Bradbury (1977), Cell, 10, 139

32. J. T. Finch, L. C. Cutter, D. Rhodes, R. S. Brown, B. Rushton,
    M. Levitt and A. Klug (1977), Nature, 269, 29

33. K. E. Van Holde, C. G. Sahasrabudde and B. Ramsay-Shaw (1974),
    Nucleic Acids Res., 1, 1579

34. H. Weintraub, K. Palter and F. Van Lente (1975), Cell, 6, 85

35. P. Oudet, C. Spadafora and P. Chambon (1977), XLII Cold Spring
    Harbor Symposium on Quantitative Biology

SCATTERING AND DIFFRACTION BY NEUTRONS AND X-RAYS IN THE STUDY OF CHROMATIN

J.F. Pardon

Searle Research Laboratories

Lane End Road, HIGH WYCOMBE, Bucks, HP12 4HL, England

INTRODUCTION

X-ray scattering (1,2) and diffraction (3,4) have been used to study the structure of chromatin since methods for the routine preparation of histone-DNA (nucleohistone) complexes were developed (5). In comparison with the earlier X-ray diffraction studies from fibres of DNA (6), the amount of information coming from these studies has until recently been relatively small. The slow progress was largely a result of the lack of detail available in the fibre diffraction patterns. Whilst it was possible to examine a variety of different models which might explain the limited data it was not possible to derive a rigorous model. Data from other physical and biochemical studies were, until recently, equally uninformative and offered little assistance in the interpretation of the diffraction results, largely, as we now know, due to the nature of the specimen preparation.

Initial scattering measurements from chromatin solutions (1,2) and diffraction (3,4) from fibres showed that the DNA is far less crystalline when combined with histones than when drawn into fibres in the absence of protein. In patterns from chromatin, the diffracted intensities corresponding to the layer lines of the DNA double helix are diffuse and only poorly oriented.

The early X-ray studies of chromatin demonstrated the presence of a series of low angle maxima (1,3) which are present both from gels and fibres of chromatin. These maxima are present at spacings greater than the dimensions of the double-helix and demonstrate the presence of regular tertiary structure. They produced the first evidence for a regular structure in chromatin other than the DNA

double-helix. Studies of chromatin gels over a wide range of concentrations showed that the intensities of the diffraction maxima
vary with concentration (1). The series of low angle rings, the
poor orientation of the DNA and the concentration dependence of
the diffracted intensities were compatible with an extendable structure such as a super-helix (4,7). It was not possible to estimate
the possible extent of superhelical structure from the limited data
available. The regularity in the structure was attributed to the
DNA and the contribution made to the scattering intensities by the
histones was largely ignored. A less regular helical structure
was proposed to explain measurements made on dilute solutions of
chromatin (2) using X-ray scattering techniques.

A subsequent re-evaluation of the diffraction from concentrated gels included an analysis of the continuous scattering
profile upon which the maxima are superimposed (8). A model to
explain the complete scattering/diffraction data included short
regions of super-helix containing a few turns - 1-3 - of helix interdispersed with non-regular super-helical DNA. This approximates to
our present concept of chromatin structure although it did not
include a histone core.

The discovery of the sub-unit structure (9,10) resulted both
from biochemical and electron microscopy studies (11,12,13), although
the 110Å X-ray spacing featured in early discussions of the spacing
between sub-units (9). A re-appraisal of the X-ray data had to
incorporate the concept of a central histone core around which the
DNA is wound. The contribution of the histones to the diffraction/
scattering maxima could no longer be neglected. When methods were
developed for isolating discrete chromatin particles (14) it was
possible to make the first unambiguous determinations of some of
the parameters of the sub-unit (15,16); also to discriminate between
those maxima which are part of the Fourier transform of the sub-unit,
as opposed to interference maxima arising from the assembly of subunits into either regular arrays or tertiary structures.

The availability of neutron scattering facilities (17,18) and
the development of the theory of contrast variation enabled a new
series of scattering and diffraction measurements to be made. These
methods, together with the availability of monodisperse solutions
of sub-units, rapidly led to the validation of the original prop-
osals for the structure. They were able to show unambiguously that
there is a protein core with DNA on the outside (15,16), and more
recently provided a model for the subunit structure to a resolution
of 25Å (19). This model has subsequently been supported by X-ray
crystallography (20). It is now possible to re-interpret the original data from gels and fibres of chromatin and make more detailed
discussions of models for the assembly of the sub-units into regular
tertiary configurations.

In this chapter the recent developments in the study of chromatin sub-units will be emphasised and the contribution made by
neutron techniques explained. In so far as it is possible, the
data from fibres and concentrated gels of chromatin will be re-
interpreted to include conclusions that can be drawn from the
structure of the sub-unit itself. Whilst much progress has been
made in studies of the sub-unit it will be apparent that the knowledge of higher levels of structure in chromatin is less complete.
The extent of regular high ordered structure in chromatin remains
to be determined.

THE LIMITATIONS OF THE EARLY X-RAY DIFFRACTION DATA

An indication of the amounts of information available in X-ray
diffraction patterns from fibres and gels of chromatin is shown in
figure 1. X-ray cameras with a low angle resolution of about 400Å
have been used to study both fibres of chromatin maintained at
various relative humidities, to give concentrations in the range
60-100% w/w, and gels at lower concentrations. Oriented gels
produced by swelling fibres in suitable buffers show that the
lowest angle maximum at about 110Å is meridionally oriented (8).
The degree of orientation obtained does not permit any distinction
between true meridional orientation and an arc apparently centred
on the meridian but resulting from two broad maxima which are
slightly off the meridian. The latter could arise from a helical
structure where the lowest angle maximum arises from the Bessel
function on the first layer line of the helix transform. Fibre
diffraction patterns demonstrate slight meridional orientation in
the 55Å and 35-38Å maxima (3) and include equatorial arcs which
vary in position in the range 35-27Å, depending upon the concentration of the chromatin in the fibre. There have been no reports
of any pronounced X-ray diffraction maxima with spacings in the
range 150-400Å, consequently the 110Å maximum is taken as the first
layer line repeat in discussions of superhelical models (4,7).
Higher angle maxima, present at spacings at submultiples of 110Å,
have relative intensities compatible with the transform from a
regular helix with pitch 120Å and diameter 110Å. At higher concentrations the intensity of the 110Å reflection decreases and at
60% w/w this reflection is no longer present. This variation in
the relative intensities was originally thought to be due to the
interference between helices as they pack together to form either
paranemic or plectanemic bundles (7). The 110Å maximum is not
observed from solutions and dilute gels where the maxima are superimposed on a strong central scattering profile (1,2,8,21,22). The
continuous central scatter was originally thought to arise either
from a non-regular helical structure (2) or from short lengths of
regular helix interdispersed with non-helical DNA (8).

With the limited data available, there was clearly no justifi-

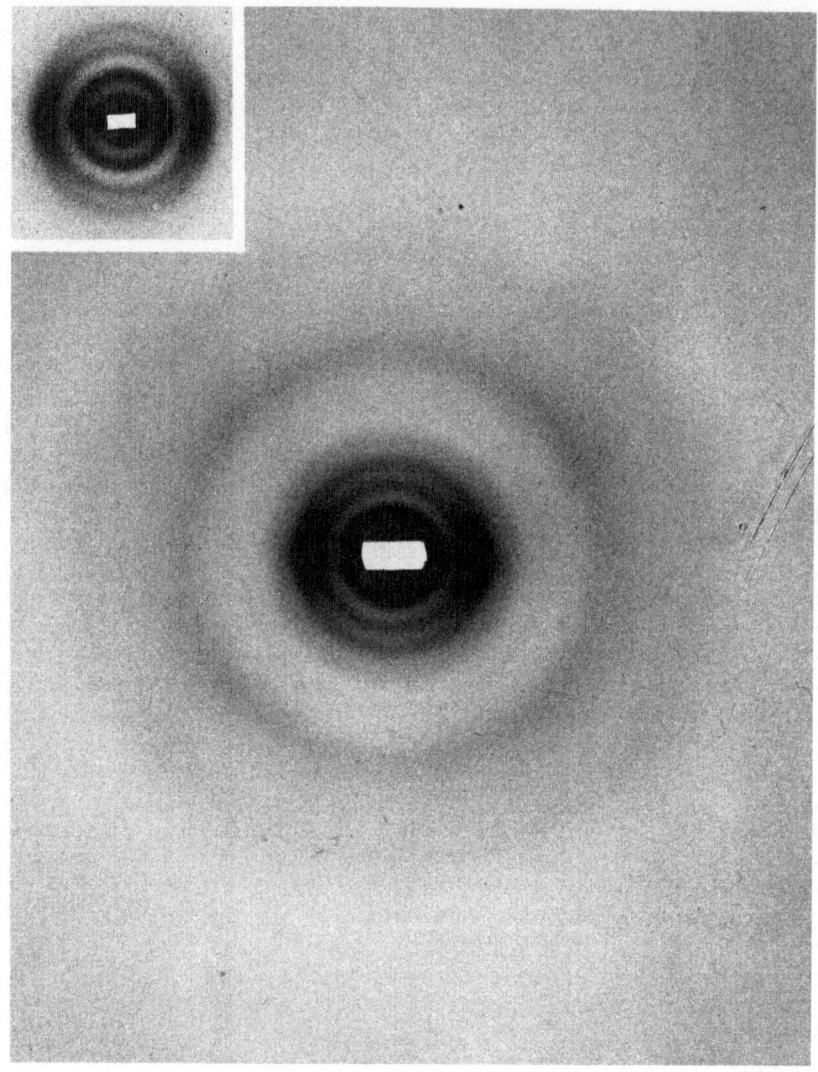

Figure 1: X-ray diffraction pattern from a fibre of chromatin
maintained at 98% relative humidity. Inner diffraction rings at
approximately 55, 37, 27 and 22Å. An equatorial reflection in the
region 27-37Å.
Inset: The central region from a less concentrated gel showing
maxima at 108Å, 55Å, 37Å, 27Å, 22Å. Equatorial maxima at 60Å and
27-37Å.

cation for more complex calculations than were made to determine
the feasibility of simple models. The assumption was made that the
diffraction was predominantly arising from a regular DNA structure
with the histone evenly distributed along the double helix (7),
possibly in the large groove (23). Contrast variation studies with
X-rays could provide no new information since it was not possible
to adjust the scattering density of the solvent to give conditions
where the histone is the dominant component. Using sucrose as the
medium to generate different contrast conditions it was possible to
obtain conditions where the solvent scattering density equalled
that of the histone. Under these conditions, where the DNA dominates, the X-ray scattering from gels is not dissimilar to that
from chromatin in water. It was not possible using sucrose to
obtain conditions where the histones dominate the scattering. Salt
solutions are inappropriate for contrast variation studies on
chromatin since histones are removed from DNA in the ionic strength
range 0.5-2.0M.

The availability of neutron beams (17,18) and the development
of the theory for analysing neutron scattering from large molecules
(reviewed in 24), especially molecules with large internal fluctuations in scattering density, provided new opportunities to put
the assumptions made in the original X-ray studies to test. Thus
it became possible to ask the following questions:-

(a) Do the scattering maxima all arise from a common struc-
     ture?

(b) Do the scattering maxima predominantly arise from regular
     DNA structures or possibly from histone complexes?

(c) How are the histones arranged relative to the DNA?

The small angle scattering camera at Harwell developed by
Haywood and Worcester (17) enabled measurements to be made covering
a similar range of spacings to those studied using X-rays. Although
more material was required, exposure times using the medium flux
source at Harwell were for fibres of chromatin similar to those used
in previous X-ray measurements. Much shorter times were required
using the small angle scattering camera D11 at Institute Laue-
Langevin at Grenoble (18).

The combination of X-ray and neutron scattering studies on
solutions of isolated subunits have provided a model for the subunit
and these data together with studies of crystals of core particles
permit a more detailed interpretation of the original X-ray data
from fibres and gels.

CONTRAST VARIATION WITH NEUTRONS

This technique utilises the large difference between the coherent scattering amplitude of hydrogen (-3.74×10^{-13}cm) and that of deuterium (6.67×10^{-13}cm). It is possible (25) to adjust the solvent scattering density using mixtures of D_2O and H_2O such that DNA and histone each have scattering densities greater than or less than the solvent or alternatively such that either the histone (60% D_2O) or the DNA dominate the scattering (40% D_2O) - see Table 1. It is important to note that the scattering density of the chromatin varies with the D_2O/H_2O ratio of the solvent since there is exchange of labile proteins.

In addition to varying the scattering density of the solvent it is also possible to vary the internal fluctuations in scattering density by deuterium labelling (26,27). This can be achieved by extracting either histones or DNA from cell cultures maintained in a medium rich in D_2O and subsequently reconstituting chromatin using a mixture of deuterated and protonated components (28). The following discussion relates to the former, namely the influence on the scattering profile of changing the solvent scattering density.

a. The Scattering Profile I(k)

Ibel and Stuhrmann (29) showed that a convenient way of expressing the excess scattering density $\rho(r)$ of dissolved particles is to divide it into two parts such that:-

$$\rho(\vec{r}) = \bar{\rho} . \, \rho_F \, (\vec{r}) + \rho_s \, (\vec{r}) \tag{1}$$

TABLE 1

Coherent Scattering Amplitude Densities

H_2O	-0.0056
D_2O	0.0638
Histone[*] (39% D_2O)	0.021
DNA[*] (60.5% D_2O)	0.036

[*]Data taken from studies of isolated chromatin core particles (15). Units of 10^{-12} cm/Å^3.

---

where $\bar{\rho}$ is the difference between the mean scattering density of the protein and the scattering density of the solvent. $\rho_F(\vec{r})$ is a dimensionless function that is unity inside regions of the particle inaccessible to solvent molecules and zero elsewhere. $\rho_S(\vec{r})$ describes the variation of scattering density inside the particles around the mean value. The scattering amplitude can therefore be written:-

$$A(\vec{k}) = \bar{\rho}.A_F(\vec{k}) + A_S(\vec{k}) \tag{2}$$

where the momentum transfer $k = 4\pi \sin\theta/\lambda$ for scattering angle 2θ and wavelength λ. Stuhrmann and Kirste (31) showed that the scattering intensity can be divided into three basic scattering functions:-

$$I(k) = \langle A(\vec{k})^2 \rangle = \bar{\rho}^2 I_F(k) + \bar{\rho}. I_{FS}(k) + I_S(k) \tag{3}$$

and $$|I_{FS}(k)| \leqslant 2\sqrt{\{I_F(k) \times I_S(k)\}} \tag{4}$$

$$I_S(\vec{k}) \text{ and } I_{FS}(\vec{k}) \to 0 \text{ for } k \to 0$$

$I_S(k)$ is observed under conditions of zero contrast, i.e. the mean scattering density of the particle equals the scattering density of the solvent. $I_F(k)$ is the intensity under infinite contrast conditions, i.e. only the shape of the particle contributes to the scattering.

This theory is only true where the particles are not affected by the exchange of atoms with the solvent. Thus for the neutron scattering of particles using $H_2O/{}^2H_2O$ mixtures:-

$$\rho(\vec{r}) = \bar{\rho}.\left[\rho_F(\vec{r}) - \rho_E(\vec{r})\right] + \rho_S(\vec{r}) \tag{5}$$

where $\rho_E(\vec{r})$ defines the positions and density of exchangeable atoms. With (5) $I(k) = \bar{\rho}^2 I_C(k) + \bar{\rho} I_{CS}(k) + I_S(k)$ (6)

By measuring the distribution of intensities as a function of k for at least three different contrast conditions it is therefore possible to solve 6 and obtain the three functions $I_C(k)$, $I_{CS}(k)$ and $I_S(k)$. $I_S(k)$ represents the internal structure function providing information of the distribution of scattering intensities about the mean. $I_{CS}(k)$ is a function of both the shape and the internal structure of the particle. $I_C(k)$ is a function mainly due to the overall shape of the particle with a small component arising from the distribution of exchangeable sites.

b. The Guinier Region

For very small scattering angles:-

$$I(k) = I(0) \exp \frac{-R_G^2 k^2}{3} \tag{7}$$

R_G is the radius of gyration where:-

$$R_G^2 = \frac{1}{\rho V} \int_V r^2 \cdot \Big(\rho(r) - \rho(s) \Big) d^3r$$

for solvent scattering density $\rho(s)$. This is Guinier's Law (30).
This law is extremely useful but can only be used for a monodisperse solution of particles where log I(k) plotted as a function
of k^2 is strictly linear. It is important to collect data within
the Guinier region to enable radii of gyration to be determined.
Thus in general the maximum value of k for a spherical particle
is given by kRg = 1.3 and for a rod-like particle by kRg = 0.7.
In most of the neutron studies of chromatin and chromatin particles
the Guinier conditions have been met, in previous X-ray scattering
this has not always been so.

The measured radius of gyration is a function of the difference
between the mean scattering density of the particle and the density
of the solvent (29). It can be expressed in the following way:-

$$Rg^2 = Rc^2 + \frac{\alpha}{\bar{\rho}} - \frac{\beta}{\bar{\rho}^2} \tag{8}$$

Rc represents the radius of gyration at infinite contrast. α and β
depend both on the solvent and internal fluctuations in the scattering density. α has the significance of a mean square distance
of fluctuations from the centre of shape and β is the square of the
mean distance of the fluctuations.

In practice, the sign and magnitude of α determines whether
the response of higher scattering density are towards the outside
or the inside of the particle. β is useful in two or more component
systems since it can be used to determine whether the centres of
shape are coincident or not. If the centres of shape are coincident
β = 0 and a plot of Rg^2 versus $1/\bar{\rho}$ is linear. Since the chromatin
subunit contains two components, DNA and protein, with different
scattering densities an analysis of the variation in Rg with contrast provides a measure of the average distribution of these components in the particle. Subsequent analysis of the entire scattering curve can be used in two ways:-

(a) By collecting data for at least three different solvent
     scattering densities and calculating the three functions

$I_c(k)$, $I_{cs}(k)$ and $I_s(k)$.

(b) By collecting data under high contrast conditions and
     calculating the scattering profile for various models
     which are constrained to comply with the known variation
     in radius of gyration for different solvent contrast con-
     ditions. It is important to include realistic variations
     in the internal scattering densities for any model con-
     sidered. Since both the structure of the DNA and the
     composition of the subunit are known, the number of models
     which can be constructed, subject to the constraints
     imposed by the radius of gyration data, is limited. Thus
     this second approach was the first to be used to investi-
     gate the structure of the core particle and is described
     in some detail in the following two sections.

RADIUS OF GYRATION MEASUREMENTS ON CORE PARTICLES AND HISTONES

     Chromatin core particles (14) were originally studied (15)
using the small angle scattering camera at Harwell (17). Measurements were made on 11 mg/ml solutions in D_2O, 75% D_2O, 25% D_2O and
H_2O. Additional data were collected for 5 mg/ml and 2.5 mg/ml solutions in D_2O. In each instance the Guinier region was linear allowing the radius of gyration to be determined and the zero angle
scattering intensity I_0 to be obtained by extrapolation. The plot
of $\sqrt{I_0}$ as a function of the mole fraction of D_2O in the buffer
provided a mean scattering density for the particle equivalent to
the scattering density of 49% D_2O (figure 2). This is the value
predicted from the composition of the particle and the mean scattering length densities of the components with 76% of the labile
protons exchanging with the solvent.

     As mentioned in the previous section, the radius of gyration of
an object with internal variation of scattering density is a function
of the scattering density and hence isotopic composition of the
solvent. Having measured the mean scattering density of the particle
from the $I_0^{\frac{1}{2}}$ versus % D_2O plot the relationship between Rg^2 and
$1/\bar{\rho}$ (equation 8) was investigated as shown in figure 3. The size of
the error bars for the data points under low contrast conditions
illustrates the limitations of using a medium flux reactor for
scattering studies of this type. From these limited data it was
concluded that β has a value close to zero. Thus the regions of
high scattering density, within the accuracy of these data, are concentric with the regions of lower scattering density. The slope of
the line in figure 3 shows that the particles are composed of regions
of very different scattering density. The value of α for the chromatin particles, 508 ($\pm$ 80) x 10^{-6} (15), is much smaller than found
for lipoprotein (2800 x 10^{-6}) (32) but significantly greater than the
value found for globular proteins (35 x 10^{-6}) (29). The sign and

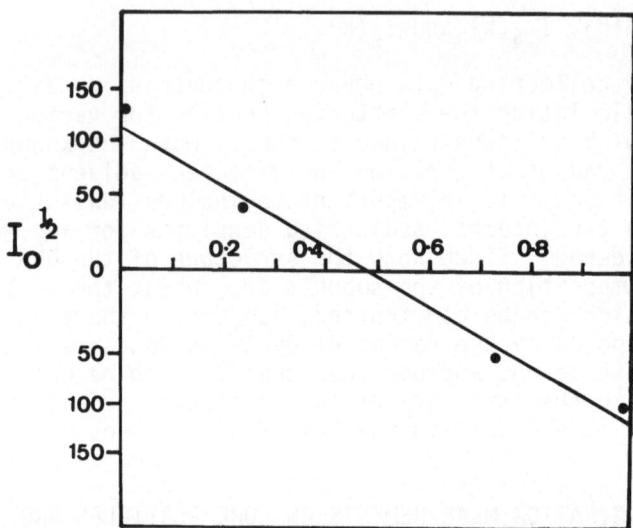

Figure 2: Plot of square root of zero angle scattering intensity
($I_0^{\frac{1}{2}}$) versus mole fraction of D_2O in the buffer for a 11 mg/ml
solution of chromatin core particles.

slope demonstrate that the material with higher scattering density
(the DNA) is at a greater radius than regions of lower scattering
density. By extrapolating from the graph of Rg^2 vs $1/\bar{\rho}$ it is possible to obtain radii of gyration where the DNA dominates the scattering and conversely where the protein dominates. Since the DNA
is towards the outside and 76% of labile protons exchange with the
solvent it was assumed that complete exchange occurs for labile
protons in the DNA. Thus the mean scattering density of the DNA is
equal to that of the solvent when it is 60.5% D_2O. Similarly
allowing for complete exchange of the protons in the DNA only 68%
of the labile protons of the histones exchange with the solvent
giving equality between the mean scattering density of the protein
and the water for 39.0% D_2O.

Using values of $\bar{\rho}$ for 60.5% and 39% D_2O and assuming $\beta = 0$, the
radii of gyration under conditions where the protein dominates and
where the DNA dominates are 30.6 ± 2Å and 50.5 ± 1.4Å respectively.
Also where $1/\bar{\rho} = 0$ (conditions of infinite contrast) the radius of
gyration is that of the particle with the same shape and dimensions
but no internal variation of scattering intensity. Under this

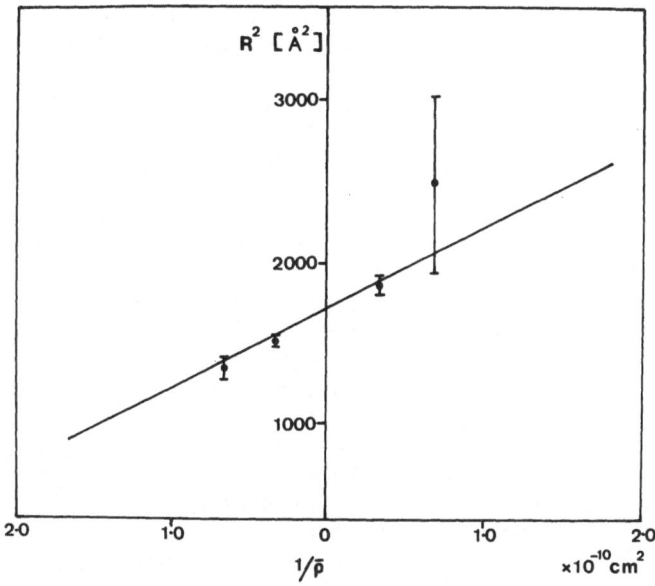

$R^2 [\mathring{A}^2]$

$\frac{1}{\bar{\rho}}$ $\times 10^{-10} cm^2$

Figure 3:   The variation of the (radius of gyration)^2 of isolated core particles as a function of $\frac{1}{\bar{\rho}}$.

condition the radius of gyration extrapolated from the plot of R_g^2 vs $\frac{1}{\bar{\rho}}$ is 41.1 (± 0.4) $\mathring{A}$.   More precise measurements using the higher flux neutron facility at Grenoble (33) have produced slightly smaller values for the radii of gyration but their conclusions are similar.   These studies unambiguously showed that the DNA is on the outside of the core particle, surrounding a protein core, with a small amount of overlap between the DNA and the histone.

Additional studies of a complex of histones H2A, H2B, H3 and H4 isolated from non-sheared chromatin using 2.0M NaCl at pH 9.0 were made over a concentration range 2.5-40 mg/ml.   The radius of gyration of this complex was not dependent on the concentration or the D_2O/H_2O ratio.   The value obtained was 30.1 ± 0.3$\mathring{A}$, similar to the value for the intact core particle under conditions where the protein dominates the scattering (34).

A MODEL FOR THE CHROMATIN CORE PARTICLE DERIVED FROM
SCATTERING CURVES FROM CORE PARTICLES AND ISOLATED HISTONE COMPLEXES

The values of the radii of gyration for the core particle provide information of the gross fluctuations in internal scattering density. However, a variety of different models can account for these radii. This choice of models is limited to those where the biochemical data dictated that the DNA is constrained to a 140 base pair segment with length about 476Å and diameter about 20Å. The number of models was further reduced by a consideration of the scattering data from the isolated core protein. Since the 'native' X-ray diffraction pattern for chromatin is obtained when DNA is added to isolated core protein this complex almost certainly retains its native conformation after isolation. Using haemoglobin, oval-bumin and human albumin as protein standards the molecular weight of isolated core protein was determined from the intensities for zero angle neutron scattering. The value obtained, 60,500 ± 5,000, is roughly equivalent to a tetramer of histones H2A, H2B, H3 and H4. The small difference between the measured radii of gyration between the isolated protein (a tetramer) and the histone core (an octamer) in the intact particle suggests that in the particle two tetrameric histone complexes stack directly on top of each other. If they are laterally spaced with respect to each other a larger variation in the radius of gyration would have been observed. Furthermore, the high angle scattering curve from isolated core protein is unlike that calculated for a spherical protein (34) but more like profiles calculated for a variety of disc-like structures (figure 4).

The maximum dimensions of the core particle and core protein were determined to within an accuracy of 10Å by calculating pair probability functions. The scattering intensities can be used to generate a function, similar to the Patterson function used in crystallography, which gives the probability for vector distances between volume elements as a function of radius. To generate this function the scattering curves have to be extended in two directions:

(a) To zero scattering angle. This can be achieved using the Guinier Law (30);

(b) To high angles using the Porod Law $I(Q) \approx Q^{-4}$ (35).

From the data shown in figure 5, and using the rules of Guinier and of Porod to extrapolate the data, the pair probability functions were calculated using the expression:

$$P(r) = \frac{1}{2\pi^2 r} \int_{0}^{\infty} Q \cdot I(Q) \, \mathrm{Sin}\,(Q \cdot r) dQ$$

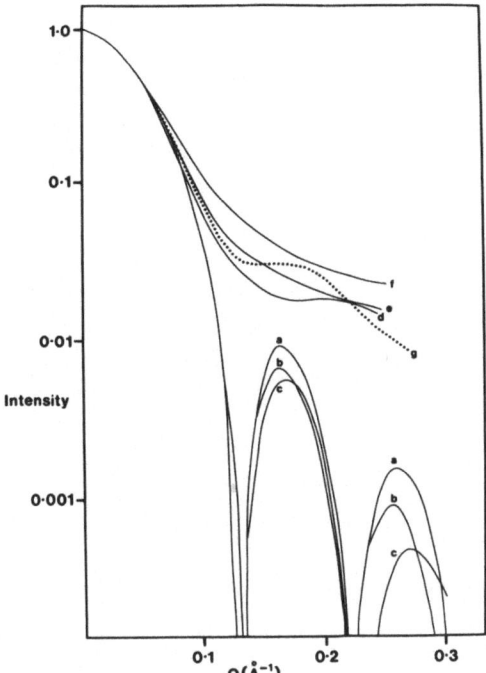

Figure 4: Experimental and calculated neutron scattering profiles for histone core protein in D_2O.

A   Spherical models. Uniform sphere radius 35Å (a) and spheres with compact centres and less dense outer domains (b) and (c).

B   More squat structures with height in the range 25-50Å and diameter in the range 82-92Å (d)-(f). More details of models in reference (34).

C   Scattering profile obtained from a solution of core protein in D_2O, 2.0M NaCl (g).

where the scattering vector Q has magnitude $4\pi \sin \theta/\lambda$ for scattering angle 2θ and neutron wavelength λ. The curves obtained are shown in figure 6. Maximum dimensions of 118 ± 10Å and 80 ± 10Å can be determined from the p(r) = 0 intercepts for the core particle and core protein respectively (36).

These combined data therefore restricted the possible models for the core particle to include:-

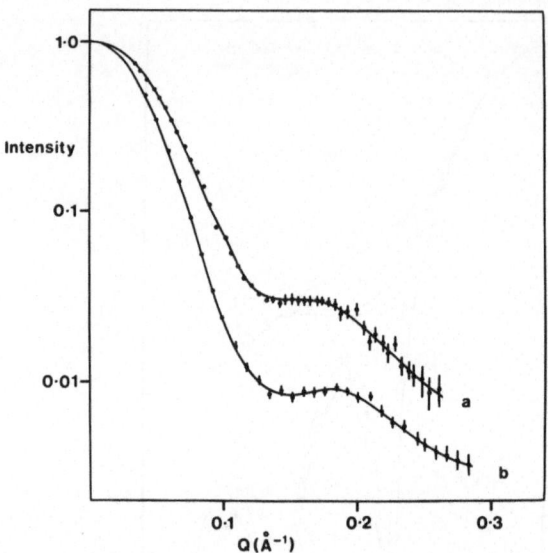

Figure 5: Neutron scattering profiles from:-

a   Chicken erythrocyte core protein at 12 mg/ml in D_2O, 2.0M NaCl;

b   Chromatin core particles from chicken erythrocyte nuclei at
     9 mg/ml in D_2O, 10mM Tris 0.7mM EDTA at pD 7.1.

Data collected using the small angle scattering camera at AERE
Harwell (17). $Q = 4\pi \sin \theta/\lambda$ where scattering angles 2θ and wavelength λ (4.7Å). The wavelength spread $\Delta\lambda/\lambda = 2\%$ FWHM.

(a)   An internal protein core diameter about 80Å composed of
      two tetrameric disc-shaped protein complexes stacked
      directly on top of each other;

(b)   A cylindrical DNA molecule of length 476Å and diameter
      20Å wound or folded on the outside of the protein with a
      small degree of protein/DNA overlap;

(c)   Radii of gyration of 30.6 and 50.5Å for the histone and
      DNA dominating respectively and 41.1Å under infinite
      contrast conditions. A radius of gyration of 41.6Å was
      obtained from X-ray scattering measurements using the
      position sensitive detector in the laboratory of Dr. D.
      Sadler;

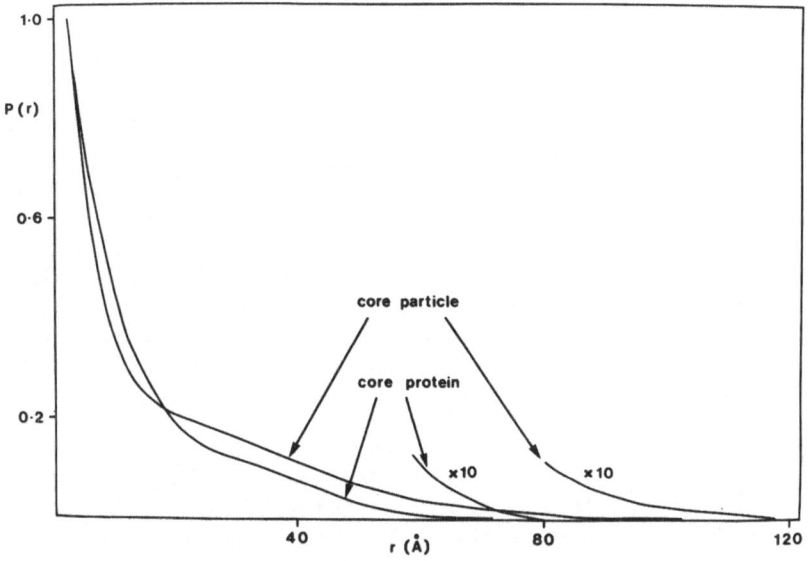

Figure 6: Probability functions derived from the scattering profiles shown in figure 5 (36).

(d) A maximum dimension for the core particle of 118 ± 10Å, suggesting a diameter of about 110Å.

These factors severly limit the choice of model. Additional data were also available from electron microscopy in at least three different laboratories. This can be summarised as follows:-

1. The DNA tends to enter and leave the particle on one side (37);

2. The DNA is concentrated at the top and bottom of the particle (38);

3. The particle is disc-shaped (39).

Initial calculations indicated that the relative neutron scattering intensities in the region of the 35Å maximum and the shape of the scattering profile were inconsistent with a roughly spherical particle (40). Therefore cylindrical structures approximating to both prolate and oblate ellipsoids were considered.

Scattering profiles were calculated using the Debye equation

(41):-

$$I(Q) = \sum_i \sum_j \rho_i \ \rho_j \ \frac{\sin (Q.r_{ij})}{Q.r_{ij}}$$

Model structures were generated by dividing the model into a number, typically 700, of scattering volumes of equal size separated by distances r_{ij}. These were assigned scattering densities (ρ) equivalent to those for histone, DNA, nucleoprotein or a mixture of one of these components and solvent. The scattering density of DNA was given the value 1.0 relative to the solvent. Calculations were also made for X-ray scattering profiles since data was also obtained using X-ray techniques. The subsequent calculations (40) can be summarised as follows:-

(a) Neither cylindrical particles 110Å high with 110Å diameter nor spherical particles with diameter 110Å explain the neutron scattering data. A series of maxima are obtained rather than the single rather broad 35Å maximum (figure 7a);

(b) Prolate structures do not explain the neutron data;

(c) Structures with a hole through their centre are unsuitable;

(d) Oblate structures with height 50-55Å and diameter 110Å give the correct type of scattering profile only when the DNA is confined to two loops at the top and bottom of the particle (figure 7b);

(e) The refined model shown in figure 7 fits the neutron data well (figure 7c) and also provides the following features of the X-ray scattering profile obtained from solutions of core particles (42):-

1. A shoulder at $\approx$ 60Å;

2. Poorly defined maxima at about 38Å and 28Å;

3. A pronounced minimum between the 38Å maximum and the 60Å shoulder (figure 7d).

These are also the main features of the scattering profile from dilute solutions of intact chromatin at low ionic strength.

This final model was subsequently validated by X-ray crystallographic work by Finch et al. (20). More recently Suau et al. (33) concluded that it can account for the shape function $I_c(Q)$ obtained from solutions of core particles - see equation 6. Thus studies of core particles both in solution and in crystals have provided very similar models to a resolution of about 25Å.

TABLE II

Oligomer size	Experimental Radius of Gyration	Calculated Separation	
		Linear assemblies	Assemblies with all particles equidistant from centre of mass
Dimer with H1 and H5	51 ± 2Å	66 ± 6Å	66 ± 6Å
Dimer without H1 and H5	72 ± 2Å	121 ± 6Å	121 ± 6Å
Trimer with H1 and H5	84 ± 1Å	91 ± 5Å	128 ± 4Å
Trimer without H1 and H5	93 ± 2Å	103 ± 3Å	145 ± 6Å
Hexamer with H1 and H5	125 ± 1Å	70 ± 2Å	119 ± 1Å

For points lying on a helix with pitch 220Å and radius 110Å: trimer has Rg 95Å and hexamer has Rg 125Å. Particle separation along helix 107Å.

RADIUS OF GYRATION MEASUREMENTS FROM SMALL CHROMATIN OLIGOMERS

It is not obvious from the structure of the core particle how the chromatin subunits assemble themselves to form the unit chromatin thread. The electron micrographs of Finch et al. (20) show that isolated core particles can, under some conditions, stack almost directly on top of each other to give a separation of about 55Å. However in chromatin these particles are connected by a length of up to 60 base pairs of DNA which may have histone H1 attached in some way that has an influence on the structure of the chromatin thread. Thus there could be up to 300Å between the centres of the particles.

Sperling and Tardieu (21) have shown from small angle X-ray scattering studies that in 0.2mM EDTA the electronic mass per unit length of chromatin is 1240 e/Å, corresponding to a separation between subunits of about 110Å. They do not record interference maxima in the 100Å region and therefore conclude that the particles are contiguous to form a more or less uniform thread.

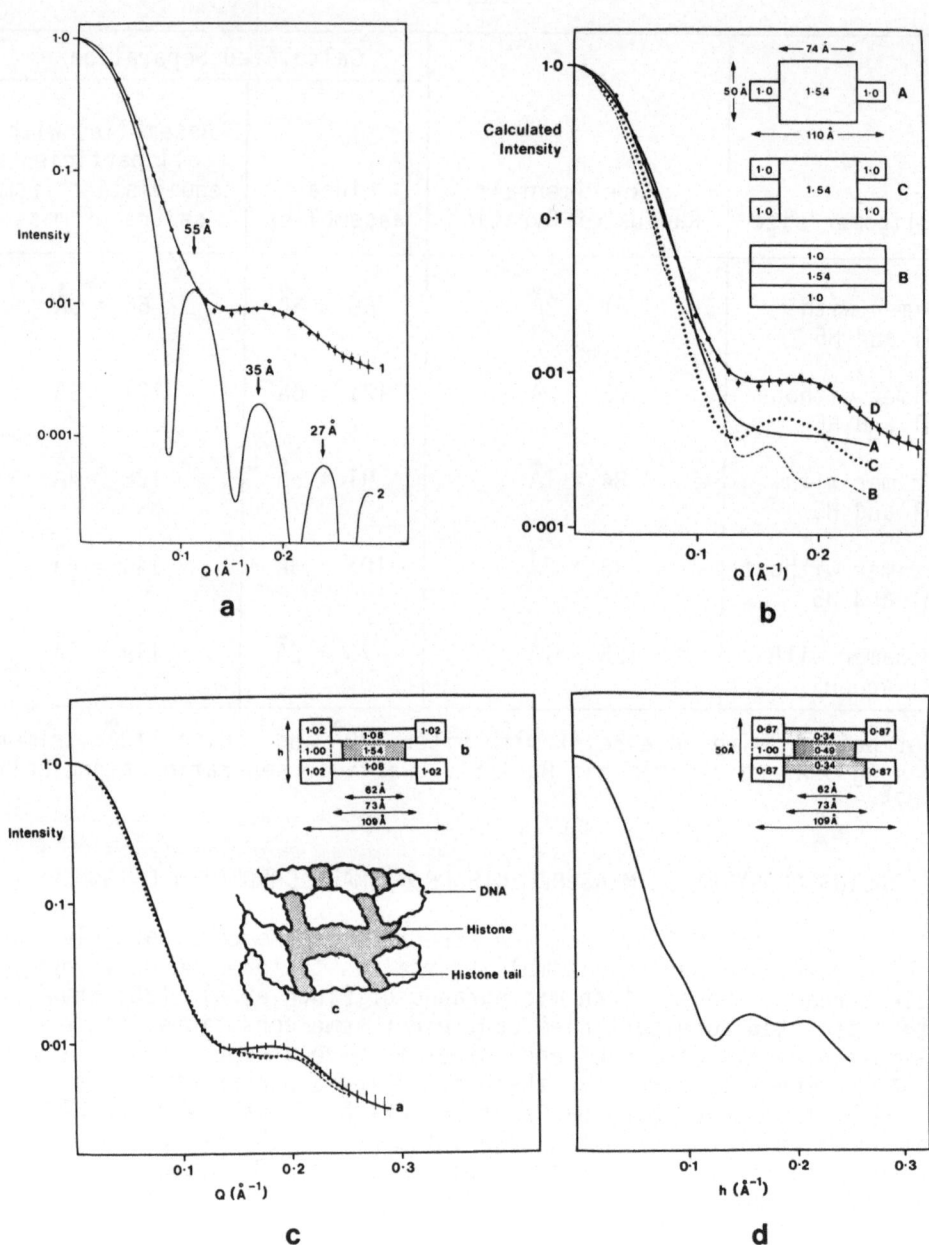

Figure 7

Figure 7:

a   Neutron scattering profile from a 5 mg/ml solution of core
        particles in D_2O, 10mM Tris 0.7mM EDTA (1) and the calculated
        scattering profile for a spherical model approximating to an
        inner sphere of protein (radius 39.5Å) surrounded by a shell
        containing DNA (radius 53Å) (2);

b   Calculated neutron scattering for oblate structures (height
        50Å, diameter 110Å) with the DNA located in domains with
        scattering density 1.0 and the histone at the centre (scat-
        tering density 1.54). Cross-sections of the models are
        shown in the inserts. Experimental scattering profile (D);

c   Calculated and experimental neutron profiles for the core
        particle. The dimensions and distribution of scattering
        intensities are shown in the schematic diagram which represents
        a section through the model along the cylindrical axis. The
        domain with scattering density 1.0 corresponds to a short
        segment of DNA joining two three-quarter filled annuli con-
        taining DNA and histone tails (scattering density 1.02). The
        protein centre has a dense centre (scattering density 1.54)
        surrounded above and below by more hydrated regions (1.08);

                    +++++   experimental profile
                    .....   model with h = 50Å
                    ------   model with h = 53Å

d   The calculated X-ray scattering profile for the model for the
        core particle showing the major features of the experimental
        X-ray scattering profile from a solution of particles.

A somewhat different approach was to isolate small oligomers
containing a limited number of subunits and to measure their radii
of gyration using the small angle neutron scattering camera D11
(18) at the Institute Laue-Langevin at Grenoble. A summary of the
results from oligomers purified by Ti14 zonal ultracentrifugation
and containing two, three and six subunits respectively is shown
in Table II. The radii of gyration were calculated from the slopes
of the Guinier plots shown in figure 8. The calculated separation
between particles is shown both for linear strings of subunits and
also for more compact clusters similar to those observed in the
electron microscope. These data indicate that the particles are
not widely spaced but have centre-to-centre separations in the range

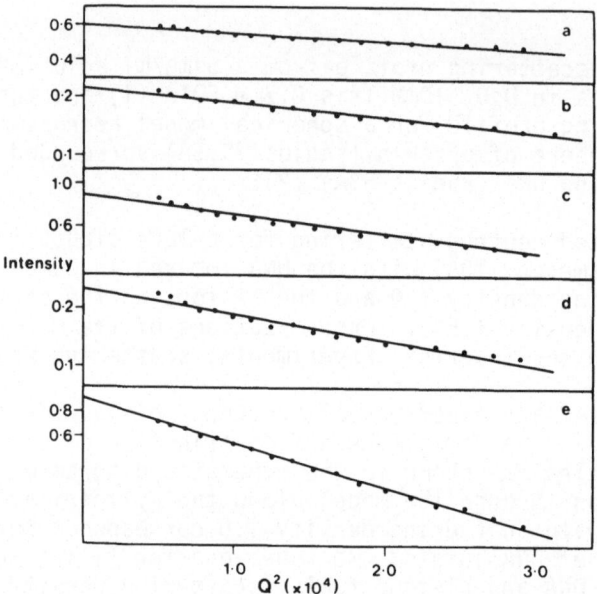

<u>Figure 8</u>:   Guinier plots from solutions of chromatin oligomers iso-
lated from chicken erythrocytes in D_2O, 10mM Tris, 0.7 EDTA pD 8.6.

a   11 mg/ml solution of di-nucleosomes containing histones H1
        and H5;

b   5 mg/ml solution of di-nucleosomes with histones H1 and H5
        removed;

c   10 mg/ml solution of tri-nucleosomes with histones H1 and H5;

d   4 mg/ml   solution of tri-nucleosomes with histones H1 and H5
        removed;

e   6 mg/ml solution of oligomers each containing six nucleosomes
        with histones H1 and H5.

Data obtained using the D11 instrument at Grenoble (18).

66-145Å.

        The higher angle scattering profiles of both dimers and trimers
(small oligomers containing two and three nucleosomes) have a single

broad maximum at 35Å similar to that obtained for single core particles. This suggests that under the low ionic strength conditions
(10mM Tris, 0.7mM EDTA) used for these studies the individual particles behave as if they are scattering independently of each other.
This would suggest a rather flexible link between particles as was
originally proposed by Kornberg (9) rather than a rigid arrangement
whereby the particles either stack directly on top of each other or
alternatively assemble themselves edge-to-edge. This may not be
the case for larger oligomers where there might be additional stab-
ilising forces which could constrain the subunits to adopt a more
regular tertiary structure. It is interesting to note that the
radius of gyration of a hexamer is 125Å. This is the value obtained
for one turn of helix containing six subunits and having a pitch of
220Å and radius 100Å. Such a helix is also able to account for
radii of gyration of larger oligomers at the same ionic strength
but studied by laser light scattering (43).

NEUTRON SCATTERING FROM FIBRES OF INTACT CHROMATIN

The position and intensity of the lowest angle diffraction maximum in X-ray patterns obtained from fibres of chromatin in the concentration range 40-70% w/w is dependent on the concentration.
Thus for a 45% w/w gel there is a strong maximum at 100-110Å (8).
At 60% w/w there is no maximum observed in the spacing range 70-120Å
(1,3,8) and in more concentrated gels a strong intensity maximum is
observed at about 75-80Å (3). An early attempt to explain the reduction in intensity and subsequent loss of the 110Å maximum in terms
of interdigitating super-helices was based on the assumption that
it arises largely from superhelical DNA (7). Studies of chromatin
fibres using neutrons were able to put this assumption to the test
and prove it to be invalid (25,44).

For chromatin in D_2O, in the concentration range 40-70% w/w, the
neutron scattering maximum at 80-110Å is present at all concentrations (25), the spacing decreasing gradually with increase in
chromatin concentration. Thus under conditions where a maximum is
not recorded with X-rays (fibres maintained at 98% relative humidity)
a strong maximum is recorded with neutrons (figure 9a). Furthermore
the variation in intensity of this maximum with solvent contrast
provides information relating to the distribution of DNA and histone
along the direction of the fibre axis. In general terms, for the
maximum to originate from a largely DNA repeat the D_2O percentage
at which its intensity goes through a minimum has to be about 60%.
For a protein repeat the D_2O percentage for a minimum intensity is
nearer 40%. For a nucleoprotein structure, such as a superhelix
evenly coated with an equal mass of histone, an intermediate value
would be expected. The minimum intensity for the 82Å maximum from
fibres at 98% relative humidity occurs at neither of these values
but at about 10% D_2O (figure 9a).

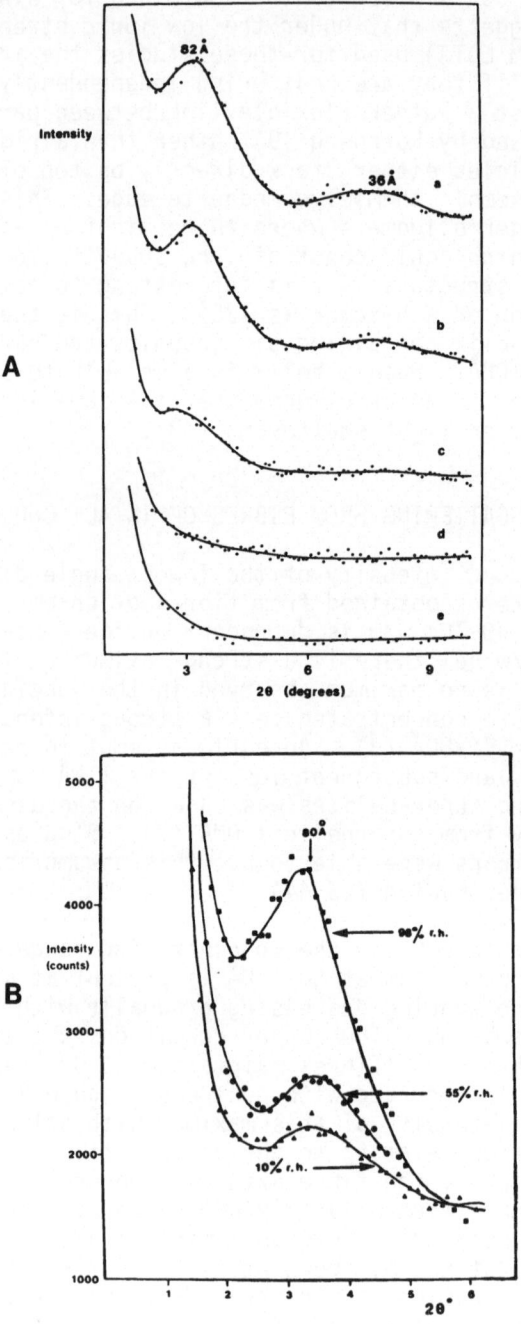

Figure 9

Figure 9:   Neutron diffraction from fibres of chromatin.   Data
obtained by scanning along the meridian using the small angle scattering camera at Harwell (17).

(A)   The effect of varying the D_2O/H_2O ratio in solutions of
      potassium chlorate used to equilibrate the fibres to a
      relative humidity of 98%; (a) D_2O, (b) 75% D_2O, (c) 50% D_2O,
      (d) 25% D_2O and (e) H_2O;
(B)   The effect of reducing the relative humidity of the fibres
      using three different salt solutions each dissolved in 100%
      D_2O.

<------

     Under these contrast conditions (10% D_2O), domains containing
hydrated DNA will scatter with an average scattering density equal
to relatively unhydrated protein domains.   The two types of domain
will not be resolved and any interference maxima arising from alternating protein rich and DNA rich regions will not be present in the
scattering profiles.   For neutron scattering in D_2O solutions the
protein domains have a higher contrast than the DNA (relative to
solvent scattering) and the interference maxima will be present at
all chromatin concentrations (figure 9a).   The densities of histone
and DNA for X-ray scattering are approximately the same as those
for neutron scattering in 10% D_2O.   Thus the absence of the 80-110Å
X-ray reflection from fibres of chromatin at 60% w/w and its presence at all concentrations of chromatin for neutron scattering in
D_2O indicates a regular protein/DNA repeat along the fibre axis
(i.e. the data were collected along the meridian).

     The intensity of the neutron 82Å maximum from fibres of chromatin
(figure 9b) decreases as the material becomes more concentrated (44).
This again supports a model where there are alternating relatively
unhydrated protein cores contrasting with hydrated DNA.   As water is
removed from the DNA containing regions the contrast between those
regions and the protein decreases and the scattering maximum becomes
less intense.   With X-rays the opposite occurs and the low angle
maximum becomes the most intense and easily recorded of all the
scattering/diffraction maxima from fibres of chromatin.

     For chromatin concentrations below 50% w/w unoriented gels have
been examined.   The position of the low angle maximum changes to 108Å
for gels with concentrations 30 and 38% w/w and to 200Å for 6% gels
(44).   Furthermore the variation in the intensity of the 200Å maximum
with D_2O percentage has a minimum at about 55% D_2O.   This indicates
that within the 6% gel there is a 200Å nucleoprotein repeat.   Such
a repeat might represent the separation of chromatin threads or
alternatively arise from a higher order of structure within the individual thread, such as a helix.   Additional studies are required to
distinguish between these two possibilities.

X-RAY FIBRE DIFFRACTION RE-EVALUATED

Recent progress resulting from the use of neutrons and studies of isolated chromatin subunits demands a review of the long established X-ray fibre diffraction data. The intensity variation with concentration of the 80-110Å maximum has already been re-interpreted in the previous section. In this section we consider the origins of the diffraction maxima and in doing so invoke information from the recent studies of crystals of core particles.

Despite early expectations, contrast variation methods in neutron diffraction studies of fibres of chromatin were unable to distinguish regorously between the various possible origins of the strongest of the diffraction maxima (25). X-ray solution scattering studies of core particles indicated that the 37Å, 27Å and 22Å maxima are present in the Fourier transform of the isolated subunit (42), which also contains a shoulder in the 55-60Å region.

The 110Å reflection is not present in profiles from dilute solutions (<20% w/w) of either isolated core particles (42) or intact chromatin (1,2,8,21,22). At higher concentrations nucleosomes (containing 165-190 base pairs of DNA and histone H1) produce diffraction patterns very similar to those recorded in diffraction patterns from intact chromatin. For instance a gel (centrifuge pellet) with concentration about 35% w/w produces maxima at 124, 62 and 38Å (34), similar to those recorded from intact chromatin. When the concentration is increased and the material maintained at 99% relative humidity the nucleosomes again behave like intact chromatin providing maxima at 55, 35 and 27Å but no maximum in the spacing range 80-120Å (figure 10). Dry samples of nucleosomes produce a strong 75-82Å maximum, again similar to intact chromatin. Thus if a higher order structure such as a helix (or solenoid) does exist in more concentrated gels of chromatin it does not depend upon the continuity of the DNA between subunits for its existence.

Studies of crystals of core particles by Finch et al. (20) validated the model for the core particle derived from neutron and X-ray scattering studies. In addition the crystal data can be used to consider the origin of some of the maxima observed in the original X-ray fibre diffraction studies. By combining this new information with the older data describing the relative orientations of the maxima in the original X-ray fibre diffraction patterns we may consider the way in which the subunits are assembled in the chromatin thread. Considering each reflection in turn the following considerations can be made from an analysis of the various reflections obtained from the crystals:

(a)   55Å maximum. The neutron fibre studies suggested (25) that this reflection arises from the transform of the DNA. However, the crystal studies show that it arises from two

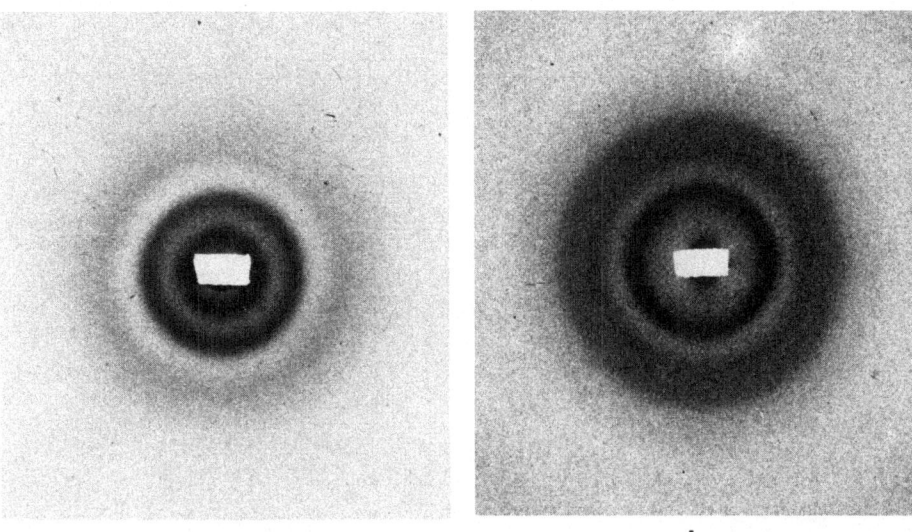

a   b

Figure 10:   X-ray scattering from concentrated gels of nucleosomes
(containing 165-190 base pairs of DNA and histones H1 and H5).

(a) Centrifuge pellet giving a concentration of ~35% w/w.   Rings
    are present at 124, 62 and 38Å.

(b) Gel equilibrated to give a relative humidity of 99% (~60% w/w).
    Rings are present at 54, 35 and 27Å.

components related to the shape of the core particle.
These are the height of the particle and its diameter.
Since these two components are at right angles it is not
possible to derive any information relating to the orientation of the subunit in chromatin based on the meridional
orientation observed in X-ray fibre diffraction studies (3).

(b)   37Å reflection.   This arises from spacings in the plane of
      the disc-like structure of the core particle most probably
      with components from both the DNA (20) and the histones
      (34,42).   The slight meridional orientation of this reflec-
      tion (3) would suggest that the subunits assemble themselves
      in the fibre with their axes aligned perpendicular to the
      fibre axis.   However, orientation in fibres is slight and
      might arise from shearing the chromatin during the prep-
      aration of the sample.   This maximum is still present in

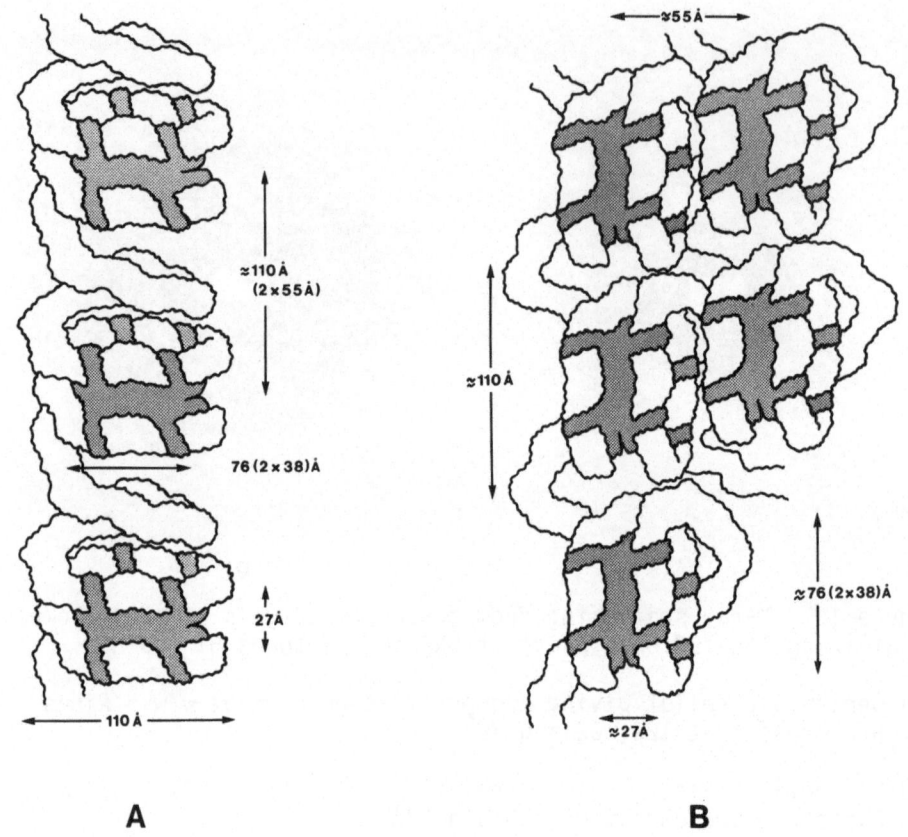

A   B

Figure 11: Models for the arrangement of the core particles in the chromatin unit thread.

A   Core particles stacking in a top-to-bottom arrangement to give a spacing of about 110Å along the fibre axis.

B   Core particles assembled edge-to-edge to give a contiguous distribution of particles spaced 110Å apart. Two chains are included parallel to each other, separated by about 60Å. With this arrangement an individual chromatin thread might form a coil-bound helix or solenoid in which the particles in success-ive gyres stack directly on top of each other. Such a helix would account for the contrast variation behaviour of the 82Å maximum described in figure 9 since there are alternating gyres rich in DNA and histone respectively.

diffraction from highly stretched fibres of chromatin (4).

(c) The 27Å reflection arising from the spacing between the
     loops or gyres of DNA in the core particle is oriented in
     a direction perpendicular to the plane of the particle.
     Fibre diffraction patterns do include an equatorial com-
     ponent at about 27Å, however this might arise from
     'stretched' chromatin molecules which pack together with a
     regular side-by-side spacing of about 27Å.

(d) The '110Å' reflection. The two possible origins for this
     reflection are illustrated in figure 11. These are:-

     i. repeat distance between subunits in a linear assembly;

     ii. the pitch of a regular helical structure such as the
          solenoid proposed by Finch and Klug (45) and Carpenter
          et al. (46).

The neutron diffraction studies of Baldwin et al. (25)
showed that this reflection moves progressively to ≈80Å as
water is removed from the fibre. Also as described above,
this maximum arises from a structure which contains a regular repeat along the fibre axis arising from the spacial
separation of DNA and histone. Such a separation would
clearly be provided by model A in figure 11 where the core
particles tend to stack on top of each other in the chromatin thread. However, it could equally well be explained
by model B (figure 11) in which the unit thread consists
of a string of subunits assembled edge-to-edge and where
the unit thread forms a helix in which particles in
adjacent gyres stack more or less on top of each other.
The latter type of structure is supported both by the
electron microscopy of Finch and Klug (45) and also by the
off meridional orientation of the 80Å reflection observed
by Carpenter et al. (46) from concentrated samples of
chromatin.

Further evidence supporting a solenoid type of structure
has been obtained by Campbell et al. (43) who studied
multimer fragments containing discrete numbers of subunits.
For molecules having between 50 and 150 nucleosomes the
radius of gyration was found to vary linearly with the
number of subunits. The slopes of the distributions
decreased with increase in ionic strength showing that the
chromatin has a more compact structure in the presence of
salt. Whilst the results at low ionic strength could be
explained by a series of helices with pitch in the range
210-440Å and outer diameter in the range 300-500Å, the

higher ionic strength data were supported only by a more
compact structure of the type proposed by Finch and Klug
(45).

FUTURE POSSIBILITIES OF USING DIFFRACTION/SCATTERING STUDIES TO STUDY CHROMATIN

A combination of several techniques has clearly made possible
the determination of the structure of the chromatin subunit (core
particle) to a resolution of about 20Å. Details of the precise
structure of this particle will emerge during the next few years
as arrangement of the histones within the particle (H2A, H2B, H3
and H4) and relative to the DNA will then be known in detail.
Interest is already beginning to be placed towards the study of
subunits modified to allow transcription. This is an area where
X-ray and neutron scattering methods are likely to provide a significant contribution.

The study of the structure of the higher order structures in
chromatin has been hampered by lack of detail in the X-ray and
neutron diffraction patterns which indicate the lack of regularity
in the structure. Whether this is inherent in the structure of
chromatin itself or a result of harsh methods used in isolating
chromatin from the nucleus is still not known. Solution scattering
from multimers of chromatin has produced a variety of different
results for the mass per unit length of the chromatin thread. An
extension of the type of experiments described by Sperling and
Tardieu (21) including a wider range of concentrations and ionic
strengths might produce a better definition of the gross features
of the chromatin thread. Any such study should take advantage of
the very low angle resolution now possible with neutron scattering
cameras such as D11 at Grenoble. It is perhaps not encouraging
that the one study of metaphase chromosomes (47), which might have
been expected to contain more regularity in their structure,
provided diffraction patterns no better than those more recently
acquired from relatively short lengths of so-called 'native' chromatin (22).

ACKNOWLEDGEMENTS

I would like to thank Dr. Brian Richards and Dr. David Lilley
for their comments on the manuscript, John Hobbs for preparing the
figures and Pat Campbell for carefully editing the final version of
the text.

REFERENCES

1. Luzzati, V. and Nicolaieff, A. (1963) J. Mol. Biol. 7, 142
2. Bram, S. and Ris, H. (1971) J. Mol. Biol. 55, 325
3. Wilkins, M.H.F., Zubay, G. and Wilson, H.R. (1959) J. Mol. Biol. 1, 179
4. Pardon, J.F., Wilkins, M.H.F. and Richards, B.M. (1967) Nature 215, 508
5. Zubay, G. and Doty, P. (1959) J. Mol. Biol. 1, 1
6. Langridge, R., Wilson, H.R., Cooper, C.W., Wilkins, M.H.F. and Hamilton, L.D. (1960) J. Mol. Biol. 2, 19
7. Pardon, J.F. and Wilkins, M.H.F. (1972) J. Mol. Biol. 68, 115
8. Pardon, J.F., Richards, B.M. and Cotter, R.I. (1974) Cold Spring Harbor Symposia of Quantitative Biology 38, 75
9. Kornberg, R.D. (1974) Science 184, 868
10. Van Holde, K.E., Sahasrabuddhe, C.G. and Shaw, B.R. (1974) Nucleic Acids Res. 1, 1579
11. Olins, A.L. and Olins, D.E. (1974) Science 183, 330
12. Woodcock, C.L.F., Safer, J.P. and Stanchfield, J. (1976) Exp. Cell Res. 97, 101
13. Woodcock, C.L.F., Sweetman, H.E. and Frado, L.L. (1976) Exp. Cell Res. 97, 111
14. Shaw, B.R., Corden, J.L., Sahasrabuddhe, C.G. and Van Holde, K.E. (1974) Biochem. Biophys. Res. Comm. 61, 1193
15. Pardon, J.F., Worcester, D.L., Wooley, J.C., Tatchell, K., Van Holde, K.E. and Richards, B.M. (1975) Nucleic Acids Res. 2, 2163
16. Hjelm, R.P., Kneale, G.G., Suau, P., Baldwin, J.P., Bradbury, E.M. and Ibel, K. (1977) Cell 10, 139
17. Haywood, B.C.G. and Worcester, D.L. (1973) J. Phys. E. 6, 568
18. Ibel, K. (1976) J. Appl. Cryst. 9, 296
19. Richards, B.M., Pardon, J.F., Lilley, D.M.J., Cotter, R.I., Wooley, J.C. and Worcester, D.L. (1977) Cell Biol. Int. Rep. 1, 107
20. Finch, J.T., Lutter, L.C., Rhodes, D., Brown, R.S., Rushton, B., Levitt, M. and Klug, A. (1977) Nature 269, 29
21. Sperling, L. and Tardieu, A. (1976) FEBS Lett. 64, 89
22. Sperling, L. and Klug, A. (1977) J. Mol. Biol. 112, 253
23. Pardon, J.F. and Richards, B.M. (1973) In Subunits in Biological Systems Vol. 6, 1, Ed. Fasman and Timasheff, Marcel Dekker
24. Jacrot, B. (1976) Rep. Prog. Phys. 39, 911
25. Baldwin, J.P., Boseley, P.G. and Bradbury, E.M. (1975) Nature 253, 245
26. Engelman, D.M. and Moore, P.B. (1972) Proc. Natl. Acad. Sci. USA 69, 1997
27. Hoppe, W. (1972) Israel J. Chem. 10, 321
28. Bradbury, E.M., Hjelm, R.P., Carpenter, B.G., Baldwin, J.P., Kneale, G.G. and Hancock, R. (1977) In The Molecular Biology of the Mammalian Genetic Apparatus (Ed. P. Ts'o), p. 53, Elsevier/ North Holland

29. Ibel, K. and Stuhrmann, H.B. (1975) J. Mol. Biol. 93, 255
30. Guinier, A. and Fournet, G. (1955) In Small-Angle Scattering of X-rays. John Wiley, New York
31. Stuhrmann, H.B. and Kirste, R.G. (1965) J. Phys. Chem. 46, 247
32. Stuhrmann, H.B., Tardieu, A., Mateu, L., Sardet, C., Luzzati, V., Aggerbeck, L. and Scanu, A.M. (1975) Proc. Natl. Acad. Sci. USA 72, 2270
33. Suau, P., Kneale, G.G., Braddock, G.W., Baldwin, J.P. and Bradbury, E.M. (1977) Nucleic Acids Res. 4, 3739
34. Pardon, J.F., Cotter, R.I., Lilley, D.M.J., Worcester, D.L., Campbell, A.M., Wooley, J.C. and Richards, B.M. (1978) Cold Spring Harbor Symposium of Quantitative Biology, Vol. 42, In Press
35. Porod, G. (1951) Kolloid Z. 2, 83
36. Lilley, D.M.J., Richards, B.M., Pardon, J.F., Cotter, R.I. and Worcester, D.L. (1978) Proc. FEBS 11th Meeting Copenhagen 1977, Vol. 43, Ed. Clark. Pergamon.
37. Olins, A.L., Breillatt, J.P., Carlson, R.D., Senior, M.B., Wright, E.B. and Olins, D.E. (1977) In The Molecular Biology of the Mammalian Genetic Apparatus (Ed. P. Ts'o) p. 211, Elsevier/North Holland
38. Varshavsky, A.J. and Bakayev, V.V.(1975) Mol. Biol. Rep. 2, 247
39. Langmore, J.P. and Wooley, J.C. (1975) Proc. Natl. Acad. Sci. USA 72, 269 1
40. Pardon, J.F., Worcester, D.L., Wooley, J.C., Cotter, R.I., Lilley, D.M.J. and Richards, B.M. (1977) Nucleic Acids Res. 9, 3199
41. Debye, P. (1915) Zerstreuung von Rontgenstrahlen, Ann. Physik. 46, 809
42. Richards, B.M., Cotter, R.I., Lilley, D.M.J., Pardon, J.F., Wooley, J.C. and Worcester, D.L. (1976) In Current Chromosome Research (Ed. Jones & Brandham) p. 7, Elsevier/North Holland
43. Campbell, A.M., Cotter, R.I. and Pardon, J.F. (1978) Nucleic Acids Res., In Press
44. Pardon, J.F., Worcester, D.L., Richards, B.M. and Wooley, J.C. (1976) Harwell Report No. MPD/NBS/29
45. Finch, J.T. and Klug, A. (1976) Proc. Natl. Acad. Sci. USA 73, 1897
46. Carpenter, B.G., Baldwin, J.P., Bradbury, E.M., Ibel, K. (1976) Nucleic Acids Res. 3, 1739
47. Pardon, J.F., Richards, B.M., Skinner, L.G. and Ockey, C.H. (1973) J. Mol. Biol. 76, 267

NUCLEAR MAGNETIC RESONANCE STUDIES OF NUCLEIC ACIDS AND PROTEINS

Paul O. P. Ts'o and Lou-Sing Kan

Division of Biophysics, School of Hygiene and Public

Health, The Johns Hopkins University, Baltimore, Md. 21205

With the advances in spectrometer instrumentation and computer technology, NMR has become a very powerful technique in the study of biopolymers, such as nucleic acids and proteins. This approach can potentially provide information about the properties and inter-actions of biopolymers at the atomic level through investigation of the magnetic properties of many nuclei, particularly those having spins of 1/2, such as ^1H, ^{13}C, ^{19}F, and ^{31}P. To most effectively utilize this approach, however, we must define (1) the type of in-formation which can be gained using NMR, (2) the type of questions to be posed about the characteristics of proteins and nucleic acids at the atomic level, and finally (3) the kinds of improvements needed in NMR studies for the future. The biochemist is mostly concerned not only with the nuclei which were mentioned previously, but also with ^{11}B, ^{15}N, and ^{14}N.

The possession of both spin and charge confers on a nucleus a magnetic moment $\vec{\mu}_N$ which is proportional to the magnitude of the spin,

$$\vec{\mu}_N = \gamma_N \hbar \vec{I} \tag{1}$$

where γ_N is the magnetogyric ratio of the given nucleus and is measured in radian $\cdot$ sec^{-1} $\cdot$ gauss^{-1}. Quantum theory demands that the allowable nuclear spin states are quantized; the component m_I, the nuclear spin quantum number, in any given direction can take up only one set of discrete values which are $+I$, $(I-1)$,..., $-I$. For the nuclei to have a nuclear spin $=1/2$, m_I may only take the values $1/2$ and $-1/2$; for the nuclei to have a nuclear spin $=1$, m_I may take three values, i.e., $1, 0, -1$. If a steady magnetic field $\vec{H}$ is applied on the nuclei, there is an interaction between the field and the magnetic moment $\vec{\mu}_N$, which may be represented in terms of a Hamiltonian

$$H = -\vec{\mu}_N \cdot \vec{H} = -\gamma_N \hbar \, \vec{I} \cdot \vec{H} \tag{2}$$

The energy corresponding to each splitting level when a magnetic field is applied will be

$$E = -\gamma_N \hbar \, m_I H \tag{3}$$

The selection rule for transitions among the energy levels is that m_I changes by ± 1; therefore

$$\Delta E = \gamma_N \hbar H \tag{4}$$

In order to induce transition between the two nuclear spin levels, an oscillating electromagnetic field must be applied to the system and the frequency ν of the oscillating field must satisfy the resonance condition $h\nu = \Delta E$; therefore

$$\nu = \frac{\gamma_N H}{2\pi} \tag{5}$$

This result clearly implies the following:

(1) For each nucleus (γ_N is a constant), the resonance frequency is directly proportional to the applied field $\vec{H}$.

(2) For a given field, nuclei with a larger γ_N will resonate at smaller magnetic fields.

GENERAL INTRODUCTION

Here, we shall first mention the principal difference existing between the NMR spectroscopy and other types of spectroscopy, such as UV absorption spectroscopy or infrared spectroscopy which is familiar to most readers. For NMR, the different states of the nucleus having varying energy levels are manifested by the applied magnetic field, but are detected by the absorption of the radiation at radio frequency. Therefore, the energy differences between these states are directly related to the strength of the applied magnetic field as revealed by changes in the frequency of radiation absorbed by the nucleus at resonance. In other words, the resolution of resonances, i.e. absorption of radiation at different frequencies, is directly proportional to the strength of the applied magnetic field. The second major difference between NMR and other conventional spectroscopy is the relaxation process. Since the energy differences between various states are relatively small, unless a proper process exists to dissipate the absorbed energy, the system will soon be "saturated" in reaching equilibrium, and hence will show no signals. This problem will be returned to briefly in a later section.

As a general introduction, the following four parameters can be derived from the NMR studies:

(1) Chemical shifts (δ, ppm)

When a molecule is placed in a magnetic field H_0, orbital currents are induced in the electron clouds. Therefore, each nucleus is, in effect, partially shielded from H_0 by the electrons, and the local magnetic field strength will be

$$H_{loc} = H_0(1 - \sigma) \tag{6}$$

where σ is the so-called screening constant (expressing the change of field). Sigma is independent of H_0 but highly dependent upon the chemical structure of the molecule and can be either positive or negative. Therefore, the resonance frequencies have to be

$$2\pi\nu = \gamma_N H_{loc} = \gamma_N H_0(1 - \sigma) \tag{7}$$

When an NMR experiment is carried out with a group or a mixture of molecules at a given magnetic field, the signals from the various nuclei are spread out in a spectrum according to their nuclear environments. Since ν is proportional to the applied magnetic field, the spacing between NMR signals corresponding to different types of nuclei is also proportional to magnetic field. Thus, in NMR spectroscopy, in contrast to optical spectroscopy, there is no absolute zero or standard reference. Hence, the chemical shift between two sets of nuclei is defined as the difference in their resonance frequencies measured at constant field. It is conveniently expressed in a field independent unit as part per million (ppm) of the constant field or frequency;

$$\delta = \frac{\nu_2 - \nu_1}{\nu_1} \times 10^6 \tag{8}$$

where δ is chemical shift in ppm, ν_2 is the measured frequency and ν_1 is the reference frequency. A standard reference substance for proton and carbon-13 spectra, tetramethylsaline (TMS), has been proposed and widely accepted.

In case of a complex molecule, such as nucleic acids or proteins, in addition to the influence of the electron clouds through covalent binding, the neighboring groups can also exert a magnetic or electric field to the atom of interest. This through-space field effect is reflected in the measurement of the chemical shifts which can now provide an indication of the spatial relationship of these groups to the atom under measurement. This is a very powerful tool in studying the conformation and interaction of biopolymers. It should be cautioned, however, that the geometrical relationships

of these field effects are not simple and are often anisotropic, as
shown in later paragraphs.

(2)   Coupling constants (J, H_2)

In addition to the lines which had different chemical shifts,
the high resolution NMR spectra of many compounds contain patterns
which reveal the interactions of neighboring magnetic dipoles.
Magnetic nuclei may transmit influence to each other indirectly
through the intervening chemical bonds. This interaction occurs by
the slight polarizations of the spins and orbital motions of the
valence electrons, and the magnitude of the interaction is expressed
in terms of a coupling constant J which is not affected by the
tumbling of the molecules and is independent of H_0. Usually one
neighboring spin (1/2) would split the resonance of a single nucleus
into a doublet with intensity 1:1; and two equivalent neighboring
spins (1/2) would split the resonance of a single nucleus into a
triplet with intensity 1:2:1. In general, if a nucleus of spin 1/2
has n equivalently coupled neighbors of spin 1/2, its resonance will
be split into n+1 peaks corresponding to the n+1 spin states.

The coupling constant for two spins is usually large when separated by one bond (^1J), smaller when separated by two bonds (^2J)
and, is even smaller when separated by three bonds (^3J). However,
the vicinal coupling constants (^3J) have been shown to be related
to the dihydral angle of a A–X–Y–B system, where the coupling is
between the spin A and spin B ($^3J_{AB}$) and the dihedral angle describes
the rotation around the X–Y bond. In the case of the couplings of
the H–C–C'–H' system in the furanose ring, the following relationship
was found by Lemieux.

$$^3J_{HH'} = J_0 \cos^2 \phi - 0.28 \text{ Hz}$$

where $^3J_{HH'}$ is the vicinal coupling constant, ϕ is the dihedral
angle between H–C–C' and the C–C'–H planes in the fragment of
H–C–C'–H in the furanose. From the experimental results, $J_0 = 9.27$Hz
for ϕ below 90° and $J_0 = 10.36$ Hz for ϕ above 90° were calibrated.

Similarly, our laboratory has adopted the equation proposed by
Govil and Smith (1973) for the coupling constant relationship in the
conformational analysis of the 3'–C–OP bond and 5'–C–OP bond in the
nucleotides, dinucleoside monophosphates, and polynucleotides,

$$^3J_{C,P} \backsim 9.5 \cos^2\theta - 0.6 \cos \theta$$

where θ is the dihedral angle between the planes ^{13}C CO and CO^{31}P,
and the values of $^3J_{trans} = 10.1$ Hz and $^3J_{gauche} = 2.1$ Hz are calculated.

(3)   The area of the resonance line or the amount of radiation

energy absorbed (relative units).

The areas of the resonance lines which represent the amount of radiation absorbed, reflect proportionally the number of nuclei contributing to these signals, subjected to the modification of the relaxation process. Therefore, if the relaxation process is properly understood, such as in a small molecule, not involving chemical exchange, spin-spin interaction, etc., then the areas of a resonance line provide the vital information as to the number of nuclei in that signal.

(4) The relaxation times (T_1, T_2, in seconds)

In the previous sections, the behavior of an isolated, spinning nucleus has been examined. When NMR is actually observed in bulk matter, the observed signal represents a large number of identical nuclei. These nuclei may interact among themselves and with their surroundings. These interactions provide the means for the dissipation of the absorbed energy. Consider an assembly of identical nuclei experiencing the same magnetic field; such an assembly constitutes a magnetically equivalent set. For a spin =1/2 nucleus, there are only two magnetic energy levels which correspond to the two alignments of the nuclear magnetic moment, i.e., either along or against the magnetic field. At equilibrium the nuclei are distributed between the two energy levels and the ratio of the number of spins in each level as given as follows

$$\frac{n+}{n-} = e^{-(\Delta E/kT)} \tag{9}$$

where n+ and n- are the populations of the upper and lower spin states respectively, ΔE is defined by Eq. (4), k is the Boltzman constant, and T is the absolute temperature. For a given ΔE, the number of lower state spins will always be larger than that of upper state spins. If the system is irradiated at a frequency $\nu=\Delta E/h$, the system absorbs energy from the radiation field with a consequent increase in the n+/n- ratio. When the system absorbs sufficient energy to equilize the population of the two states, it is said to be "saturated". A saturated or partially saturated spin system itself will tend to return to a thermal equilibrium when the radiation field is lifted. Two simultaneous processes are involved in the return of a saturated system to equilibrium: (1) The absorbed energy is given up from the spin system to the lattice; this process is called spin-lattice relaxation. A time period is required to accomplish this relaxation and is denoted by T_1, the spin-lattice relaxation time; (2) Redistribution of the absorbed energy among the nuclei by mutual exchange of nuclei between the higher and lower states; this process is called spin-spin relaxation. T_2 represents the spin-spin relaxation time. These two processes, having different mechanisms, do not necessarily occur at the same rate.

In general, the relaxation processes in biopolymers are related to their freedom of motion and rotation as a means for dissipation of energy and to their interaction of neighboring spins. As an example for the first case, large and rigid DNA molecules do not show narrow resonance lines because of the lack of motion, and as an example for the latter case, ^{13}C resonances of ^{13}C bonded to proton (i.e. ^{13}C-H) are effectively relaxed by the dipole-dipole interaction.

Application

With the instrumentation currently available, under optimal conditons the individual nuclei in ^{1}H NMR can be detected at 10^{-4}M (or 30 μmole in 0.3 ml), which is about the same for ^{19}F NMR, while ^{31}P NMR can be detected at 10^{-3}M and ^{13}C NMR at 10^{-2}M (at natural abundance, ^{13}C-enriched compounds can be detected with proportionally greater sensitivity). The sensitivity of detection is also dependent on the various relaxation processes of the sample.

Owing to the intrinsic capabilities and limitations of this technique, the NMR investigation of the biopolymers can be formulated from two different vantage points: (1) for static information, and (2) for dynamic information.

In the first type of inquiry, the investigator wishes to obtain the description of a three-dimensional structure at the atomic level. Such a study is not unlike a study using the x-ray diffraction technique, except that it is carried out in solution. The information is provided mostly from the chemical shifts and the coupling constants. The molecule under investigation is often assumed to have only one predominant conformation; occasionally a dynamic distribution between two conformational states can be ascertained with some confidence. When the conformation of the molecule is distributed dynamically into three or more major states, then the population of the conformational states of this molecule cannot be readily described. While such a study may or may not produce a precise description of the conformation, it has been most useful in detecting changes in conformation due to interaction or perturbation. Considerable success in this approach was achieved in the conformational studies on small molecules (mol. wt. less than 5000), including cyclic oligopeptides and nucleic acid short helices. Often a comparison is made between the conformation in the crystalline state determined by x-ray diffraction and that in solution determined by NMR. The most important quantitative equations in this study are the calculation on ring-current magnetic anisotropy (particularly for the bases in nucleic acids) made by B. Pullman, C. Giessner-Prettre and coworkers concerning the through-space effect on chemical shifts of ^{1}H, and the relationship between the dihedral angle and the three-bond coupling

constant, ^{3}J, particularly between ^{1}H and other nuclei, as proposed
by M. Karplus. Examples for the application of these theoretical
considerations to the experimental data will be presented.
Currently, the effects of the neighboring magnetic environment and
the conformation on the chemical shifts of ^{13}C, ^{31}P, and ^{19}F are
not understood in quantitative terms as well as the interpretation
of the coupling constants of these nuclei. Both theoretical and
experimental work are urgently needed in this area. In order to
facilitate this investigation, a computer graphics program has
been constructed in our laboratory so that the three-dimensional
coordinates of every atom of nucleic acid of any sequence and
conformation can be displayed numerically and graphically. The
main thrust of the NMR-computer technology is to extend the investigation on nucleic acids and possibly protein-nucleic acid interaction to the atomic level. The present challenge is to focus on
tRNA and interaction of tRNA with its cognate synthetase.

1. A Study of Short Nucleic Acid Helices in Solution. In
1975, we reported the comprehensive study of the proton magnetic
resonance on the non-exchangeable protons and on the hydrogen-
bonded NH-N protons of the ribosyl ApApGpCpUpU helix in solution
(Borer et al., 1975, Kan et al., 1975). All the resonances of the
base proton, ribose-H$_1$, and the hydrogen-bonded NH-N protons have
been assigned with a great deal of certainty. The effects of helix-
coil transition on these twenty resonances have been carefully
recorded and analyzed in terms of changes in chemical shifts, coupling constants, and line width. The effect of salt concentration
and oligonucleotide concentration on the helical coil transition
process has also been investigated.

Confronting the challenging task of making spectral assignments
for the 17 C-H resonances and the three hydrogen-bonded NH-N
resonances, we have adopted three major procedures, in addition to
the conventional approach of relying on chemical shifts, coupling
constants, and the spin lattice relaxation time, T$_1$.

The first procedure compares the spectra of a sequence-related
series of oligonucleotides in which each member of the series is
incremented one nucleotide unit from its predecessor. Hence, this
aspect is termed "incremental assignment". The shortest member of
this series must be previously assigned by other standard methods
and the longest member is obviously the molecule of interest. In
the present case, ApA, A$_2$G, A$_2$GC, A$_2$GCU, and A$_2$GCU$_2$ comprise the
series. The spectra of the series of oligomers are usually recorded in low salt, at high temperature, and, if possible, at low
strand concentration. In this environment the inter- and intra-
strand interactions are greatly reduced as indicated by the
similarity of the oligomer and monomer spectra, and also the
resonances are narrow and usually better resolved. Each of the
oligomers in the series exists as a substantially unstacked single

strand; therefore, any major change in the resonance pattern from
an oligomer to the one incrementally longer is due to the resonances of the new nucleotide and its shielding effects on the
previously present protons. The magnetic field effect of the newly added nucleotide acts principally on its immediate 5' neighbor.
The procedure is illustrated in Figure 1.

The second assignment procedure compared the effect of temperature variations as demonstrated in Figure 2. The entire spectrum
of the hexamer and those of many oligomers in the series have been
recorded over a 0-90°C range. The interval between temperature
points was 2-4°C in regions where δ changes rapidly with temperature or several protons resonate very close to each other. The
interval was ~10°C in temperature regimes of little overlap or
change in δ. From such data the spectral assignment established
at one temperature can be transferred to that at another temperature.

The third procedure involves the assignment of NH-N resonances
which depends on the line-width measurements. The line widths of
the NH-N resonances of the helical duplex are related to the rates
of the following exchange reactions:

$$NH_{helix} \xrightleftharpoons[k_{ch}]{k_{hc}} NH_{coil} \qquad (1)$$

$$NH_{coil} \xrightleftharpoons[k_{wc}]{k_{cw}} H_2O \qquad (2)$$

Based on various analyses of the known rate of exchange and of
helix-coil transition, we reached the conclusion that the line-
widths of the NH resonances in the current experiment are primarily
determined by the lifetime of the helix.

The above conclusions readily lead to the assignment of the
NH resonances based on the linewidth data. The NH resonance which
has the largest linewidth and the highest sensitivity to thermal
effects on the line-broadening and line-shifting is assigned to
the NH-N of the two terminal A(1)·U(6) pairs; the NH resonance
which has the smallest linewidth and the lowest sensitivity to
thermal effects is assigned to the NH-N of the two middle G(3)·C(4)
pairs; and the NH resonance which has an intermediate linewidth
and sensitivity is therefore assigned to the two interior A(2)·U(5)
pairs.

In analyzing the chemical shift data, a helical duplex of
Kendrew's molecular model of A_2GCU_2 was constructed from the
coordinates of A'-RNA and from the coordinates of B-DNA. Based on

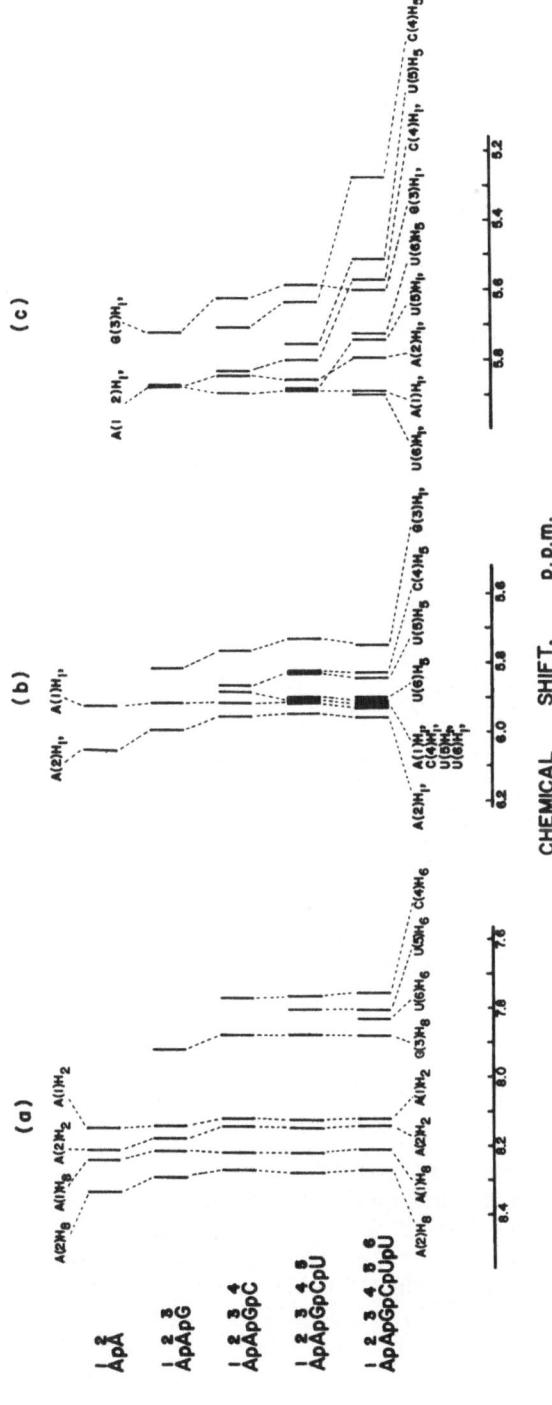

Fig. 1. The incremental assignment scheme for all the base and H_1' resonances of A_2GCU_2. The thickness of the lines represents the number of signals contained in a resonance envelope. (a) The assignment of H2, H6, and H8 resonances at $\sim 65^\circ C$. (b) The assignment of H1, and H5 resonances at $\sim 65^\circ C$. Four resonances are clustered at this temperature. (c) The assignment of H1, and H5 resonances clustered at $\sim 65^\circ$ are resolved at this temperature and can now be assigned. $[Na^+]=0.02$ M and $C_S=1$ mM for each of the spectra except for the hexamer spectrum shown in the 5-6 ppm region where $C_S=10$ mM and $[Na^+]=0.07$ M. Chemical shifts are expressed in reference to DSS. (From Ref. 4).

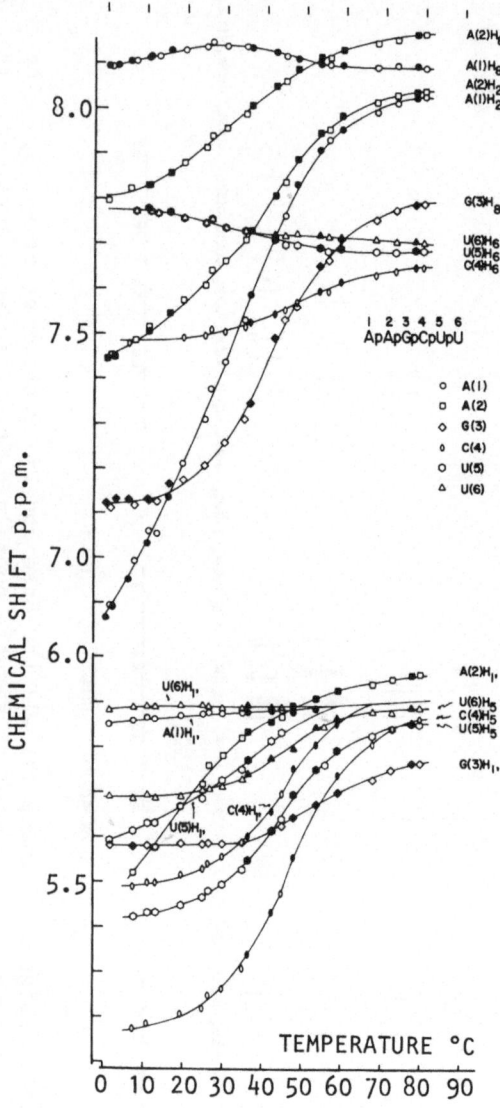

Fig. 2. The plot of the chemical shifts of base and $H_{1'}$ protons of A_2GCU_2 in D_2O (10 mM in strand concentration, 0.01 M sodium phosphate buffer, pD = 7.0, 0.07 M Na^+) versus temperature. All chemical shifts are expressed in reference to DSS. The solid symbols represent the data from the 100 MHz spectrometer and the open symbols represent the data from the 220 MHz spectrometer. (From Ref. 4).

these models, projections of neighboring protons onto the plane of each base were made from the Kendrew model, such as that for the A_2GCU_2 helix in A'-RNA geometry in Fig. 3. These projections now permit determination of the distance of a proton from the base ring.

Additional and valuable information can also be obtained from the coupling constants data on the H_1' resonances during the helix coil transition and the temperature variation profile. The low $J_{1'-2'}$ values of the H_1' resonances in the helical conformation indicate that the furanose of the helical duplex is most likely in the C3'-endo conformation. It is known from the x-ray diffraction studies that the C3'-endo conformation belongs to the A and A'-RNA, while the C2'-endo conformation belongs to the B form of nucleic acids, such as B-DNA. Therefore, the data based on $J_{1'-2'}$ clearly favors the notion that the A_2GCU_2 duplex in solution assumes an A or A' conformation.

A comparison between the predicted chemical shifts of these twenty proton resonances and the observed resonances were made. The predicted chemical shifts of formation for the A_2GCU_2 sequence in both the A'-RNA and B-DNA geometries are reported in Table IA. The computation used ring current isoshielding contours which were mapped in planes parallel to the bases at the distances shown in Figure 3. In a similar manner, the measured chemical shifts of the three NH-N resonances are compared with the computed values based on A'-RNA and B-DNA model, as shown in Table IB.

In summary, conformational details of the A_2GCU_2 helix in solution can now be ascertained by NMR on the basis of the following self-consistent information: (i) The $J_{1'-2'}$ values of all the residues in the helix indicate that each furanose is in a 3'-endo conformation. Model building based on the x-ray diffraction data of nucleic acid fibers reveals that the furanose conformation in the A'-RNA (or A-RNA) is 3'-endo, while the furanose in B-DNA is 3'-exo or 2'-endo. (ii) The chemical shifts of 17 C-H resonances of the A_2GCU_2 helix in solution agree with the computed values based on the geometry of the A'-RNA much better than those based on the geometry of B-DNA. (iii) The chemical shifts of three sets of NH-N resonances representing six base pairs of the A_2GCU_2 helix also agree with the computed values based on the geometry of A'-RNA significantly better than those based on the geometry of B-DNA. Thus, these data strongly support the conclusion that the A_2GCU_2 helix in solution must assume a conformation closer to that of the A'-RNA than to that of B-DNA. It would be of great interest to synthesize a short DNA helix containing the A_2GCU_2 sequence and to study the conformation of this short DNA helix in solution by PMR following the approach outlined above.

The NMR data on the helix-coil transition of this short helix $(A_2GCU_2)_2$ contains both thermodynamic as well as structural infor-

Table IA. The Chemical Shifts of Formation of 17 Non-Exchangeable Protons in (1) The A_2GCU_2 Duplex as Compared with Calculated Ring Current Shifts from Two X-Ray Crystallographic Models and (2) The A_2GCU_2 Coils at 90°C (δ in ppm from DSS)

	δ^a_{obs}	$\Delta\delta^b_{obs}$	$\Delta\delta$ calc^c		$\Delta\delta_{obs}-\Delta\delta$ calc		δ^d	$\Delta\delta^e$
			A'	B	A'	B	(COIL FORM)	
G(3)H8	7.20	0.93	0.85	0.33	0.08	0.60	7.91	0.13
C(4)H6	7.56	0.48	0.42	0.21	0.06	0.27	7.77	0.15
C(4)H5	5.14	0.97	1.07	0.66	-0.10	0.31	5.89	0.19
A(2)H2	7.50	0.76	0.64	0.44	0.12	0.32	8.16	0.13
A(2)H8	7.84	0.67	0.79	0.25	-0.12	0.45	8.28	0.27
U(5)H6	7.95	0.07	0.15	0.18	-0.08	-0.11	7.79	0.11
U(5)H5	5.43	0.51	0.46	0.31	0.05	0.20	5.87	0.06
A(1)H2	6.89	1.36	1.05^f	1.10^f	0.31^f	0.26^f	8.16	0.12
A(1)H8	8.16	0.22	0.00^f	0.02^f	0.22^f	0.20^f	8.20	0.16
U(6)H6	7.95	0.07	0.03	0.03	0.04	0.04	7.81	0.09
U(6)H5	5.69	0.25	0.20	0.09	0.05	0.16	5.90	0.03
G(3)H1'	5.59	0.35	0.08	0.17	0.27	0.18	5.77	0.13
C(4)H1'	5.48	0.50	0.04	0.07	0.46	0.43	5.92	0.04
A(2)H1'	5.45	0.69	0.10	0.27	0.59	0.42	5.98	0.15
U(5)H1'	5.60	0.38	0.02	0.05	0.36	0.33	5.92	0.08
A(1)H1'	5.86	0.27	0.00	0.22	0.27	0.05	5.92	0.19
U(6)H1'	5.86	0.12	0.00	0.00	0.12	0.12	5.92	0.08

Table 1A. Footnotes

a   These are the low temperature plateau values of δ at 10 mM strand concentration, pD 7.0, 1.07 M Na$^+$. If the 0.17 or 0.07 M Na$^+$ δ vs. T($^\circ$C) profiles level off at low temperature, their plateau values were averaged with the 1.07 M Na$^+$ number to generate the δduplex value.

b $\Delta\delta$duplex=δduplex-δ^{0-5°mononucleotide is the chemical shift of formation for a proton in the duplex. Appropriate mononucleotide values were selected from the following list (δ in ppm from DSS): pG-H8, H1', 8.126, 5.936, respectively; pC-H6, H5, H1', 8.038, 6.110, 5.979; pA-H2, H8, H1', 8.257, 8.512, 6.141; pU-H6, H5, H1', 8.015, 5.935, 5.982; and Ap-H2, H8, H1', 8.252, 8.381, 6.132, measured in 0.01 M sodium cacodylate buffer, pD 5.9+0.1 in D$_2$O, 1 mM in the appropriate 5'-mononucleotide (a 2.0 mM sample of 3'-AMP was also measured). Little or no association of the mononucleotides is expected at these low concentrations. Temperatures were 0-5°.

c   Based on the A'-RNA and B-DNA geometries (Arnott et al., 1972; Arnott and Hukins, 1972) and ring current isoshielding contours provided by B. Pullman (private communication, see discussion).

d   These are averages of the 90°C chemical shifts of protons in 0.07, 0.17, and 1.07 M Na$^+$ at 10 mM strand concentration, pD 7.0.

e $\Delta\delta$coil=δcoil-δ^{90°mononucleotide is the chemical shift of formation for a proton in the high temperature coil form. These mononucleotide values were used (see note b for buffer and nucleotide concentrations): pG-H8, H1', 8.041, 5.890; pC-H6, H5, H1', 7.921, 6.084, 5.960; pA-H2, H8, H1', 8.287, 8.547, 6.130; pU-H6,H5, H1', 7.905, 5.930, 6.001; and Ap-H2, H8, H1', 8.282, 8.359, 6.109.

f   These calculated values are subject to a correction of $\sim$0.15 ppm due to shielding in end-to-end aggregates.

Table 1B. The Chemical Shifts (in ppm) and the Linewidth at Half-
           Height (in Hz) of the Hydrogen Bonded NH-N Resonances
           of Guanine and Uracil in $(A_2GCU_2)_2$ at 1°C.[a]

	observed	δ_{calcd}[b] (A'-RNA)	δ_{calcd}[c] (B-DNA)	$\delta_{obsd} - \delta_{calcd}$ A'-RNA	B-DNA	linewidth $(\omega_{\frac{1}{2}})$
G(3)NH	13.5[d]	13.3	12.8	0.2	0.7	30
U(5)NH	14.2	14.1	14.1	0.1	0.0	44
U(6)NH	13.2	13.5	14.3	-0.3	-1.1	80

[a] The intrinsic values of the chemical shifts of the NH resonance
     in the A·U pair and in the G·C pair is taken to be 14.7 and
     13.6 ppm respectively (Kearns and Shulman, 1973), with the A·U
     pair in the B-DNA geometry given the same value, 14.6 ppm, as
     derived for the A·T pair (Patel and Tonelli, 1974).

[b] Calculated values based on the geometry of the A'-RNA helix
     (Borer, et al., 1975).

[c] Calculated values based on the geometry of the B-DNA helix
     (Borer, et al., 1975).

[d] All negative signs are omitted.

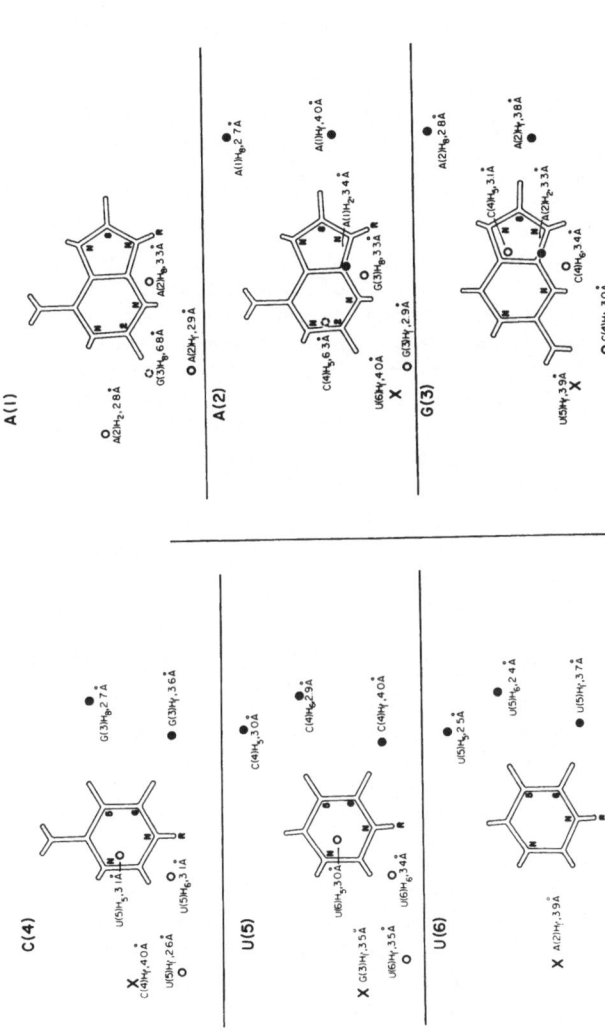

Fig. 3. The projections of neighboring protons onto the plane of each base of the A2GCU2 helix (A'-RNA geometry). Projections and distances were taken directly from the Kendrew model. Open circles (O) indicate the protons positioned below the base plane; filled circles (●) indicate the protons positioned above the plane; broken circles (◌) indicate protons from a distant neighbor about 6.8 Å away; and crosses (X) indicate H1' atoms in the other strand. The views are normal to the base planes with the free 5'-OH terminus above and the 3'-terminus below the base planes. Vertical distances of the protons to the base planes are indicated in parentheses. (From Ref. 4).

mation. These helix-coil transition profiles are reports of 20
atoms at separate locations within the helix (including the three
NH-N profiles) about their magnetic properties during the thermal
transition; therefore, in principle, they can provide useful data
to assign statistical weights to the various partially bonded
states. However, each transition curve is a reflection of the
changes in the local magnetic environment of each proton rather
than a direct report on the helix-coil populations. In the
partially formed duplex the δ value of a proton is determined by
its δ value in each microstate, weighted according to the population of the state. Moreover, the value in a particular microstate
is determined by shielding influences which are anisotropically
distributed through space.

Thus, it is a very challenging task to disentangle the various
factors in order to obtain meaningful information without more
accurate knowledge of the application of NMR theory to nucleic acid
research. However, two general conclusions can be reached at
present. First, the NMR data follow the expected patterns derived from the optical studies carried out at lower concentrations
concerning the effect of concentration and ionic strength on the
helix-coil transition of this short helix (Borer et al., 1975).
The "average T_m" increases with increasing concentration, showing
a linear relationship in a $1/T_m$ vs. -log conc. (strand) plot,
and also with an increase in ionic strength. Second, the melting
of this short helix clearly does not reflect an all-or-nothing
pattern. Both C-H resonances and NH-N resonances reveal that,
with respect to temperature change, the $G_3 \cdot C_4$ pair in the center
is more stable than the $A_2 \cdot U_5$ pair, which in turn is more stable
than the $A_1 \cdot U_6$ pair at the end of the helix. These results are
equivalent to the fraying of the ends of this short helix.

It should be noted that valuable studies on short helices have
also been made by others (Patel and Tonelli, 1974, 1975; Arter
et al., 1974, Patel, 1976, and Early et al., 1977).

2. Theoretical computation of proton chemical shifts. It is
self-evident that a quantitative application of the NMR theory is
urgently needed. We have taken two approaches in our research.
In the first approach, the spatial dependence of the ring-current
magnetic anisotrophy of the nucleic acid bases has been calculated
over a cylindrical domain of 10 Å radius, which extends 8 Å above
and below each ring of these bases (Giessner-Prettre, et al., 1976).
As shown in Fig. 4, separating the bicyclic purine base into two
individual cyclic rings allows this spatial dependence to be
presented in a series of graphs in cylindrical domains. An example of this graphic approach for the adenine ring is shown in
Fig. 5. By this approach, the through-space ring-current effects
on the chemical shifts of any resonance for atoms of nucleic acids
 with known and fixed conformations can be calculated by this first

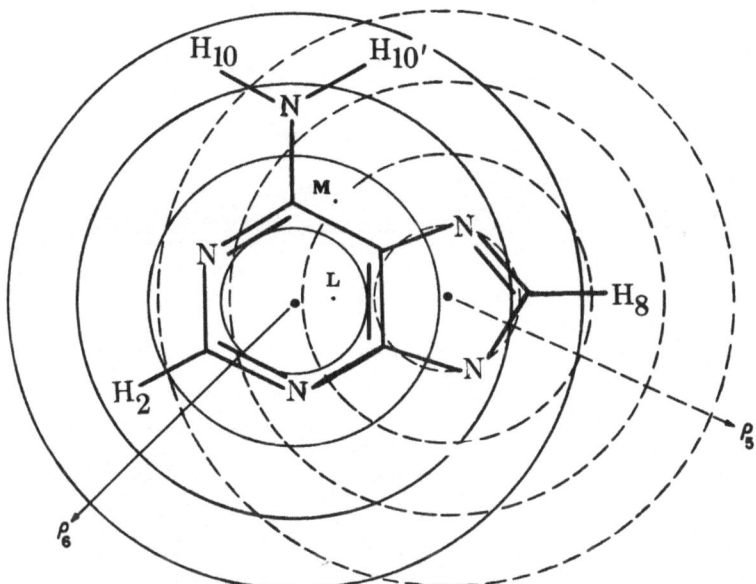

Fig. 4. Concentric contours for each ring of adenine base spaced at 1.0 Å intervals. The large dot in the center of each ring is the origin of the cylindrical coordinate system for that ring; ρ_5(→) represents the radius for the five-membered ring and ρ_6(→) represents the radius from the six-membered ring. (From Ref. 9).

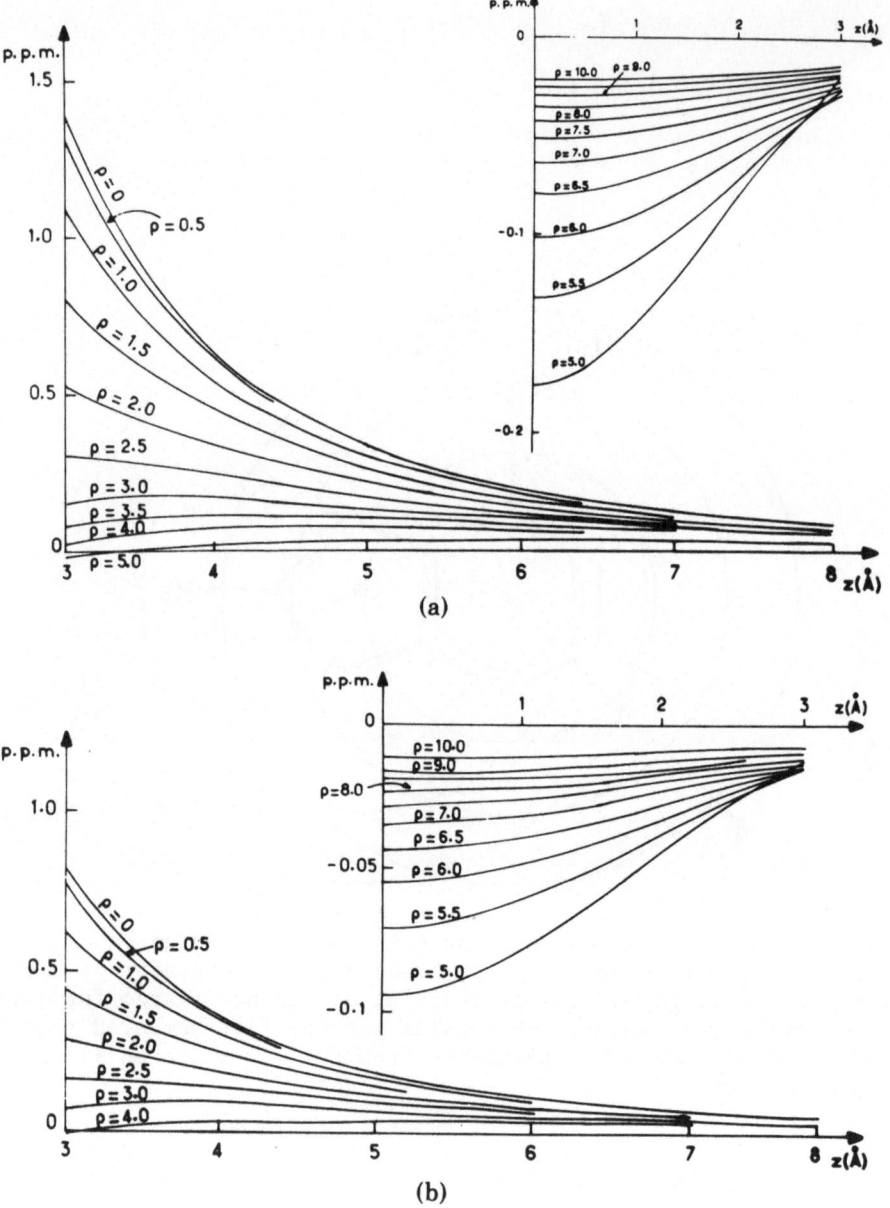

Fig. 5. Adenine, (a) $\Delta\delta_6$ vs. $|z|$, the vertical distance from the base plane and ρ_6, the radius from the center of the six-membered ring, (b) $\Delta\delta_5$ vs. $|z|$ and ρ_5, the radius from the center of the five-membered ring. (From Ref. 9).

approximation and by the independently determined atomic coordinates. While this graphic approach does not replace the more precise computer–computation approach, it does provide a simple method for manipulation of geometric parameters in order to predict the probable mode of base-base or even base-drug interactions.

In the second approach (Kan, Kast, Ts'o, and Ts'o, unpublished data), we have just completed a computer program using Fortran-10 (a DEC-10 computer) which will print out the precise Cartesian coordinates and cylindrical coordinates of all atoms of nucleic acid duplexes having three helical structures (A form, A' form, and B form) of any specified sequences up to 30 base pairs in length. This program can also calculate the distance between atoms. Programs for other helical structures can be readily added. Thus, the precise coordinates of all atoms of any nucleic acid duplex up to 30 base pairs (such as that shown in Fig. 3) can be ascertained in three or more conformations for use in NMR calculations. We hope now we will be able to use the computation program established by Professor B. Pullman and Dr. C. Giessner-Prettre in conjunction with this computer program for coordinates for the calculation of chemical shifts of all of the atoms of known coordinates. Various through-space, electromagnetic effects exerted on the atoms, such as ring current anisotrophic effect, the hydrogen-bonding effect, and the electrostatic effect of the charged group (such as the phosphate), can now be considered individually and collectively in various combinations (Giessner-Prettre et al., 1977) in a quantitative fashion.

3. Quantitative NMR Studies on the Yeast Transfer RNA^{phe}. We have attempted to apply the above systematic approach at least in a trial fashion to the study of yeast tRNA^{phe} in solution (Kan et al., 1977). The two-dimensional structure of yeast tRNA^{phe} in the proposed clover-leaf model is shown in Fig. 6.

Recently, the structure of yeast phenylalanine transfer ribonucleic acid (tRNA^{phe}) in crystalline state has been clearly elucidated by x-ray diffraction studies. Furthermore, the three-dimensional coordinates of all atoms (except hydrogen) in this tRNA molecule have been reported by several laboratories (Quigley et al., 1975; Ladner, et al., 1975b; Sussman and Kim, 1976; Stout et al., 1976; See review by Rich and RajBhandary, 1976). The tRNA^{phe} structures determined from these two crystal forms are very similar to each other. Therefore, a very crucial question arises: Does the common conformation of tRNA^{phe} determined in the crystalline state also exist in aqueous solution? At present, information obtained from nuclear magnetic resonance (nmr) studies can provide a direct and defined answer to this question.

Two spectral regions in the ^{1}H NMR spectrum of tRNA^{phe} can be investigated for quantitative conformational information. The

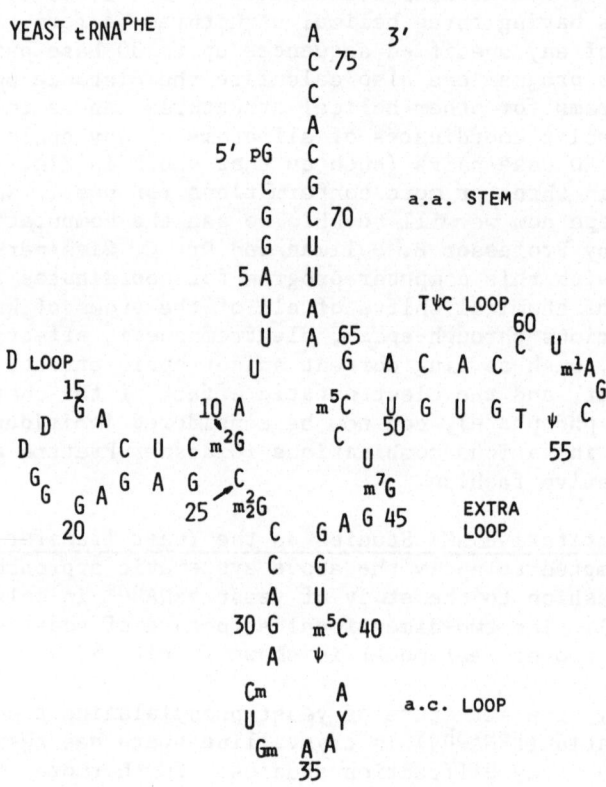

Fig. 6.   The cloverleaf structure of Baker's yeast tRNA^{phe}.

first region, which will be discussed later, contains the very low field NH-N hydrogen-bonded proton resonances in H_2O. The second region is described in this section, in which the high field (1-4 ppm) resonances of the methyl/methylene protons from the minor nucleosides in yeast tRNAPhe in D_2O are investigated. The emphasis in this study is not on base pair regions which can be studied by the NH-N hydrogen bonded proton resonances but rather on the loop regions and tertiary structure of tRNA molecule (Kan, et al., 1974; Kastrup and Schmidt, 1975; Reid and Robillard, 1975; Daniel and Cohn, 1975 and 1976), especially the D, anticodon, and TψC loops on tRNAphe.

In addition to the intact tRNA^{phe} molecule, four important fragments have also been investigated. These fragments are not only very useful in the assignment of the methyl/methylene resonances of the intact tRNAphe, but also provide information on the conformation of the whole molecule of tRNAphe.

First, by taking advantage of recent developments in the distance dependence of the ring-current magnetic effect of the bases described above (Giessner-Prettre, et al., 1976) and in the refined atomic coordinates of yeast tRNAphe in the crystalline state (A. Rich, private communication), a quantitative comparison between the calculated shielding effects ($\Delta\delta$) and the observed shielding effects was made. This study suggests that the conformation of yeast tRNAphe in aqueous solution is grossly, but not totally, identical to that determined in the crystalline state, especially in the TψC and D regions. The initial step for assignment of these high field resonances from tRNAPhe and its fragments is to compare their spectra to those from monomers at high temperature. All modified mononucleosides (-nucleotides) found in tRNAphe molecules were investigated separately at high temperatures. This knowledge from the mononucleosides (-nucleotides) together with the sequence data provides the basis for the unambiguous assignment of the methyl, methylene resonances in tRNAPhe fragments, which were then used for the assignment of the spectra for the whole tRNA molecule. The intact tRNAphe in the presence of Mg^{++} was investigated at 360 MHz frequency, as shown in Fig. 7.

The calculated ring-current shielding/deshielding effects of the methyl or methylene groups from their neighboring bases in yeast tRNAphe has been made (Kan et al., 1977). These calculations of shielding or deshielding effects were based on the structure of yeast tRNAphe in orthorhombic crystal as described in atomic coordinates (A. Rich, private communication) and the graphic approach to the computation of the ring current effects (Giessner-Prettre, et al., 1976). A comparison between the calculated ring-current effect (in terms of ppm) is made to the observed shielding/deshielding effects which are defined by the differences in ppm of the chemical shift values from the intact tRNA versus those from

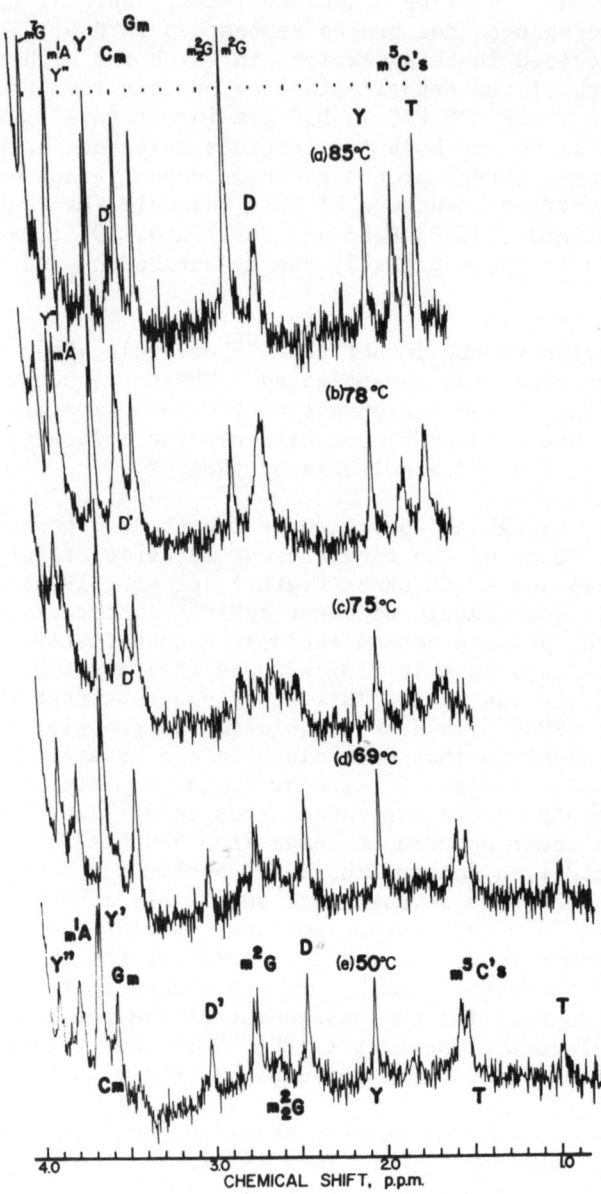

Fig. 7. The 360 MHz nmr spectra of high field proton resonances, region of Baker's yeast tRNA^{phe} in a temperature range of 50–85°C. These spectra were taken under the following conditions: acquisition time, 1 second, 512 transients for spectrum a, 1024 transients for spectra b, c, d, and e. (From Ref. 16).

the mononucleotides (-sides) determined at 40-50°C, i.e. the lower plateau region with respect to the temperature perturbation.

The results in this comparison can be classified into four categories. The first category contains 7 resonances, Cm_{32}, Gm_{34}, Y_{37} (all four methyl resonances) and m^5C_{40}. The observed and the calculated $\Delta\delta$ values of these seven resonances are in agreement with each other within 0.1 ppm. Since these four modified residues are either in anticodon loops or in the anticodon stem, the results suggest no difference in the conformations of anticodon stem and the anticodon loop of this tRNA in aqueous solution versus that in the crystalline state can be found by this approach. In addition, this category also contains one unusual case which is $m_2^2G_{26}$. There are two ring-current effects predicted for the two methyl groups in $m_2^2G_{26}$ (0.82 and 0.13 ppm) based on a static structure of tRNA in crystal, but only one broad signal was seen on ^1H nmr spectrum at ~65°C (Fig. 6) having an observed $\Delta\delta$ value of 0.46 ppm. This observed value of $\Delta\delta$ is very close to the average of the two calculated values. As expected, the $C_2-N(CH_3)_2$ bond in m_2^2G rotates at a sufficiently fast rate so that only one broad peak is observed. Therefore, it is possible that the average conformation of the two methyl groups from m_2^2G in aqueous solution may also be no different from the average positions from its crystalline form. Since $m_2^2G_{26}$ is located on top of the anticodon stem, this reasoning again reinforces the preceding conclusion, i.e. the conformation of the anticodon region of yeast $tRNA^{phe}$ in aqueous solution is similar to that in the crystalline state.

The second category contains m^2G_{10}, $D_{16,17}(C_5)$, $D_{16,17}(C6)$ and m^5C_{49}. The observed $\Delta\delta$ of these resonances in this category are much higher (more than 0.1 ppm) than the calculated $\Delta\delta$, indicating these protons are _more_ shielded than the predicted values based on crystal structure and ring-current effects. While there exists some doubt about the assignment of $D_{16,17}(C_6)$ resonance in the native tRNA spectrum, this conclusion is most likely correct, since it is supported by the results on $D_{16,17}(C5)$.

The third category contains m^7G_{46} and m^1A_{58}. The observed $\Delta\delta$ of these two resonances are much lower (0.15-0.20 ppm) than the calculated $\Delta\delta$, indicating these protons are _less_ shielded than the predicted values based on crystal structure and ring-current effects.

Finally, the fourth category contains a T residue. The methyl resonance from T has only one predicted ring-current effect value but two $\Delta\delta$ values are observed (Kan et al., 1977). However, both experimental values are not in good agreement with the predicted value (interestingly, one was too high and the other too low).

In summary, the comparison indicates that no differences between observed and calculated $\Delta\delta$ values from resonances in the

anticodon stem and loop can be found, but differences in the TψC
stem/loop and D stem/loop have been uncovered. The nature of the
results, with both agreement and disagreement involving values that
are too high or too low, suggests that the difference may indeed be
due to the difference in conformation. It should be noted that the
regions showing the differences in conformation are the regions of
the tertiary structure of the molecule which are more readily
influenced by packing and also are the areas which are less defined
by the x-ray diffraction data. The segment involving $m_2^2G_{26}$ residue,
which connects the D stem and the anticodon stem may have the same
conformation in aqueous solution as in the crystalline state except
that the $C_{(2)}-N(CH_3)_2$ bond in m_2^2G is rotatable in aqueous solution
but fixed in the solid state.

A similar approach is being used for the hydrogen-bonded
NH-N resonances of yeast tRNA^{phe} (Kan and Ts'o, 1977). The difference in this case is that the resonances have not been assigned
with certainty.

Figure 8 shows the 360 MHz spectrum of the ^{1}H nmr resonances
of the hydrogen-bonded NH from yeast tRNA^{phe} sample at 23°C. Under
this condition, the spectrum is essentially insensitive to temperature variation within +10°C and can be considered as a reliable
representation of the hydrogen-bonded NH resonances of yeast tRNA^{phe}
in native conformation. This spectrum has a good signal-to-noise
ratio, and contains 15 well-resolved peaks plus a shoulder (k') in
the region of 11 to 15 ppm from DSS. This observed spectrum closely
resembles the published and unpublished spectra of the same tRNA
obtained under slightly different conditions (Kearns et al., 1974;
Robillard et al., 1976).

Based on the experimental spectrum in Figure 8, the estimation that the spectrum contains 25 NH's, a simulated spectrum was
constructed (Fig. 9b). This simulated spectrum is considered to
be the representative spectrum of the hydrogen-bonded NH resonances
of yeast tRNA^{phe} in the native state as observed in Fig. 8. This
simulated spectrum is used subsequently in a comparison with the
computed spectrum (Fig. 9).

In the calculation of the chemical shifts of these hydrogen-
bonded proton resonances, two parameters are needed: First, the
shielding (or deshielding) effect on the resonance, and second, the
intrinsic chemical shift of the resonance of this proton in different base pairs. For the NH-N resonance in base pairs of Watson-
Crick type, we adopted the values of Kearns and Shulman (1974),
i.e., 14.7, 13.4, and 13.6 ppm for AU, Aψ, and GC base pairs, respectively. As for the tertiary hydrogen bonds, there are many
types, most of which are not Watson-Crick base pairs such as $G_{15}C_{48}$
(Table IIIB). Recently, Kallenbach et al. (1976) have examined
the (U)N_3H-(A)N_7 hydrogen-bonded proton resonance of reverse

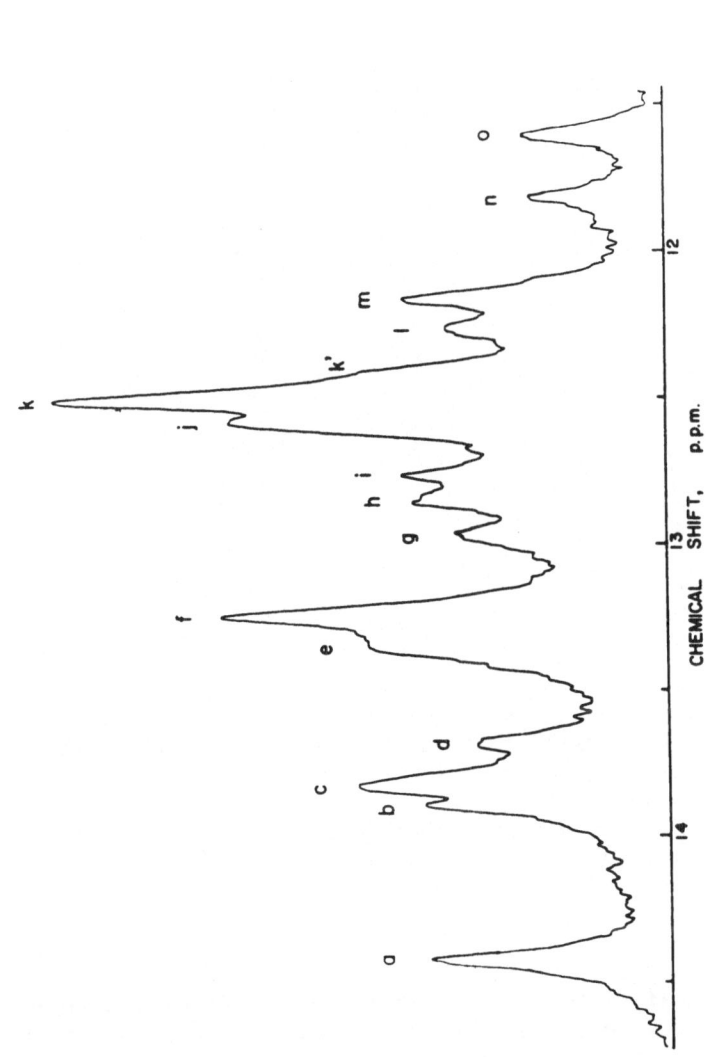

Fig. 8. A 360 MHz spectrum of the yeast tRNAPhe at the 11.5-14.5 ppm region from DSS showing the hydrogen-bonded NH resonances at 23°C. The sample contained 25 mg/ml of tRNAPhe, dissolved in 0.01 M MgCl2, 0.15 M NaCl, 0.002 M EDTA and 0.01 M potassium phosphate buffer, pH 7.0. (From Ref. 13).

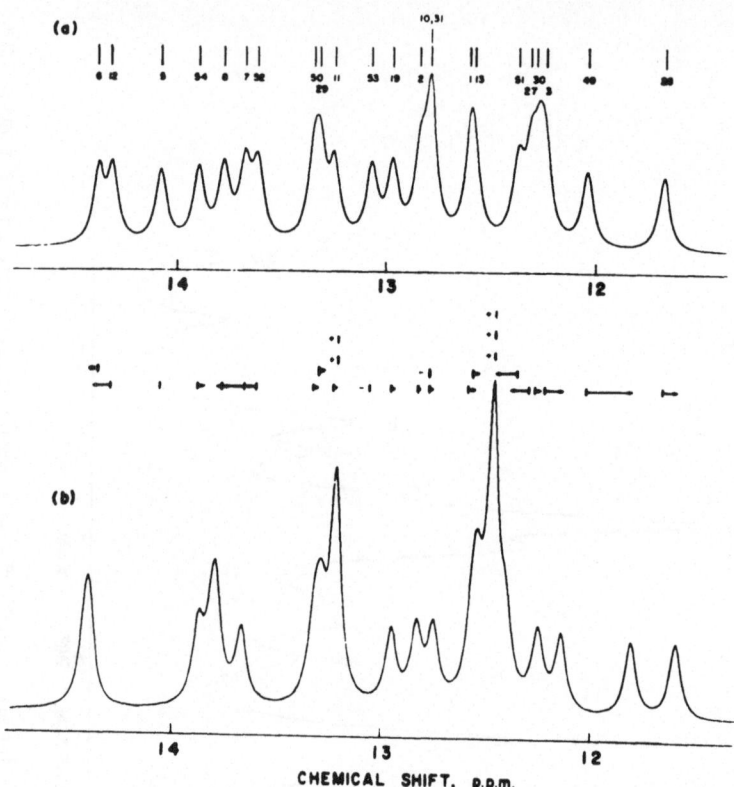

Fig. 9. (a) The computed spectrum of NH–N hydrogen–bonded proton resonances in yeast tRNA^{phe} based on calculated chemical shifts (from refer. 13). The number of these NH resonances represent the following base pairs:

6: U_6A_{67}, 12: $U_{12}A_{23}$, 5: A_5U_{68}, 54: $T_{54}m^1A_{58}$, 7: U_7A_{66}, 8: U_8A_{14}, 52: $U_{52}A_{62}$, 50: $U_{50}A_{64}$, 29: $A_{29}U_{41}$, 11: $C_{11}G_{24}$, 53: $G_{53}C_{61}$, 19: $G_{19}C_{56}$, 2: C_2G_{71}, 10: $m^2G_{10}C_{25}$, 31: $A_{31}\psi_{39}$, 1: G_1C_{72}, 13: $C_{13}G_{22}$, 51: $G_{51}C_{63}$, 27: $C_{27}G_{43}$, 30: $G_{30}m^5C_{40}$, 3: G_3C_{70}, 49: $m^5C_{49}G_{65}$, and 28: $C_{28}G_{42}$. Every computed peak has an equal linewidth at half–height; 36 Hz in 360 MHz scale.

(b) A simulated spectrum for the observed spectrum shown in Fig. 8. Every peak has a 36 Hz linewidth at half–height in 360 MHz scale. The symbols between (a) and (b) represent the adjustments needed to transform the computed spectrum (a) to the simulated spectrum (b); (–) represents removal of resonance peaks, (+) represents addition of resonance peaks, and (→ or ←) represent moving the chemical shifts to high field or low field, respectively.

Hoogsteen type in an U·A·U triple stranded helix by ^{1}H nmr and
recommended an intrinsic chemical shift of 14.1 ppm. From their
data, together with a careful evaluation of the geometry and
shielding effect, we estimated a value of 14.3 ppm as the intrinsic
chemical shift of a reverse Hoogsteen (U)N$_3$H-(A)N$_7$ hydrogen-bonded
proton resonance in an A'-RNA conformation (Arnott, et al., 1973).
As for the resonances of hydrogen-bonded NH-O in G·U, G·ψ, or G·C
base pairs, as well as that of NH-N in G·A base pairs, no intrinsic
values of their chemical shifts have been determined experimentally.
Therefore, no calculations of the resonances of these five base
pairs were made.

The interatomic/intermolecular magnetic field experienced by
the NH-N resonances in yeast tRNA[phe] was then calculated based on
the ring current effect, as evaluated from the coordinates derived
from the x-ray diffraction data (Kan and Ts'o, 1977). This result
was plotted by the PDP-10 computer as shown in Figure 9. The
adjustments needed to transform the computed spectrum to the
simulated experimental spectrum are shown between Figs. 9A and 9B.
This computed spectrum contains 23 resonances, 6 resonances less
than the total recommended by the three dimensional structure of
tRNA determined by x-ray diffraction.

Calculation of hydrogen-bonded NH resonances have been made by
others (Robillard, et al., 1976, Kearns, 1976 , and Geerdes and Hilbers, 1977). A discussion has been made in comparison with these
calculations (Kan and Ts'o, 1977). Special attention should be
given to the interesting results by Geerdes and Hilbers (1977).
Their work showed that the calculated NMR spectra are very sensitive
to slight changes in structure.

In conclusion, despite some uncertainties in the theoretical
treatment, this quantitative comparison between the simulated
experimental spectrum and the calculated spectrum based on the
atomic coordinates of the tRNA in crystal and on ring-current
effects clearly indicates that the native conformation of yeast
tRNA[phe] in solution is fundamentally similar to that in the
crystalline state. The minor difference is probably in the tertiary
structure involving the folding of the TψC loop and stem to the D
loop and stem. This conclusion is reinforced by the ^{1}H nmr studies
on the methyl/methylene resonances of the minor bases reported in
the preceding section. In addition, some of the hydrogen-bonded
base pairs existing in the tRNA in the crystalline state may not
be detectable in solution.

4. Concluding Remarks. The above description indicates that
the [1]H-nmr research on nucleic acids has been highly informative
and sophisticated. In comparison, the research on ^{13}C-NMR and [31]P-
NMR properties of nucleic acid is still in its infancy but has a
promising future (see General References for publications on this

subject). However, so far the investigation has been focused
mainly on the static conformational properties of nucleic acids,
which was treated as the first type of inquiry in this section.

In the second type of inquiry, the investigator wishes to
obtain information about the kinetic parameters of the relaxation
process. From this information, the specific and general interaction of these nuclei with the environment may be known, particularly about their freedom in motion. The studies of histone
interaction are mostly in this category. Specific information on
conformation can also be obtained if the relaxation process is
predominantly influenced by a paramagnetic center near the nuclei
of interest. The interatomic distance between the paramagnetic
source and the relaxed nuclei can be deduced from the relaxation
data. Also, soon the relaxation process will be more readily
investigated by spectrometers of different frequencies. The
frequency-dependent data provide a better estimate of the mechanism
of the relaxation process. The ability to study the relaxation
process of these nuclei is an unique feature of NMR. In order to
utilize the investigative power of NMR fully, both static and
dynamic data about the sample should be obtained and integrated
into a coherent conclusion.

Looking into the future, in order to derive more quantitative
information from the NMR data, we have an urgent need to develop
theory, followed by experimental verification. Also, NMR studies
should be extended to complex mixtures, including possibly living
cells. This can be achieved by the use of NMR label containing
^{19}F or enriched ^{13}C, nuclei normally not found in the biological
system. Under special conditions, the hydrogen-bonded NH——N
resonances and the ^{31}P resonances can also serve this purpose.

ACKNOWLEDGMENTS

This research was supported in part by grants from the
National Science Foundation and from the National Institute of
General Sciences as well as by various fellowships granted to the
individual investigators over the past 10 years. The valuable
assistance of Cathryn Alden and Christine Dreon in the preparation
of the manuscript is gratefully acknowledged.

GENERAL REFERENCES ON NUCLEAR MAGNETIC RESONANCE AND ITS
APPLICATION TO RESEARCH ON NUCLEIC ACIDS AND CHROMATIN

1. J.D. Roberts, ed. Nuclear Magnetic Resonance: Applications to Organic Chemistry, McGraw-Hill Book Co., 1959.

2. J.A. Pople, W.G. Schneider, and H.J. Bernstein, eds., High-Resolution Nuclear Magnetic Resonance, McGraw-Hill Book Co., 1959.

3. F. Bovey, ed. Nuclear Magnetic Resonance Spectroscopy, Academic Press, 1969.

4. M. Boublik, E.M. Bradbury, C. Crane-Robinson, and E.W. Johns. "An Investigation of the Conformational Changes of Histone F2b by High Resolution Nuclear Magnetic Resonance", Eur. J. Biochem. 17, 151 (1970).

5. V.M. Clark, D.M.J. Lilley, O.W. Howarth, B.M. Richards and J.F. Pardon, "The Structure and Properties of Histone F2a Comprising the Heterologous Group F2a$_1$ and F2a$_2$ Studied by ^{13}C Nuclear Magnetic Resonance", Nucleic Acids Research, Vol. I, No. 7, 865 (1974).

6. P.O.P. Ts'o, ed., Basic Principles in Nucleic Acid Chemistry, Vols. I and II, Academic Press, 1974.

7. L. Jackman and F.A. Cotton, eds. Dynamic Nuclear Magnetic Resonance Spectroscopy, Academic Press, 1975.

8. T.L. James, ed. Nuclear Magnetic Resonance in Biochemistry: Principles and Applications, Academic Press, 1975.

9. D. Fitzsimons and G. Wolstenholme, The Structure and Function of Chromatin, Ciba Foundation Symposium 28, London, on April 3-5, 1974, Elsevier/North-Holland Biomedical Press, 1975.

10. E.M. Bradbury, P.D. Cary, C. Crane-Robinson, H.W.E. Rattle, M. Boublik and P. Sautiere, Biochemistry 14, 1876 (1975).

11. P.N. Lewis, E.M. Bradbury, and C. Crane-Robinson, "Ionic Strength Induced Structure in Histone H4 and its Fragments", Biochemistry 14, 3391 (1975).

12. D.G. Gorenstein, "Dependence of ^{31}P Chemical Shifts on Oxygen-Phosphorous-Oxygen Bond Angles in Phosphate Esters", JACS, 97, 4, 898 (1975).

13. D.G. Gorenstein and D. Kar, "^{31}P Chemical Shifts in Phosphate Diester Monoanions. Bond Angle and Torsional Angle Effects", BBRC 65, 3 (1975).

14. D.J. Patel, "Proton Nuclear Magnetic Resonance Studies of the Helix-Coil Transition of d-ApTpGpCpApT in D$_2$O Solution", Biochemistry 14, 3984 (1975).

246   P. O. P. TS'O AND L.-S. KAN

15. D.J. Patel and A.E. Tonelli, "Nuclear Magnetic Resonance Investigations of the Structure of the Self-Complementary Duplex of d-ApTpGpCpApT in Aqueous Solution", Biochemistry 14, 3990 (1975).

16. M.A. Young and T. Krugh, "Proton Magnetic Resonance Studies of Double Helical Oligonucleotides. The Effect of Base Sequence on the Stability of Deoxydinucleotide Dimers", Biochemistry 14, 4841 (1975).

17. C.W. Hilbers and D.J. Patel, "Proton Nuclear Magnetic Resonance Investigations of the Nucleation and Propagation Reactions Associated with the Helix-Coil Transition of d-ApTpGpCp-ApT in H_2O Solution", Biochemistry, 14, 2656 (1975).

18. D.M.J. Lilley, O.W. Howarth, V.M. Clark, J.F. Pardon, and B.M. Richards, "An Investigation of the Conformational and Self-Aggregational Processes of Histones Using ^{1}H and [13]C Nuclear Magnetic Resonance", Biochemistry, 14, 4590 (1975).

19. A.E. Pekary, S.I. Chan, C.-J. Hsu, and T.E. Wagner, "Nuclear Magnetic Resonance Studies on the Solution Conformation of Histone IV Fragments Obtained by Cyanogen Bromide Cleavage", Biochemistry, 14, 1184 (1975).

20. T. Moss, P. Cary, C. Crane-Robinson, and E.M. Bradbury, "Physical Studies on the H3/H4 Histone Tetramer", Biochemistry 15, 2261 (1976).

21. P.D. Cary, C. Crane-Robinson, E.M. Bradbury, K. Javaherian, G.H. Goodwin and E.W. Johns, "Conformational Studies of Two Non-Histone Chromosomal Proteins and Their Interactions with DNA", Eur. J. Biochem. 62, 583 (1976).

22. D.G. Gorenstein, J.B. Findlay, R.K. Momii, B.A. Luxon, and D. Kar, "Temperature Dependence of the [31]P Chemical Shifts of Nucleic Acids. A Probe of Phosphate Ester Torsional Conformations", Biochemistry 15, 17 (1976).

23. P.J. Cozzone and O. Jardetzky, "Phosphorus-31 Fourier Transform Nuclear Magnetic Resonance Study of Mononucleotides and Dinucleotides. I. Chemical Shifts", Biochemistry 15, 22 (1976).

24. P.J. Cozzone and O. Jardetzky, "Phosphorus-31 Fourier Transform Nuclear Magnetic Resonance Study of Mononucleotides and Dinucleotides. II. Coupling Constants, Biochem. 15,4860 (1976).

25. D.J. Patel, "Proton and Phosphorus NMR Studies of d-CpG(pCpG)$_n$ Duplexes in Solution. Helix-Coil Transition and Complex Formation with Actinomycin-D", Biopolymers 15, 3 (1976).

26. J.E. Coleman, R.A. Anderson, R.G. Ratcliffe and I.M. Armitage, "Structure of Gene 5 Protein-Oligodeoxynucleotide Complexes as Determined by ^{1}H, ^{19}F, and [31]P Nuclear Magnetic Resonance",

Biochemistry 15, 25 (1976).

27. D.J. Patel, "Proton and Phosphorus NMR Studies of d-CpG(pCpG)$_n$ Duplexes in Solution. Helix-Coil Transition and Complex Formation with Actinomycin-D", Biopolymers 15, 533 (1976).

28. T.R. Krugh, J.W. Laing, and M.A. Young, "Hydrogen-Bonded Complexes of the Ribodinucleoside Monophosphates in Aqueous Solution. Proton Magnetic Resonance Studies, Biochemistry, 15, 1224 (1976).

29. N.R. Kallenbach, W.E. Daniel, and M.A. Kaminker, "Nuclear Magnetic Resonance Study of Hydrogen-Bonded Ring Protons in Oligonucleotide Helices Involving Classical and Nonclassical Base Pairs", Biochemistry 15, 1218 (1976).

30. D.J. Patel and L. Canuel, "Nuclear Magnetic Resonance Studies of the Helix-Coil Transition of Poly (dA-dT) in Aqueous Solution", PNAS 73, 3 (1976).

31. The Molecular Biology of the Mammalian Genetic Apparatus, Vol. I, ed. by P.O.P. Ts'o, Elsevier/North Holland Publishing Co., 1977.

32. W. Egan, H. Shindo, and J.S. Cohen, "Carbon-13 Nuclear Magnetic Resonance Studies of Proteins", Ann. Rev. Biophys. Bioeng. 6, 383 (1977).

33. D.G. Gorenstein, B.A. Luxon, and J.B. Findlay, "The Torsional Potential for Phosphate Diesters. The Effect of Geometry Optimization in CNDO and AB Initio Molecular Orbital Calculations, BBA 475, 184 (1977).

34. D.R. Kearns, "High-Resolution Nuclear Magnetic Resonance Studies of Double Helical Polynucleotides", Ann. Rev. Biophys. Bioeng. 6, 477 (1977).

35. D.M.J. Lilley, J.F. Pardon and B.M. Richards, "Structural Investigations of Chromatin Core Protein by Nuclear Magnetic Resonance", Biochemistry 16, 2853 (1977).

36. P.G. Hartman, G.E. Chapman, T. Moss, and E.M. Bradbury, Studies on the Role and Mode of Operation of the Very-Lysine-Rich Histone H1 in Eukaryote Chromatin. The Three Structural Regions of the Histone H1 Molecule, Eur. J. Biochem. 77, 45 (1977).

SPECIFIC REFERENCES FOR THE MATERIALS IN THIS CHAPTER

1. J.L. Alderfer and P.O.P. Ts'o, Biochemistry 16, 2410 (1977).

2. S. Arnott, D.W.L. Hukins, S.D. Dover, W. Fuller, and A.R. Hodgson, J. Mol. Biol. 81, 107 (1973).

3. D.B. Arter, G.C. Walker, O.C. Uhlenbeck, and P.G. Schmidt,

Biochem. Biophys. Res. Commun. 61, 1089 (1974).

4. P.N. Borer, L.S. Kan, and P.O.P. Ts'o, Biochemistry 14, 4847, (1975).

5. W.E. Daniel, Jr., and M. Cohn, Proc. Natl. Acad. Sci. USA 72 2582 (1975).

6. W.E. Daniel, Jr., and M. Cohn, Biochemistry 15, 3917 (1976).

7. T.A. Early, D.R. Kearns, J.F. Burd, J.E. Larson, and R.D. Wells, Biochemistry 16, 541 (1977).

8. H.A.M. Geerdes, and C.W. Hilbers, Nucleic Acids Res. 4, 207 (1977).

9. C. Giessner-Prettre, B. Pullman, P.N. Borer, L.S. Kan, and P.O.P. Ts'o, Biopolymers 15, 2277 (1976).

10. C. Giessner-Prettre, B. Pullman, and J. Caillet, Nucleic Acid Res. 4, 99 (1977).

11. G. Govil and I.C.P. Smith, Biopolymers 12, 2589 (1973).

12. N.R. Kallenbach, W.E. Daniel, Jr., and M.A. Kaminker, Biochemistry 15, 1218 (1976).

13. L.S. Kan and P.O.P. Ts'o, Nucleic Acids Res. 4, 1633 (1977).

14. L.S. Kan, P.O.P. Ts'o, F. von der Haar, M. Sprinzl, and F. Cramer, Biochemistry 16, 3143 (1977).

15. L.S. Kan, P.N. Borer, and P.O.P. Ts'o, Biochemistry 14, 4864, (1975).

16. L.S. Kan, P.O.P. Ts'o, M. Sprinzl, F. von der Haar, and F. Cramer, Biochemistry 16, 3143 (1977).

17. R.V. Kastrup and P.G. Schmidt, Biochemistry 14, 3612 (1975).

18. D.R. Kearns, Prog. Nucleic Acids Res. Mol. Biol. 18, 91 (1976).

19. D.R. Kearns and R.G. Shulman, Acc. Chem. Res. 7, 33 (1974).

20. J.E. Ladner, A. Jack, J.D. Robertus, R.S. Brown, D. Rhodes, B.F.C. Clark, and A. Klug, Nucleic Acids Res. 2, 1629 (1975).

21. D.J. Patel, Biopolymers 15, 533 (1976).

22. D.J. Patel and A. Tonelli, Biopolymers 13, 1943 (1974).

23. D.J. Patel and A. Tonelli, Biochemistry 14, 3990 (1975).

24. G.J. Quigley, N.C. Seeman, A.H.-J. Wang, F.L. Suddath, and A. Rich, Nucleic Acids Res. 2, 2329 (1975).

25. B.R. Reid and G. Robillard, Nature (London) 257, 287 (1975).

26. A. Rich and U.L. RajBhandary, Annu. Rev. Biochem. 45, 805 (1976).

27. G.T. Robillard, C.E. Tarr, F. Vosman, and H.J.C. Berendsen, Nature (London) 262, 363 (1976).

28. C.D. Stout, H. Mizuno, J. Rubin, T. Brennan, S.T. Rao, and M. Sundaralingam, Nucleic Acid Res. 3, 1111 (1976).

29. J.L. Sussman and S.-H. Kim, Biochem. Biophys. Res. Commun. 68, 89 (1976).

TECHNIQUES FOR CYTOCHEMICAL STUDIES OF THE NUCLEUS AND ITS SUBSTRUCTURES

Torbjörn Caspersson

Karolinska Institutet, Inst.for Tumor Pathology
Karolinska Hospital
S 104 01 Stockholm 60, Sweden

Qualitative methods for determining the chemical composition of cellular structures go quite far back in time. Even in the early days the main interest was the cell nucleus. Overenthusiastic use of these staining techniques brought, however, the field into disrepute especially in the 1920-ies and this came to be one of the main reasons for developing quantitative optical microscopy for cytochemical research.

With passing years and the rapid development of modern biology and medicine, the main fields of interest in experimental cell biology have changed so that a considerable number of different types of instruments are available to the cytochemist of today.

The first quantitative optical cytochemical work concerned nuclear structures. The central problem at that time was to localize and determine the amounts of nucleic acids by ultramicro-spectrography (UMSP). This photometric work was done in the ultraviolet using UV-microscope objectives developed already at the turn of the century by von Rohr and Köhler. In the very beginning, photographic photometry was used, but this was soon replaced by photoelectric spectrophotometry.

To your eyes, these early instruments may look primitive but they were quite powerful in precision and resolution.

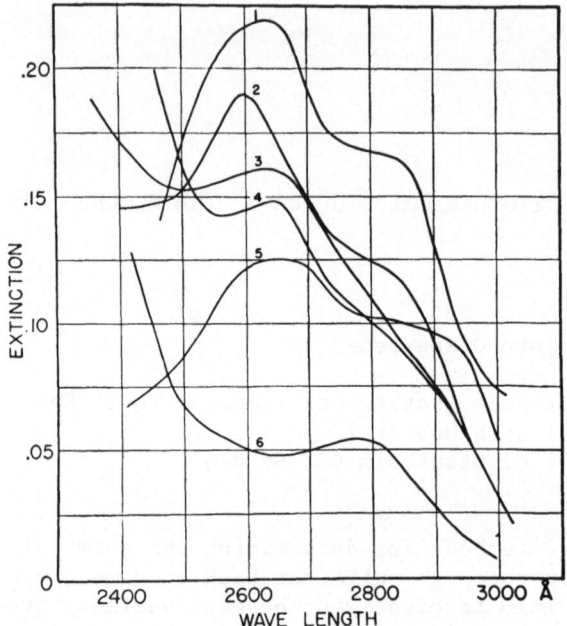

Fig 1. Absorption spectra from different parts of a Drosophila salivary gland nucleus: (1) chromocenter, (2) chromosome insertion on chromocenter, (3 and 4) euchromatic bands, (5) nucleolus, (6) interband space.

Figure 1 gives examples of ultraviolet absorption spectra of small nuclear structures measured in 1940 by the first photo-electric UMSP. The size of the measuring spot is 0.3 μ. These UMSPs also gave <u>semiquantitative</u> information on the amount of protein in cellular structures. The data won in the early forties indicated a clear relationship between protein synthesis and nucleic acids.

Later, improved models of the original UMSP were built. Figure 2 shows an instrument built for spot measurements as well as integrating measurements, the "Universal UMSP". Together with Carl Zeiss Company of Oberkochen this model was further developed (figure 3) and Zeiss began commercial production of that instrument in the early sixties.

This instrument is still the most advanced instrument available for high resolution spectrophotometric analysis of nuclear structures and other small elements inside the cell and because of that, a short description of its function will be given. Basically it is a double beam instrument with identical pathways for

Fig 2. Universal-ultramicrospectrophotometer (UMSP) for spot measurements and for integrating work.

the measuring beam and the comparison beam. It works in the wavelength range from 220 to about 700 nanometers. The special "Ultrafluar" microscope objectives developed by Zeiss for the instrument work well over this entire range. Recordings of absorption spectra can be made of spots within the cells, with the smallest spot size determined only by the resolution of the optics (up to numerical aperture 1.25, even in UV). In addition automated scanning measurements can be made across an object. During the scan the running transmission values taken as the running integral of the extinction values are recorded on the same paper. The distance between each of the subsequent scanning lines can be chosen arbitrarily down to 0.5 μ.

For instance, if a whole free lying cell is scanned, the endpoint of the extinction-integral curve gives the total extinction of the whole cell directly at the wavelength used one can correlate different points in the object with the corresponding points on the recorded curves and it is thus possible to determine the total extinction of any intracellular element, such as nucleoli or chromatin. The instrument is also equipped for photography and with an

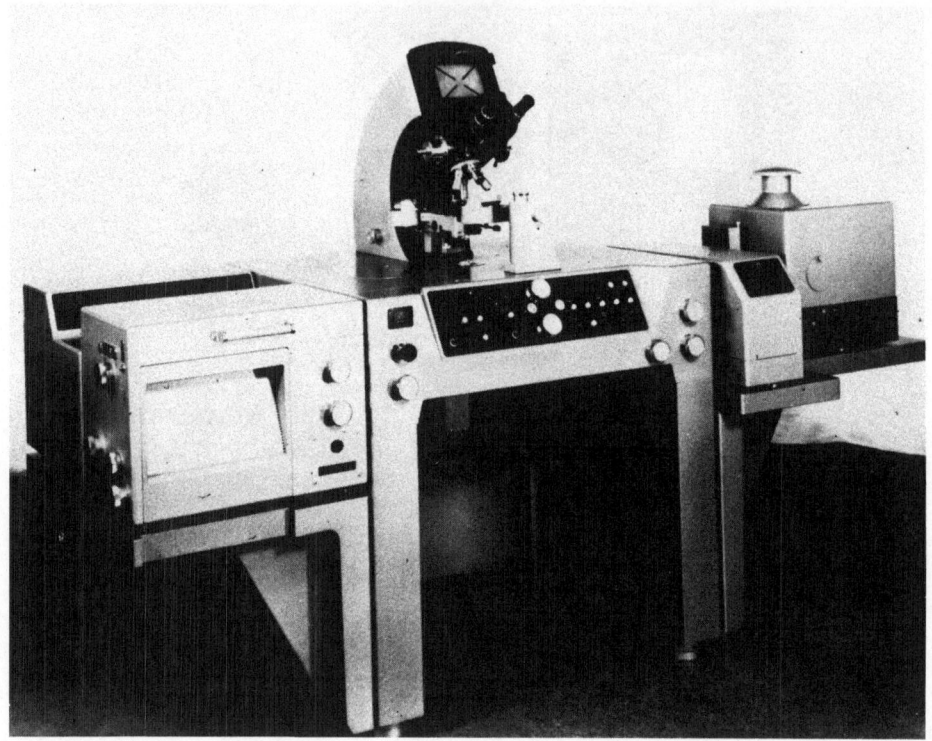

Fig 3. The UMSP built by C. Zeiss.

image converter which visual observation in the UV is quite easy.
This instrument has been used and is being used by many groups for
different kinds of studies of cells and their internal structures
during the course of growth and differentiation and during different types of functions.

The instrument permits work with extreme optical resolution
and is suitable for work on very small structures. For example,
Figure 4 shows recording UV-measurements along rye metapahse chromosomes. Rya has seven chromosomes of practically the same length
which are almost indistinguishable from each other in the microscope. The recordings show fine patterns in the DNA-distribution,
the details of which lie very close to the theoretical limits for
the UV microscope resolution. The reproducibility of the measurements is so good that all seven of the individual chromosome types
can be recognized through their individual absorption profiles.

In the late forties and early fifties several UMSP-model were

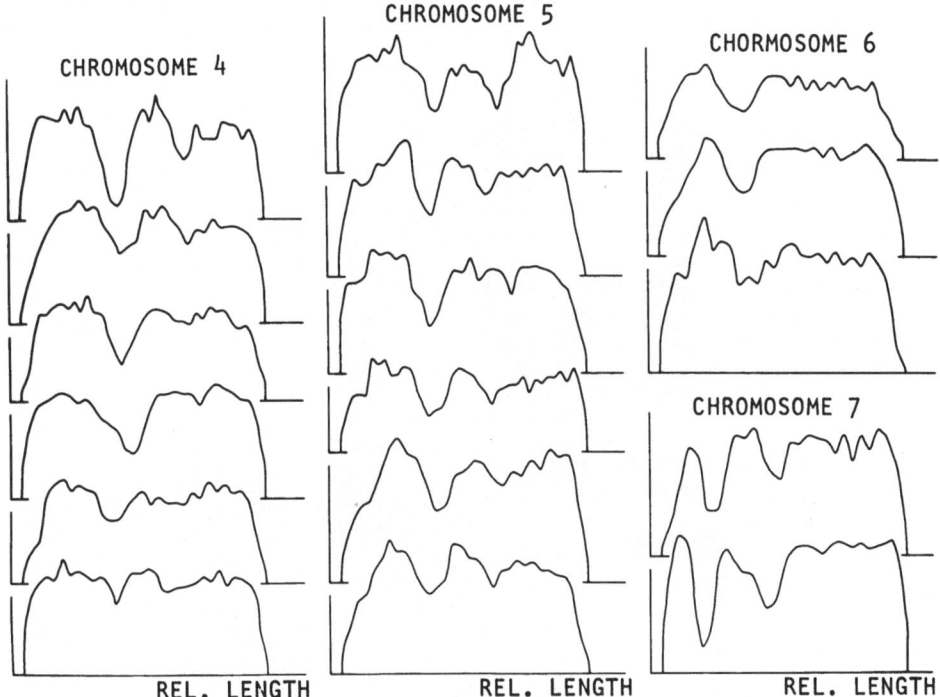

Figure 4.   High resolution recordings of DNA-distribution along
four rye chromosomes.   The absolute length is 7-9 μ.

built in different laboratories, a few for UV but most for use in
the visible region. The main aim was DNA determination by Feulgen-
UMSP and this work contributed greatly to our present view of the
DNA constancy in the nucleus. Most of the instruments permitted
only "plug" - measurements and not high resolution integration
scanning work. This gives low accuracy in the determinations, even
if such "tricks" as the "two-wavelength procedure" are used. Similar instruments are now available from several of the large optical companies.

Several rather complex instruments have been built combining
microspectrophotometers with computers and are now commercially
available. Some of these incorporate scanning stages, permitting
steps down to 0.5 μ range. One use for such instruments is to scan
slides for selected cell types in studies aiming at "automated cytology", which, however, falls outside the frame of this presentation. A field of application which is of considerable interest for
chromatin analysis, is the determination of the substance distribution within an area, for instance the nucleic acid component in
a cell nucleus.

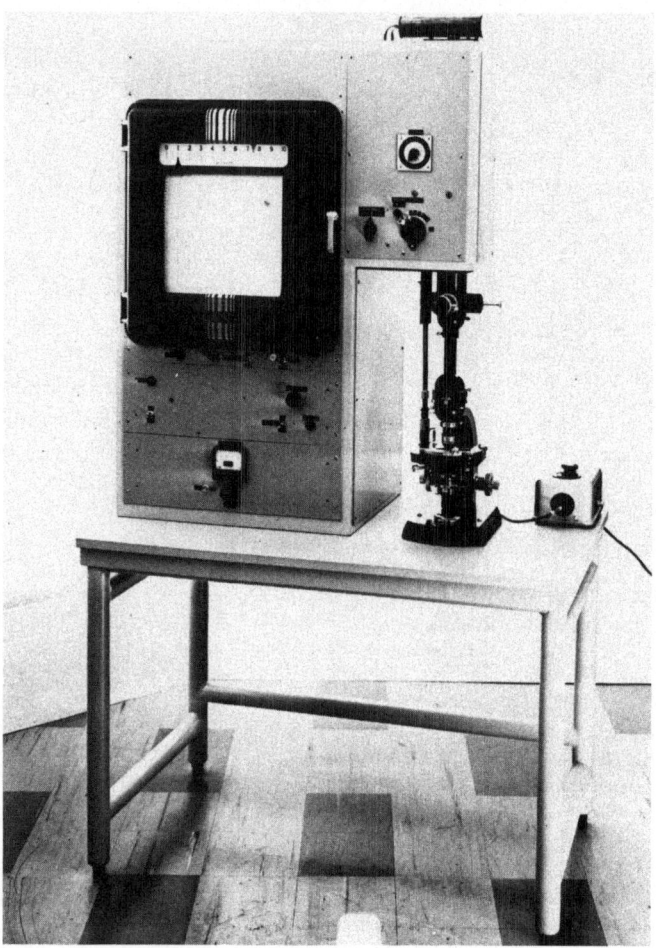

Fig 5. A recording and integrating ultramicrointerferometer.

        Out of our early cytochemical work on the role of the cell
nucleus in cell growth and on its interplay with the cytoplasm, a
desire ensued for better methods for protein and/or cellular dry
mass determinations. In 1948 Engström and Lindström developed a
method based on X-ray absorption, a technique which Leon Carlson
then carried to a resolution comparable to that of microscopy in
the visible range. The technique works very well, but was eventually replaced by the more convenient interferometric techniques. The
first dry mass determinations by interference microscopy were made
in 1950 by Davies and Maurice Wilkins with the Dyson interference
microscope. Ever better interference optics became available
during the following years.

In our group we have built several models of microinterferome-
ters for different purposes. Figure 5 shows a scanning-recording
instrument using the Jamin,Lebedeff type of optics. Its main use
is to measure the distribution of cell dry mass in different compartments of the cell, but it can also perform integrating measurements, even of intracellular details. Interferometry has not
proved to be very useful for analysis of small intranuclear details
because of certain limitations in resolution when these optics are
applied to conventional biological specimens. On the other hand, it
has proved to be an important tool in studies of the nucleus-cyto-
plasm interaction.

It should be mentioned here that the limitations of interferometry at high resolution led to the development in 1959 of a method
for dry mass determination in small cellular elements by measurement of electron scattering in the electron microscope (Bahr, Bloom,
Carlson, Zeitler). This method has later on by Bahr and Zeitler
been proved to be a very valuable tool for the analysis of nuclear
details in general, and especially of chromosomes.

Microfluorometry, a logical outgrowth, developed rather slowly despite efforts in many laboratories. This was due mainly to
lack of suitable optics and adequate light sources. Around 1960, the
situation in this field changed. Figure 6 shows an example of an
instrument for the study of nuclear elements in which both fluorescence, excitation spectra and absorption spectra can be measured
one after the other without the object having to be moved out of
the field of vision of the microscope system.

The instruments referred to as yet were all the result of
efforts to analyze fine nuclear details and reached their present
stage of development in the early sixties. They are quite a potent
array of instruments for cytochemical chromatin studies of different kinds. These instruments are fast, efficient and are easy to
use accurately on a routine basis. As to further improvements,
they have shown little reason for future development.

In the early sixties as a result of the interest in cancer
research there was a desire for instruments more suitable for kinetic studies of protein and nucleic acid synthetic processes during
growth. At that time, tissue culture material became more easily
available for studies of both normal and pathological growth. In
such studies, whole cells must be analyzed and measurements made
on large numbers of cells. Tissue culture offered good opportunities for preparation of cell suspensions in which entire cells
could be measured directly without the complications of tissue em-
bedding and sectioning. The instruments described above, in spite

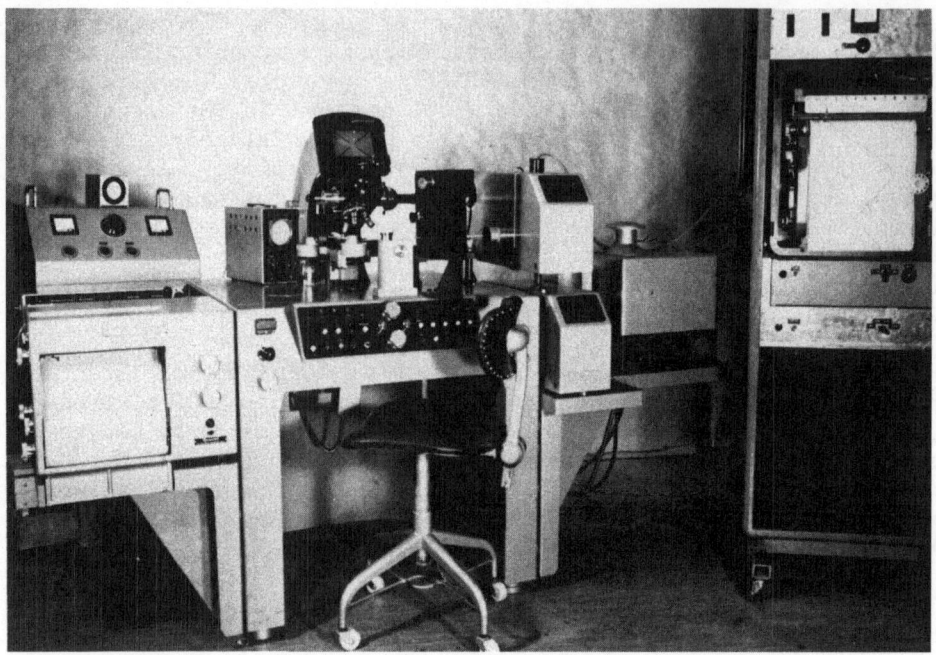

Fig 6. Fluorometer for high resolution work. In the instrument recordings can be made of fluorescence and excitation spectra and
also of the absorption spectrum of small cell details.

of their having been built primarily for analysis of fine structural details, also permit whole cell measurements, although they
are hopelessly slow in kinetic studies where one must analyze populations of hundreds of cells in each experiment.

    This led us to the development of a series of instruments for
whole cell measurements by UMSP, interferometry and fluorometry
all working many times faster than was possible in the instruments
predecessors. Different instrument-models were built between 1960-
1970 and were used by several research groups for growth and dif-
rerentiation studies of many different kinds. As most of these
studies did not specifically concern chromatin questions, I need
not describe them here. Figure 7 illustrates the type of data obtainable with such instruments for cell population studies. The
object is a rapidly growing tissue culture and the histograms describe the distribution of DNA, RNA and total dry mass in the cell
population.

Nuclear cytochemistry can be said to have gotten an extra stimulus around 1970 from the general work in carcinogenesis. A need
was felt for better ways to study the interplay between nucleus
and cytoplasm during cell growth and transformation processes.
Furthermore, considerable attention was being paid to chromosomal
changes during "transformation" and there was a need for better
ways to analyze metaphase chromosomes. Both these fields needed
considerable improvement of the then available equipment and procedures. During the last few years our group has tried to develop
better techniques for study of nuclear-cytoplasmic interaction as
well as for chromosome analysis. New equipment for high resolution
scanning and integrating UMSP and interferometry was built in
such a way that the nucleus could be measured separately from the
cytoplasm. For fluorometry, an instrument was built with which one
can, with great precision, outline the field of the cell to be
measured. Figure 8 shows the appearance of an UMSP equipped with
an arrangement for "preselection of the measuring field". The instrument works in the 220 - 700 nanometers range and permits work

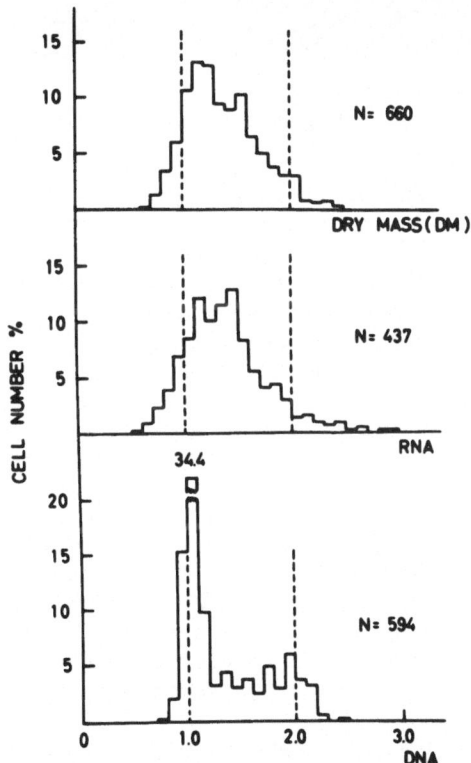

Fig 7. Histograms of the distribution of dry cellular mass (largely proteins), total cellular RNA and DNA in a rapidly growing normal tissue culture.

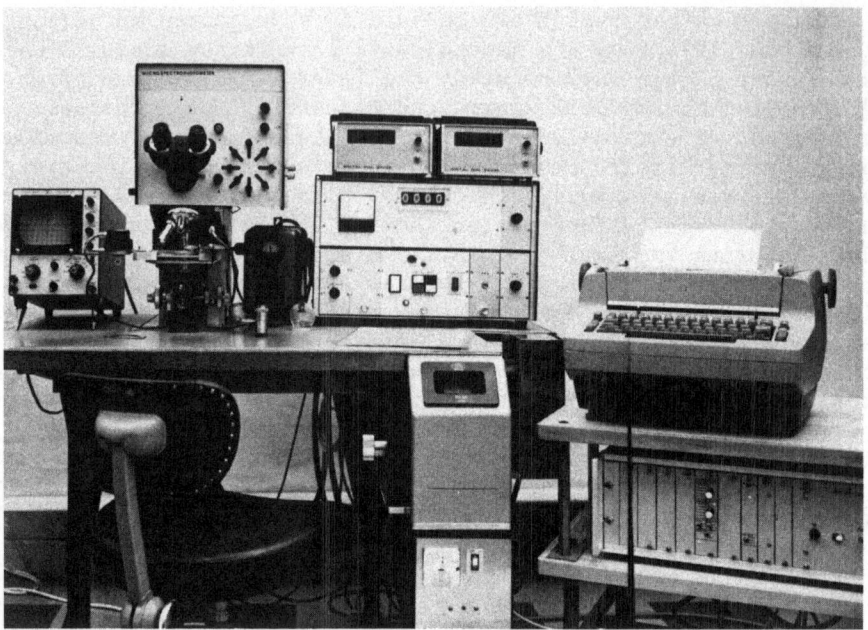

Fig 8. Rapid scanning integrating UMSP for UV and the visible regions, equipped with arrangements for preselection of field.

close to the theoretical limit given by the lens resolution for the wavelength used. The scanning is by mechanical movement of a high precision mechanical stage. Long experience has shown that this type of scanning gives much higher precision than scanning by movement of optical elements, especially in the ultraviolet. The underlying reasons are the difficulties in producing an absolutely homogeneous illumination of a large enough field for the specimen. These difficulties are especially great in the UV. The chosen mode of scanning makes the arrangement for preselection of field somewhat complex mechanically, but the design chosen has proved to work in a satisfactory fashion even for large scale routine work. The instrument is built so as to permit measurements of selected fields in the cell. In routine work, the size of the measuring spot is generally set at 0.25 μ and the distance between the scanning lines at 0.5 μ. For an average mammalian cell this gives a measuring time of about 5 seconds. This value has been chosen as it is about one magnitude less than the average time needed for a trained operator to go from one cell to the next.

Provided the nuclear elements can be spread in the specimen so that there is no overlying material, the instrument can be used for measurements of nucleoli, chromosomes and chromosome details. Large series of measurements of individual mammalian chromosomes have, for instance, been made in conventional metaphase spreads.

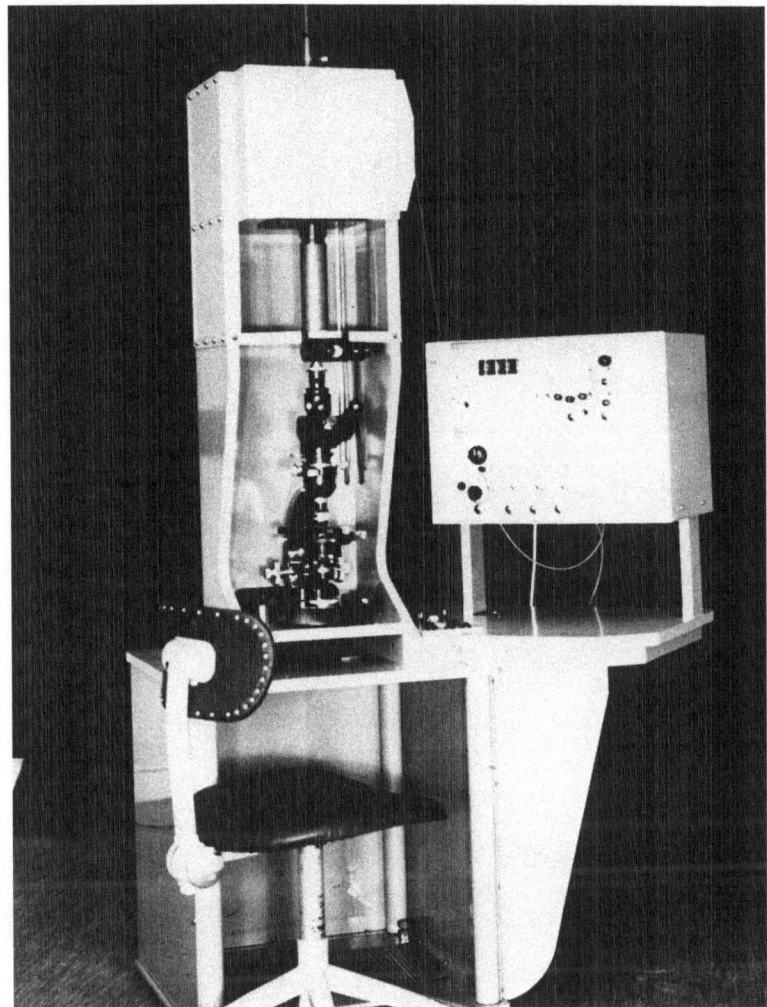

Fig 9. Rapid integrating interferometer with preselection of field.

Figure 9 shows a scanning, integrating interferometer with an arrangement for field preselection. In this case it is not necessary to use stage scanning, which somewhat simplifies the situation. In routine work, a series of interchangeable diaphragms of different shapes are used as field limiters. This instrument has also stood the test of large scale routine work well. The main value of such an instrument in chromatin work is the possibility it gives for measuring changes in nuclear mass (largely proteins), which can be very great during different stages of cell growth, especially during blastogenesis.

Fig 10. High resolution fluorometer.

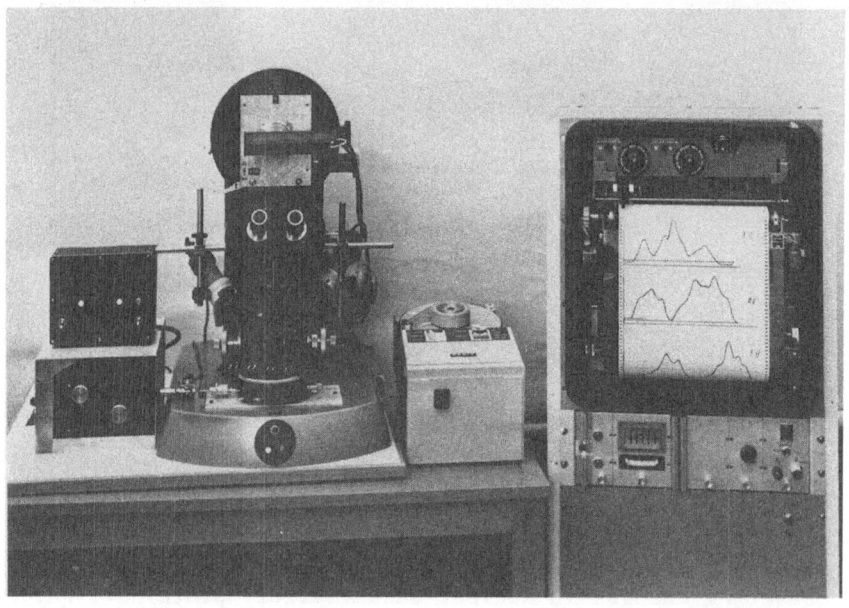

Fig 11. Arrangement for recording photographic fluorometry in chromosomes.

A fluorometer with arrangements for preselection of field has also been constructed (figure 10). It is built so as to permit choice of the cell to be measured and setting of the preselection device without any UV-irradiation of the specimen, which is thus exposed to short wave irradiation only during the time of measurement.

During the last few years, growing interest in making several different kinds of measurements in one and the same cell has led to development of systems for determination of the location of individual cells on a slide so that one cell can later be measured in different instruments. In our laboratory a computer driven stepping microscope stage is used for cell search. By pushing a button the observer, after having selected a cell, automatically transfers its coordinates to the memory of a PDP 8 computer together with a number, defining the cell. After a suitable number of cells has been selected, the teletype writes out the coordinates in millimeters as a list or presents them as a punched tape. The individual cells can then be found and measured in the different measuring instruments, either by aid of digital gauges directly connected to the microscope stage (visible in figures 8 and 10) or by aid of a teletype regulated stepping stage identical with that in the search microscope.

It is often desireable to use both phase microscopy as well as staining procedures for the determination of the cell type in complex cell populations such as those met in cytopathology. Most conventional nuclear stains greatly impair subsequent spectrophotometric and interferometric work because of the high light absorptions in the stained specimens and it is rarely possible to extract the stains without losing cellular substances. Some help in these matters can be obtained by the use of extremely weak staining. By combining TV-microscopy with electronic image enhancement the microscope image can be "lifted up" to a clearly observable level on the TV monitor.

The second field referred to above for which interest has grown conspicuously during the last few years concerns the endonuclear structures, particularly the chromosomes. Ever since the beginning of quantitative cytochemistry, chromosomes have been under study, but these studies have mainly concerned species with very large chromosomes or with peculiar chromosome organization. The rapid progress of cancer research, however, has put the very small metaphase chromosomes from higher animals and plants in the center of interest. The length of such chromosomes, for instance those of man, lies around $3 - 8$ μ and the thickness of the individual chromosome is around 1 μ, that is about two wavelengths of green light. In spite of their small size they can be worked on in the UMSP and microfluorometric instruments described in the beginning of this article. This is illustrated in figure 4, where the size of chromosomes corresponds to that of the larger human chromosomes.

However, before any reasonable cytochemical analytic work on mammalian karyotypes could be done it was necessary to have procedures for identifying individual chromosomes and preferably also regions within these chromosomes. In our laboratory this led to the quinacrine banding procedure for metaphase chromosomes. I mention that here mainly because it is a <u>direct</u> derivative of quantitative cytochemistry in that the whole work establishing the constancy of the chromosome patterns and the differences between chromosomes rested on fluorometric measurements of fluorescence profiles. These were first recorded in the instrument depicted in figure 10. Later supplemented by a faster working arrangement with photographic fluorometry (figure 11), With the latter such a large number of fluorescence profile measurements were collected that a comprehensive statistical analysis could be made by computer. This showed banding pattern analysis to be a reliable tool for chromosome identification.

For most routine applications of the chromosome banding techniques as yet there is no need for cytochemical measuring devices. They were, however, necessary requisites for the development of the method. In the work aiming at automation of the chromosome analytic work, measurements of banding patterns is an important link.

In summary it can be said that quantitative optical cytochemistry offers a considerable number of tools for study of chromatin structure. Microspectrophotometers for ultraviolet and the visible, as well as microfluorometers permit analysis of nuclear structures down to the size range of 0.5μ (in ultraviolet 0.3μ). Integrating measurements can routinely be made of nuclear details, whole nuclei and/or whole cells. Microinterferometry, used for dry mass determinations can be applied to objects of over 1μ in size. Microinterferometric mass determinations - as spot measurements or integrating measurements - can also be applied to cellular elements larger than 1μ.

For the study of chromatin function, the new instruments with preselection of measuring field can give information about the exchange between nucleus and cytoplasm. In this connection it is also an advantage to have access to the arrangements for multiparameter measurements which have recently been developed.

CHROMATIN STUDY IN SITU:

I - IMAGE ANALYSIS

F. Kendall, F. Beltrame and C. Nicolini

Division of Biophysics
Temple University Graduate School
Philadelphia, PA 19128

INTRODUCTION

In all physico-chemical and functional characterization of
isolated chromatin, there is of course a large degree of artifact
introduced during the isolation process. In this respect abundant
evidence exists in the literature that the structural and functional properties of native chromatin are modified by physical alterations introduced during chromatin preparation (1,2,3). In order
to bypass such limitations, one approach has been to analyze the
*geometry and densitometric textures of the nuclear images produced
by differential staining of chromatin*-DNA *in situ (such as by a
Feulgen reaction)*. For a detailed account of the most recent findings during image analysis on the study of chromatin during cell
cycle, visus transformation and cell proliferation, the readers
should refer to recent publications (4-7). Here we intend only to
outline a critical overview of the state of the art with advantages and limitations.

AUTOMATED IMAGE ANALYZER

An image analyzer for densitometric-geometric studies is
usually equipped with a plumbicon scanner (that, having a light
transfer characteristic of 0.99, warrants an overall linear calibration between 0.0 OD and 2.00D) and densitometer module (see
Fig. 1) on line with a digital computer. The scanned area is
divided into 880 x 588 picture elements, each of which can be
digitized into 64 gray levels where the area of each picture
point is (0.09 x 0.09) μm^2 under the given optical conditions
(4-7). Fig. 2 shows how, by applying the proper optical threshold

Automated Image Analysis

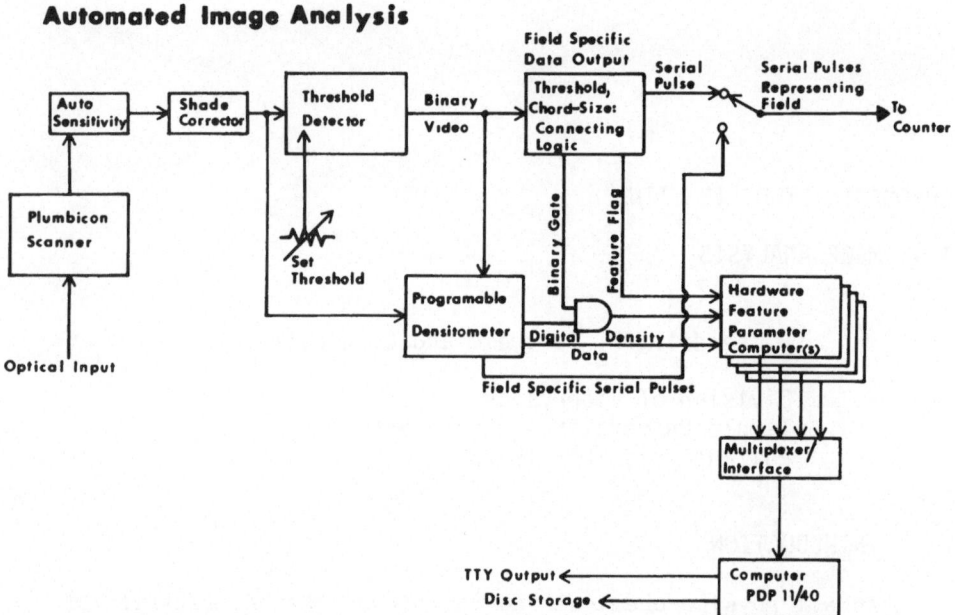

Fig. 1. Block diagram of inage analysis system which utilizes hardware real-time processing of morphometric and densitometric data in accordance with chord-size and programmable optical density criteria. The system can be expanded to four feature parameter computers for simultaneous real-time computation.

to eliminate background, the nuclear image can be properly filtered for subsequent geometric-densitometric analysis.

Nuclear morphometry itself is a technique that readily exploits existing hardware to preprocess image data. Thus threshold and minimum chord size criteria can be used in real time to define the border of a feature during scanning, and such measures as integrated optical density (proportional to DNA content for Feulgen staining) and nuclear area can be output at the end of a scan through a feature. Distributional error is negligible using oil-magnification. This technique places a premium on such global descriptors as the area and perimeter which are independent of the state or orientation of the cell. Such measures as Horizontal and Vertical Feret diameters (shadow projections) and complex projections (sums of shadow projections of lagging edges) can be made substantially independent by employing the Euclidean norms of orthogonal projections, thus yielding global descriptors that are sensitive to both shadow diameters and re-entrance. Table I shows the dramatic improvements in coefficient of variation and total range which are realized when the Euclidean norms, and horizontal and vertical complex projections are averaged for various re-entrant figures as they are rotated 90 degrees.

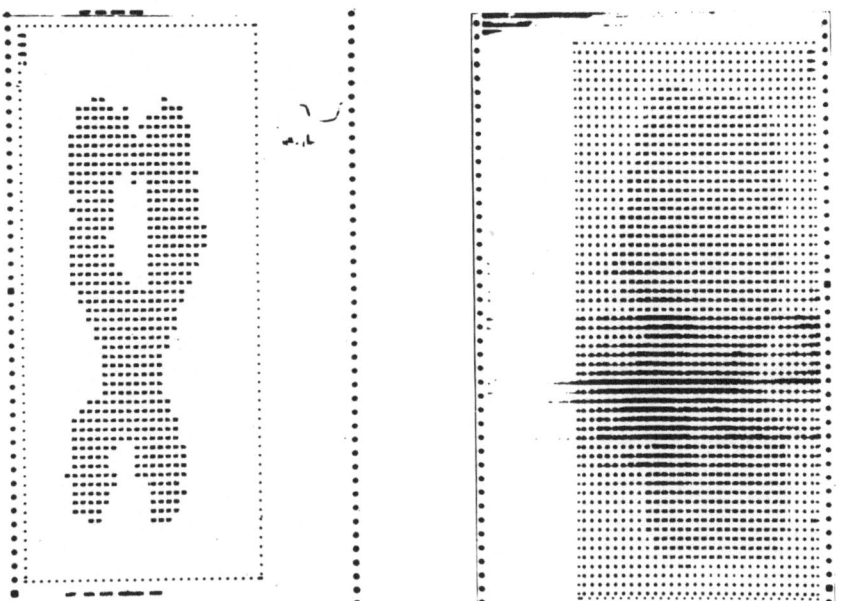

Fig. 2. The left panel shows a line printer listing of discretized
optical density values unloaded point-by-point from a given area of
a microscopic field into the core of a minicomputer. Application
of the detection logic using analog thresholding and chord-sizing
results in the "cleaned" image which appears in the right panel.

Speed can be enhanced by *parallel information processing*.
This becomes obvious to the users of analyzers which are capable
of providing a point-by-point image transfer into a computer. A
considerable amount of subsequent computational time is required
to establish connectivity in order to measure the perimeter of the
feature. In contrast, the perimeters of 10 or 20 objects can be
computed and released with existing hardware in 200 msec. A block
diagram of such a system is shown in Fig. 1. This system computes
one geometric parameter (e.g., area) for all the features (e.g.,
nuclear images) in the field per feature parameter computer per
field scan. Two different parameters for each feature can be computed simultaneously in a single field scan if 2 computers are
used. Permissible regions are established by *analog* optical density settings of the threshold detector and chord-sizing logic
which establishes connectivity of the resulting border and the
x-y coordinates of the lowest point (feature flag). Feature parameter data are only computed by the feature parameter computer on
the basis of picture elements which fall within the permissive

TABLE I

GLOBAL INVARIANCE OF THE EUCLIDEAN NORM
OF THE COMPLEX PROJECTIONS OF THREE RE-ENTRANT FIGURES

Letter	Parameter	N	Co-efficient of Variation Percent	Percent (1) Spread	Percent (2) Range
Y	Horizontal	10	15	41	53
	Vertical	10	15	39	47
	Norm	10	1.2	3.9	4.0
X	Horizontal	10	3.8	10	11
	Vertical	10	3.2	8.7	9.1
	Norm	10	0.58	1.5	1.5
S	Horizontal	10	8.6	25	28
	Vertical	10	6.8	21	24
	Norm	10	0.54	1.8	1.8

(1) Percent spread is defined by the range divided by the mean.
(2) Percent Range is defined by the range divided by the minimum
    value.

The test figures used were opaque letter "Y", "X", and "S" which
were affixed to a glass slide, magnified 30x, and rotated through
90°.

regions and whose discrete optical density values equal or exceed
those which are established by the densitometer. This configuration permits disconnected geometric data (such as the total area
of several separate islands of high optical density) to be associated with the feature to which they belong, rather than generating
additional features as functions of optical density threshold.
The feature parameter computer and the multiplexer/interface asso-
ciate the geometric data for each feature with the corresponding
feature flags. When a scan line crosses a feature flag, the geometric and coordinate data associated with the flag are unloaded
directly into the core of a minicomputer. The densitometer can
transform analog picture element luminances into logarithmic optical density values on a scale of 63 discrete digital values. The
feature parameter computer can thus compute integrated optical
density (IOD-proportional to DNA content in Feulgen-stained nuclei)

for individual features. It is also possible to load the digitized
optical density values for each element in a feature directly into
the computer core memory. With parallel processing this system
can compute up to 4 different parameters of many features simultaneously. It is also possible to automatically program optical
density threshold, which may be programmed in equal density steps
at least .01 OD to obtain 7 or more thresholds.

We have found that a satisfying compromise can be made between
the desirable speed of parallel processing (and its attendant hardware costs), the necessity of insuring that only desired images
are collected, and the necessity that resulting data sets are as
free of artifact as possible. The greatest saving in time is
effected by on-line computation of any particular parameter for as
many features as may exist in a field. Restricting the number of
parallel processors to one increases the number of scans required
to compute the specified number of parameters at each specified optical density threshold. This increase in the time necessary to
make all of the necessary measurements is advantageous because
artifacts which depend on electrical noise in the image circuits
and in the analog threshold circuitry will be afforded opportunities to add or delete features to the field and to shift the positions of others. Software coordinate matching at each optical
density threshold deletes features which cannot be matched on successive scans. Other software checks the total number of features
at the completion of all scans at each optical density threshold.
If the total number of features at all thresholds is invariant, a
software check then uses two orientationally independent global
descriptors, to compute a form factor (area divided by perimeter
squared) which cannot exceed $\frac{1}{4}\pi$.

As an added precaution the numerical index and integrated
optical density (IOD) value of each image are displayed next to
the image, and the parameters acquired are printed as hard copy
in a procedure which was designed to slow the acquisition process
to a point where a human observer would be able to manually select
a single feature for deletion from the acquired data set. As a
practical matter, this complimentary use of hardware and software
has routinely resulted in the lowest coefficients of variation compatible with various parameters of images of nuclei of cells in such
functionally homogeneous states as may be expected several hours
after synchronization by selective mitotic detachment. Inasmuch
as the total acquisition cycle for an artifact-free field of 5
images is of the order of 30 seconds, the additional 2 second (7
percent) burden imposed by lack of parallel processing is a relatively small price to pay for the resulting quality of data.

The absence of serious distributional error leaves the questions of gray scale size, stability (repeatability) and field uniformity (including distortion and shading error) to be answered.

The first case represents a practical tradeoff between various
costs and usefulness. There is little merit in achieving even 6
or 7 Bit gray level resolution unless it can be justified for a
specific application. State of the art equipment may approach
this resolution capability in terms of signal-to-noise ratio if
signal-averaging is employed. Stabilities of scanners, analog circuits, and illumination systems can be made acceptable for periods
of one hour between recalibration. We have made 110 measurements
of integrated optical density (IOD) and area over a one hour period.
The coefficient of variation of nuclear IOD was 1.8%, while the coefficient of variation for area was 0.9%. Some of this variation
was induced by the bleaching of the Feulgen-stained specimen. We
routinely achieve field uniformities (estimated by measuring the
same image at nine locations through the top, center, and lower
portions of the scanned area) sufficient to yield a *coefficient
of variation less than 3% for nuclear IOD and less than 1% for area.*

Cell populations which are in relatively homogeneous functional
conditions (shortly after synchronization by selective mitotic detachment or in confluency) will easily produce larger coefficients
of variation for these parameters. The only instrumental error
directly comparable to flow systems performance is IOD and the
validity of comparison can be questioned on the basis of the physical nature of the analog quantities measured and the inherent influences of random artifacts common to each system. As with most
exotic systems with many optical and electronic degree of freedom,
error is also a function of methodology or lack of it.

GENERAL CONSIDERATIONS

State of the art hardware processing adds desirable speed to
the necessary transformation of monumental volumes of data in
pattern space (which can be acquired with great ease by image scanning systems) to feature space so that the result can be applied to
description space. This is also a necessary process of image
classification (8) because it is required in order to produce results which are physically comprehensible to humans. The point of
departure between the classical use of image analysis for purposes
of pattern recognition and classification and its uses for an analytical determination of the structure-function relationships of
chromatin resides in the viewpoints of the two approaches: in the
former the image is to be described in a fashion whereby it can
be recognized and/or classified; in the latter case the image
is treated as an incidental necessity in the same manner as a
hologram or other optical transform.

A frequent complaint that was voiced during Automated Cytology
V in 1976 was that financial support for image analysis investigations was small compared to that available for investigations involving flow systems. This state of affairs is readily understand-
able when the literature of biological (specifically medical) image

analysis is reviewed: generally the approach taken is that of *image classification* (9,10). The overriding purpose is to supplant slow, error-prone and expensive human classifiers with fast and efficient automatons that have suitably low error rates. In short, the objective is all too frequently to achieve *sophisticated, rapid and economical quantitative evaluations* of geometric and densitometric data in order *to duplicate uncertain, qualitative, and frequently subjective visual value judgements* that rely heavily upon experience, less upon stoichiometry, and least upon precise geometric or densitometric measurements. Although the basis of human discrimination of texture has been modeled by Markovian processes and satisfying agreement with human judgement has been achieved by image analyses based on this model, the actual process is probably not fully described (11), and no systematic body of quantifiable data exists which relates this optical property in rigorous fashion to either chromatin structure or function. The apparent morphological integrity of fixed and stained cells has subtly beguiled countless workers into a reduced awareness of the fact that they are observing a standardized artifact which in both microscopic and molecular definitions is not synonymous with the original living system.

   Contrary to most workers, we attempt to assign physical boundary conditions to image analysis in the manner of flow systems where we made every attempt to relate stoichiometry and fluorescence as well as interference *to structure and structural interactions.* Perhaps accurate geometric and densitometric measurements are such exotic entities that habit does not suggest that they be systematically applied to understanding the physical differences which exist in "artifacts" obtained from well defined relatively homogeneous cell populations in different functional states.

CHROMATIN CHARACTERIZATION

   Recent interest in nuclear morphometry based on the Feulgen reaction seeks to exploit biological *image analysis as a tool to quantitatively measure differences in the resulting organization of nuclear DNA spaces from cells in different functional states.* These studies were prospectively designed to determine if physical differences in DNA spaces probably reflected structural differences which had been observed to exist at the cellular and molecular levels. The advantage of this approach when contrasted to the classification approach, is that the former method usually makes it easier to design experiments in accordance with testable hypotheses. Even within its limitations, the Feulgen approach may permit highly specific and stochiometrically accurate (12,13) (under proper optimal acid hydrolysis (14)) identification of chromatin DNA organization *in situ*. Indeed, convincing evidence exists that such functional properties as template activity of

isolated chromatin and such physical properties (which are related to the tertiary-quaternary structure of chromatin) as
ethidium bromide binding sites are directly related to the morphometric and densitometric properties of the DNA space of Feulgen-
stained nuclei.

In order to focalize on the meaning of "form factor" in the
study of chromatin *in situ*, we may analyze its relationship to
average optical density (Fig. 3) and to the number of primary
binding sites of chromatin isolated from the same cell lines
(Fig. 4 bottom): these data clearly indicate that *increased chromatin convolution* (lower form factor, accompanied by a decreased
mean bound path in the face of increased area) *corresponds to
an increased chromatin condensation* (higher degree of DNA superpacking) *in situ* and to *an increased number of sites for inter-*

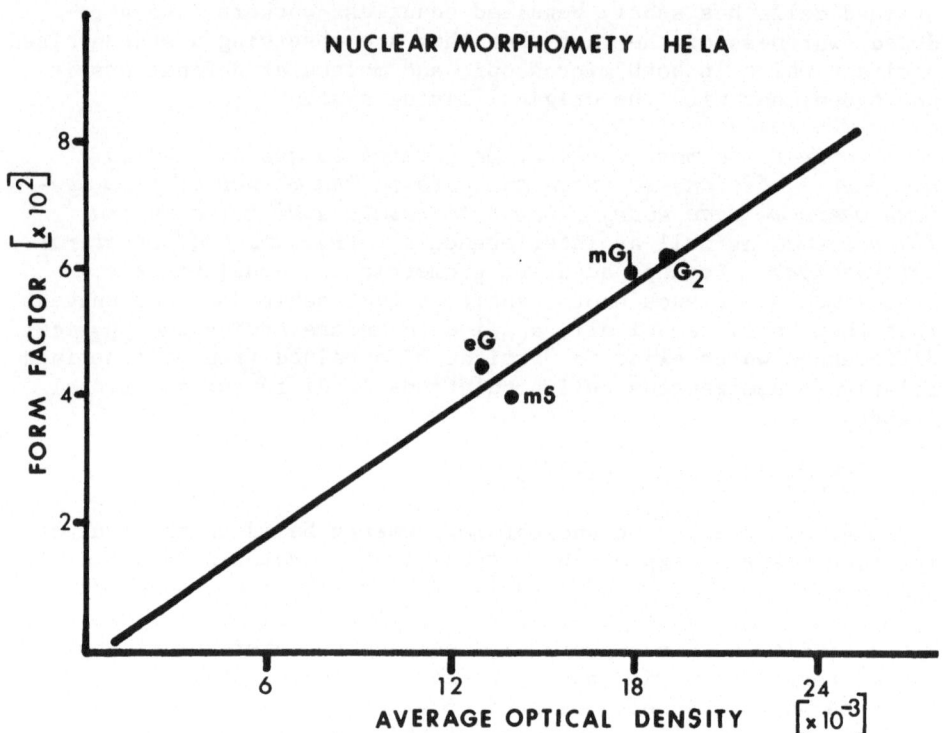

Fig. 3. Relationship between mean values of form factor and
average optical density of nuclear images of Feulgen-stained
HeLa cells synchronized by selective mitotic detachment and
observed at 3 hours (early G1-eG), 5 hours (mid G_1-mG1), 12 hours
(mid-S-mS) and 15 hours (G2) under overall magnification of
1000X at a wavelength of 540nm.

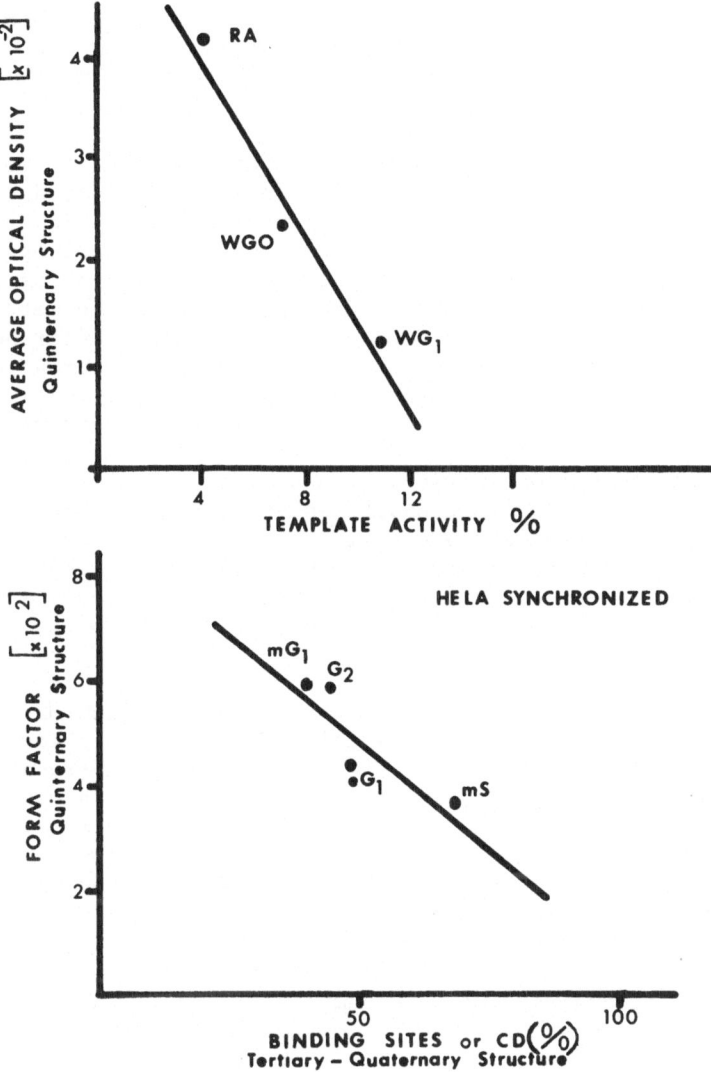

Fig. 4.   (top panel)   Relationship of Average Optical Density of
nuclear images of Feulgen-stained 7-day confluent WI-38 cells
(WGO) and stationery 2RA cells (RA) and of 7-day confluent WI-38
cells three hours after receipt of a nutritional stimulus (WG1) to
the percentage of template activity of chromatin extracted from
similar cell populations.

      (bottom panel)   Relationship between measured form fac-
tor and percentages of available binding sites of chromatins from
synchronized HeLa cells in various stages of the cell cycle
(early and mid G1, mid S, and G2).

calating dyes or macromolecules such as RNA polymerase. Separate
experiments have shown that *an inverse relationship exists between
the average optical density and the measured template activities
of cells in different functional states* (Fig. 4 top).

The interpretation of chromatin morphometry, and its modulation
during various functional states of the cells is confirmed by the
threshold dependence of the geometric parameters (see Fig. 5 for
synchronized HeLa nuclei) which for instance may uniquely prove the
existence of a drastic increase of chromatin dispersion during late
G1 (5 hours) – early S (8 hours). Indeed, while the geometric
parameters of Feulgen-stained HeLa cells are quite similar between
5 and 8 hours after selective detachment, the nuclear area (upper
panel, Fig. 5) or perimeter(lower panel go to zero for early S
(8 hours) with increasing optical density threshold, but maintain
a large positive value for late G1 (5 hours). This is compatible
with an increased chromatin dispersion permitting objective identification of the two distinct subphases even if geometric and densitometric (DNA) parameters are similar at the base threshold.
Similar *objective identifications* on the basis of threshold dependence, are possible on other *cell cycle phases and subphases* (see
ref. 4-5 and Fig. 5) to a level *up to now impossible to any human
observer.*

Even a conservative a priori binary classification of functional
state should be based on some physical evidence that at least one
discrete state is involved. The study of the morphometry of the
DNA space has revealed that, in those instances where a binary state
classification is warranted, the geometric and densitometric changes
in the DNA space are consistent with alterations in chromatin structure which have been observed by other means (6,7). And the specifi-
cation of functional state of a cell population in accordance with
a *"magic recipe"* will probably prove to be inadequate to the task
of definition. Definition of functional state in terms of the
physico-chemical analysis of *bulk* samples of chromatin may itself
be misleading inasmuch as all *distributional information is lost;*
measurable differences in physical properties can be due simply to
changes in percentage composition of the contributors. This was
the case in the particular instance of the conformational change
of chromatin associated with the G0-G1 transition. The discrete
nature of the transition could only be reasonably inferred from
flow system measurements which yielded enough distributional data
to warrant the assumption that conformation-dependent changes in
dye content were not part of a continuum. In this instance the flow
system was indispensible, but subsequent image analysis studies have
provided support for the physical interpretation of the results of
the previous work (6,7). Similarly, geometric-densitometric analysis
has itself provided sufficient distributional information to show
that it is possible to *unequivocably differentiate populations of
Feulgen-stained confluent WI-38 cells and*

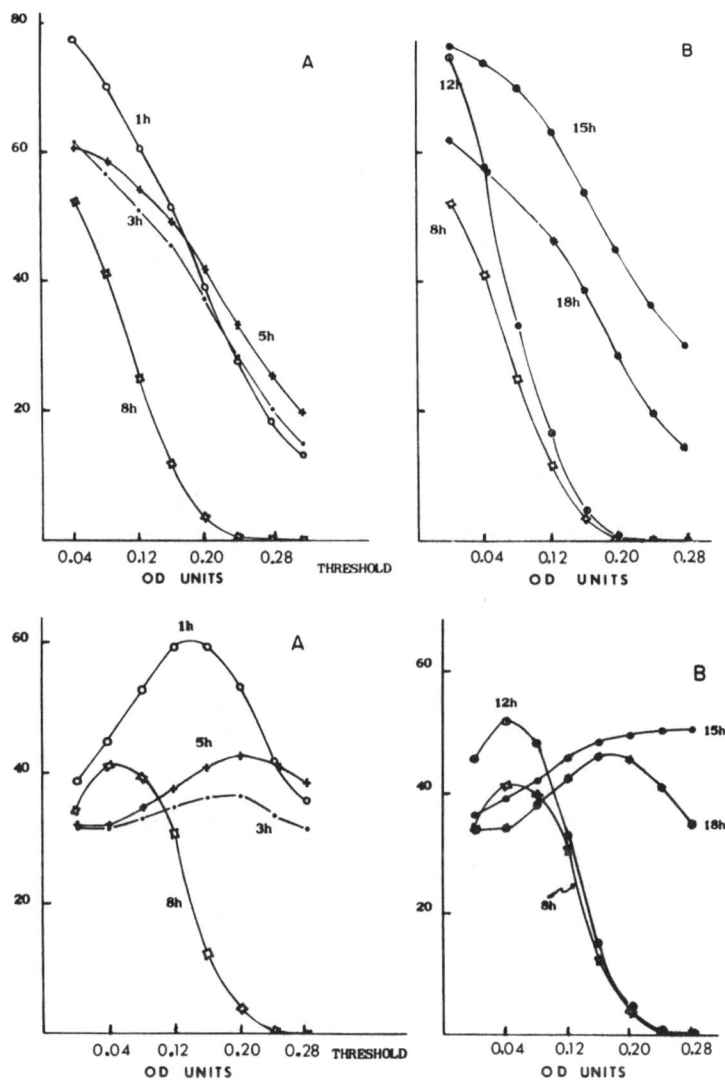

Fig. 5. (upper panel) Mean values of the area (μm²) of Feulgen-
stained nuclei of S3 HeLa cells at seven different times after mitotic detachment (1,3,5,8,12,15, and 18-h levels) versus threshold
from 0.04 OD (defining the nuclear border of each image) to a maximum value of 0.32 OD with equal spacing of 0.04. The average
points versus threshold are obtained on the same constant number
of cells (about 100), including zero values at high threshold.
(lower panel) Perimeter mean values (μm) of Feulgen-stained nuclei
from HeLa cells at 1,3,5,8,12,15 and 18 h after mitoses, measured
at eight different optical density levels (0.04 OD-0.32 OD) (ref.
10).

*their stationary SV-40 transformed counterparts (2RA cells) on the
basis of average optical density.* The IOD (Fig. 6 top) and area
(Fig. 6 center) of the 2RA nuclear images show increased disper-
sions, which are not surprising inasmuch as chromosome numbers can
vary considerably from cell to cell with reported values of 68% of
metaphases containing 61-80 chromosomes and 19% exceeding 120
chromosomes. A correlation between IOD and area of the transformed cells is to be expected on the elementary basis that nuclei
with more chromosome will probably be larger. The striking degree
of correlation (r = 0.98) suggests that a very systematic relationship may exist in the particular case of quiescent 2RA cells
which is sufficient to be clearly seen through the biological and
instrumental "noise" which are causes of dispersion of measured
values of IOD and nuclear area of the WI-38 cells (the majority
of which have 46 chromosomes). Indeed, while the frequency distributions of directly measured parameters, such as area and IOD,
show a substantial overlap between WI-38 and 2RA cells, a derived
parameter (IOD per unit area) yields two distinctly separated
distributions, which permits a unique objective identification of
individual SV-40 transformed cells *regardless of heterogeneity of
chromosome numbers* (Figure 6 bottom).

For the present, the full realization of the usefulness of
rigorous multiparametric analysis of image data for analytical
purposes is hampered by a lack of adequate physical definition of
the functional state of the target population that can never be
overcome by the percentage of agreement among the subjective
visual judgements of a panel of *"experts."* N-dimensional cluster
analysis will not in itself provide any physically meaningful information relating structure to function in the absence of adequate experiments and models to explain the possibly non-linear
connection.

FUTURE DEVELOPMENTS

In spite of the high hardware costs involved it is our
opinion that biological image analysis systems will evolve (with
certain specialized exceptions) in favor of image-plane scanning
systems employing an amalgam of analog video technology and
digital logic and which will be interfaced to small computers for
purposes of control and data acquisition. The current state of
the art is such that major advances in technology are *not* required
to provide the demand necessary to generate sales volume that will
lower with costs à la pocket calculators. Demand is predicated
on the acceptance of utility and acceptance is the factor that is
lacking even because of the *primitive suspicion or superficial
attitude* of most life scientists toward advanced automated technology.

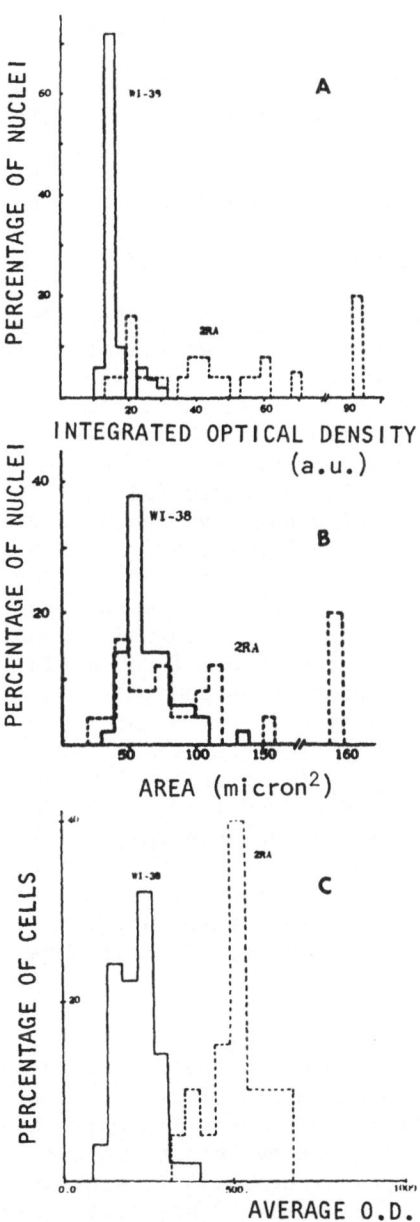

Fig. 6. Histograms of Integrated Optical Density (top), Area (center) and Average Optical Density (bottom panel) of nuclear images of Feulgen-stained confluent WI-38 cells (solid lines) and stationary 2RA cells (broken lines).

The recent systematic application of morphometric analysis
to cell population will receive added impetus from recent advances
in multiparameter cell sorting. Under proper conditions (acridine
orange concentration of 10^{-4}M and low molar ratio R (dye/DNA)
fluorescence-"sorted" cells have sufficient quantum yields to be
detected by a plumbicon scanner and consequently permit the anal-
sis of *chromatin differentially stained* by a supravital dye. It
is then realistic to assume that this technological marriage will
eventually lead to a systematically developed body of knowledge
which will better connect the morphometry of DNA space to the
physical reality from which it was created.

In order to reconstruct the three-dimensional organization
of chromatin-DNA at the *micron* level (*quinternary* structure), we
are presently transferring the nuclear image point by point into
the core of a mini computer by direct memory access under software
control. The image is then stored on a mass storage device and is
available for display on any display device connected to the
Unibus. (See Appendix A) The image (Fig. 7) is presented in
terms of discrete optical density values in a properly calibrated
scale of 63 possible levels for analysis of the patterns of
optical density distribution within each nucleus relative to its
center of mass. By means of one and two dimensional Fast Fourier
transforms, we are searching for eventual periodicities in the
geometric-densitometric properties of chromatin-DNA in situ which
are not detectable by other means and that would be indicative of
an ordered structure at the higher "quinternary" level of chromatin
organization. Preliminary results (Figs. 7-8) indicate indeed
that reproducible periodicity, a unique function of cell cycle
phases, does exist around the nuclear DNA border within a
5000-8000A° band (readily detectable only in mid-G1 but not in
S or M phases) and on the central sections of the same nuclei,
(for every cell cycle phase) compatible with a model previously
outlined for the "quinternary" organization of chromatin-DNA
in situ (see pg. 660). Any periodic variations in optical
density due to a drapery-like periodic structure around the
nuclear DNA boundary (as detected in G1) would tend to be
obscured in most central areas by squashing the nucleus and
would most likely be enhanced in the region lying along the image
border (where absorbance would be larger). Analysis of several
nuclei in the same cell cycle phase indicate the existence
of a highly reproducible chromatin-pattern and DNA periodicity,
suggesting that interphase chromatin is organized in a quite
ordered and isotropic "quinternary" structure (see Appendix A)
which modulates along the cycle. Preliminary studies using crosscorrelation of parallel averaged scans obtained at intervals
well spaced with respect to optical resolving power indicate
that DNA periodicities can be detected even in central sections
of synchronized HeLa nuclei, in early S and mid-G1 phases
(Fig. 9).

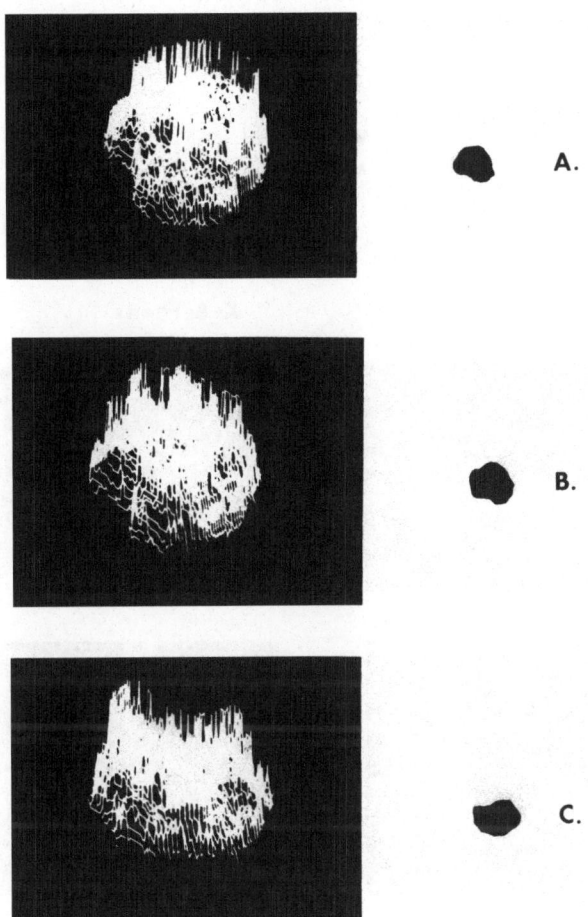

Fig. 7. Perspective optical density displays of three images
chosen at random of Feulgen-stained HeLa nuclei from a smear made
of cells selectively detached 3 hours earlier. An annular group-
ing of higher optical density values is apparent from the displays
but is not obvious in the photographs. (middle G1 nuclei)

Fourier Analysis of Nuclear Section from HELA G₁ Cells

$$X(k) = \sum_{n=0}^{N-1} x(n) W_N^{kn}$$

$0 \leq k \leq N-1$

$W_N = e^{-j(2\pi/N)}$

$N = \#$ of Samples $= 64$

Y- Sections   **X- Sections**

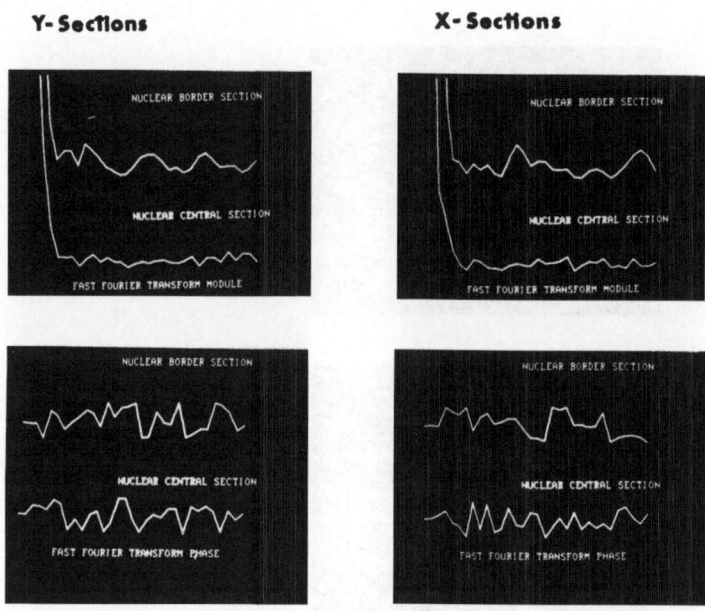

Fig. 8. Power (top panels) and phase (lower panels) spectra of optical density values acquired along the borders and through the center of the nuclear image of middle Gl Feulgen-stained HeLa cells. Comparison between data obtained in two orthogonal axes indicates that the similarity of results is dependent upon the portion of the nucleus scanned rather than the direction of scan. Identical results are obtained in other nuclei from middle Gl HeLa cells.

Auto Correlation of
Two Periodic Functions

Cross Correlation Between Two
Central Sections of Nuclear Images
(256 p.p./section)

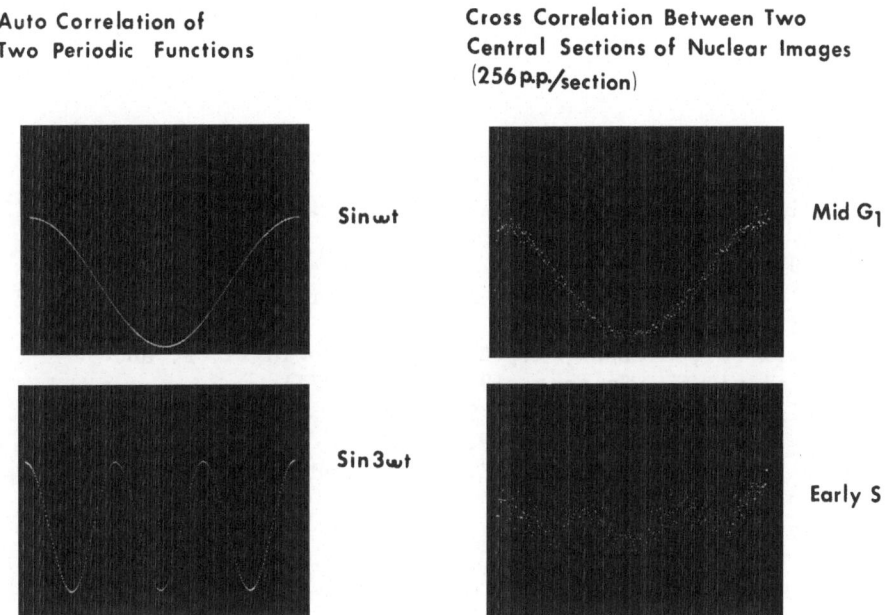

Sin ωt

Mid G_1

Sin 3ωt

Early S

Fig. 9 (left panels) 256-points auto correlations of two 256-points periodic functions, with the period being three times smaller in the bottom respect to the top function.

(Right panels) Cross correlations between monodimensional parallel scans (averaged over 1000 times to enhance signal to noise ratio) of a synchronized HeLa nuclei, at early S (bottom) and mid-G1 (top) phase.

In these studies we have utilized linear cross correlation algorithm (see Appendix).

Identical cross correlations were found, when parallel scans were separated by either 10, 20, 40 or 60 picture points.

These data show the repeatibility of the same optical pattern along one direction, as function cell cycle phase.

The correlation was computed always via Fast Fourier Transform (see Appendix A).

ACKNOWLEDGEMENT

This work was supported by Grant CA20034 from the National Cancer
Institute.

REFERENCES

1. Nicolini, C., Baserga, R. and Kendall, F., Science, 192, 796
   (1976).
2. Nicolini, C., and Kendall, F., Physiol. Chem. and Phys. (1977).
3. Noll, M., Thomas, T. and Kornberg, A., Science, 187, 796
   (1975).
4. Nicolini, C., Kendall, F., and Giaretti, W., Biophys. J., 19,
   163 (1977).
5. Kendall, F., Swenson, R., Borun, T., Rowinski, R. and
   Nicolini, C., Science, 196, 1106 (1977).
6. Nicolini, C., Kendall, F., Desaive, C., and Giaretti, W.
   Exp. Cell. Res., 106, 119, (1977).
   Nicolini, C., Linden, W., Zietz, S. and Wu, S., Nature,
   270, 607 (1977).
7. Kendall, F., Wu, S., Giaretti, W., and Nicolini, C., J.
   Histochem. Cytochem., 25, 724 (1977).
8. Andrews, H., Introduction to mathematical techniques in
   Pattern recognition, Wiley-Intersci., New York, Ch. 1-2
   (1972).
9. Bartels, P.H., Bahr, G.F., Jeter, W.S., Olson, G.B.,
   Taylor, T., Jr., and Wied, G.L., J. Histochem. Cytochem.,
   22, 69 (1974).
10. Bacus, J.W., IEEE Trans. Sys. Man. Cyber, SMC-2, 2, 513 (1972)
11. Pickett, R.M., J. Exp. Psychol., 68, 13, (1961).
12. Garcia, A.M. Stoichiometry in Dye Binding versus Degree of
    Chromatin Coiling, Wied, G.L. and Bahr, G.F. (eds),
    Introduction to Quantitative Cytochemistry-II, Academic
    Press, New York, 153 (1970).
13. Dijndam, W.A.L. and Van Duijn, P. The influence of chroma-
    tin compactness on the stoichiometry of the Feulgen-Schiff
    procedure studied in model films II. Investigations on films
    containing condensed or swollen chicken erythrocyte nuclei.
    J. Histochem. Cytochem., 23:891 (1975).
14. Linden, W., Zietz, S., Fang, S. and Nicolini, C., in "Chroma-
    tin Structure and Function," ed. Nicolini, C., Plenum Co.,
    pubs. (1978).

APPENDIX

Although the pattern of chromatin morphology studied by image analysis is essentially a two-dimensional one, much information can be obtained by analyzing one dimensional cross sections of the image. In particular, such monodimensional techniques can help in identifying isotropy and repeatibility of certain optical density patterns acquired column by column.

One of our approaches to trying to discover periodicities in chromatin morphology is to use cross correlation techniques between different scan lines of the image, either parallel or orthogonal to each other. To enhance the signal to noise ratio, we first average each scan line of 256 pp 100 times. We then attempt to find existing periodicities in the 256 pp average lines.

To aid in this search as well as to calculate certain correlation functions, it is extremely useful to utilize techniques from Fourier analysis. For the reader, we will review some of its theory below. Although our problem is inherently discrete, for sake of compactness of presentation, we will use the continuous analog where no confusion can develop.

If f(x) is a real function of position, the Fourier transform F(w) of f(x) is defined as:

$$F(w) = \int_{-\infty}^{\infty} f(x) \, e^{-jwx} dx \qquad \text{where } j = \sqrt{-1} \qquad (1)$$

It is a basic fact that the original function f(x) can be obtained from its Fourier transform by the following inverse formula

$$f(x) = \frac{1}{2\pi} \int_{-\infty}^{\infty} F(w) e^{jwx} dw \qquad (2)$$

The precise hypothesis that permits us to define (1) and (2) and show their equivalence can be found in reference 1.

What is to be noted is that although f(x) is a real function, F(w) is a complex valued function. We take advantage of this by writing F(w) in its polar form.

$$F(w) = A(w) e^{j\psi(w)} \qquad (3)$$

where A(w) and ψ(w) are real functions. A(w) is known as the amplitude and (w) is the phase.

With this representation, we then can interpret equation (2) as representing f(x) by a sum of elementary periodic functions (in sine and cosine), the contribution of the period w to the function f(x) is given by the amplitude function A(w).

CORRELATION FUNCTION

If two functions $f_1(x)$ and $f_2(x)$ are given, this *cross correlation* R(t) is defined as

$$R(t) = \lim_{T \to \infty} \frac{1}{2T} \int_{-T}^{T} f_1(x) \, f_2(x+t) \, dx \qquad (4)$$

In the case that f_1 and f_2 are periodic functions of period T the cross correlation becomes

$$R(t) = \frac{1}{T} \int_{-T/2}^{T/2} f_1(x) \, f_2(x+t) \, dt \qquad (5)$$

If f_1 and f_2 are the same function, then the function R is called the *autocorrelation* of f.

RELATIONSHIP BETWEEN FOURIER TRANSFORM AND CORRELATION FUNCTIONS

The correlation function R(t) can be expressed in terms of the Fourier transform. Let us define the *power spectrum* S(w) of a function f(x) as the square of the Amplitude function A(w) i.e.,

$$S(w) = |A(w)|^2$$

Then the autocorrelation function R(t) can be shown to be equal to the inverse Fourier transformation of the power spectrum, i.e.,

$$R(t) = \frac{1}{2\pi} \int_{-T}^{T} A^2(w) e^{jwt} \, dw.$$

This relationship is useful in computing correlation functions.

APPLICATION OF FOURIER TECHNIQUES TO DISCRETE MEASUREMENTS

In applying the above theory to help search for regular structures in the chromatin geometry, we must first consider how the discrete nature of the data affects the problem. Recall that a row of the nuclear image consists of 256 picture points. We

wish to ascertain whether there are periodic regularities to such
a column. However, it is impossible to ask whether a finite
record is periodic. Instead, what is usually done is to arbitrarily
make a periodic function by repeating the record. We then search
for periodicities in the new function, remembering that we have
arbitrarily introduced a period T = length of record = 256 pp.

An important result of the Fourier analysis of discrete
signals is the Nyquist Sampling theorem which states that if
$f(x)$ is a function such that its Fourier transform $F(w) = \emptyset$
for $|w| > w_c$ then f is uniquely determined from its values

$$fn = f(\frac{nT}{w_c})$$

at a sequence of equidistant points.

In the case of time varying signals this condition requires
that the sampling frequency be at least twice as great as the
highest frequency which is present in the signal which is
being sampled in order to avoid the phenomenon of *aliasing*
(in which the high frequency components become "folded" into the
lower frequency components by the sampling process itself). In
practice, the sampling frequency is usually set to about three
times the highest significant frequency in the incoming signals.
The sampling process itself is entirely oblivious to the nature
of the incoming "signal". If the "signal" contains high-frequency
noise components of greater than one half the sampling frequency,
these components will be aliased and will appear as lower-frequency
spectral amplitudes which are intermixed with those of the true
"signal". Consequently, most sampling systems contain one or
more *low pass filters* which attenuate frequencies higher than a
value determined by the sampling frequency.

In the case of image-plane scanning systems, the picture
information is "read" by an electron beam which is swept across
the target face at constant velocity. The signal is sampled at
equal discrete distance intervals by means of clock circuitry.
This space-domain signals are mapped into time domain signals and
the corresponding frequency domain representation of these signals
has its counterpart in the "spatial frequency" (or inverse distance)
domain of the image itself. For any scanning velocity the maximum
spatial frequency is dependent on both amplifier band width and
optical resolution. Several techniques are utilized to reduce
noise, and prevent aliasing.

Amplifier noise can be reduced by reducing amplifier band
width. Spatial resolution can be preserved if slow-speed
scanning techniques are employed to keep the highest frequency
components of the signal within the reduced band width. The

optical system itself can be used as an effective low-pass filter. Low distributional error of sampled optical density values requires that the dimensions of a single scanner picture element correspond to dimensions of the external object which are below the limits of resolution of the optical system. Inasmuch as the limit of optical resolution of a 100x, 1.25 NA objective is about 0.3μ at 540 nm and the dimensions represented by an individual picture element in our system is about 0.7μ, the minimal resolvable distance is about 1/3 that which can be resolved by the optics. In other words, the spatial sampling frequency is above three times the optical band width, so aliasing will not be a problem. However, discretized luminance signals contain random electrical noise which can contain some components corresponding to high spatial frequencies. Inasmuch as this noise is random, it can be removed by the simple process of sampling the same field 100 times and averaging each individual sampled value.

Making the computation more explicit and keeping in mind the relationship between "time" and "frequency", for example for a periodic signal of period T, i.e.

$$T = 1/f = \frac{2\pi}{\omega_o} \quad , \quad \omega_o = 2\pi f$$

we can interpret the transformed domain as an inverse-distance space, and attribute to ω the dimensionality of distance^{-1} with unit $[1/\mu]$ $(1\mu = 10^{-6}m)$

Optical Density or Transmittance Values

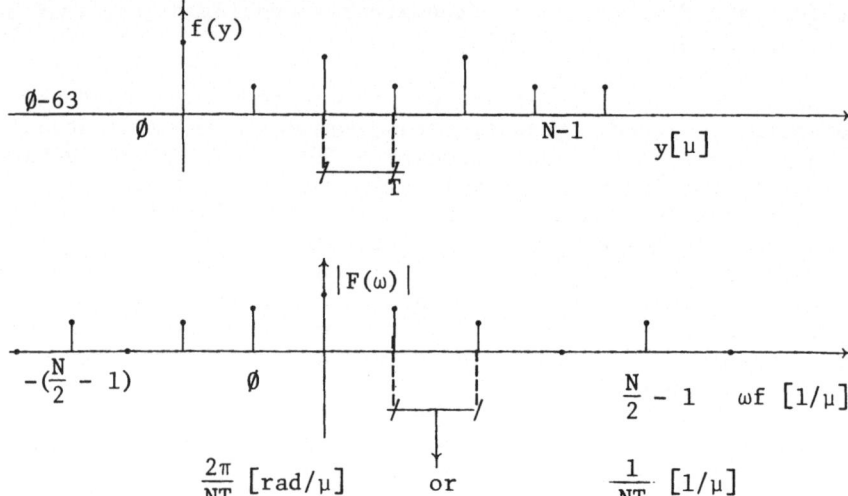

Suppose T = .05μ, N = 256

Resolution in frequency = $\dfrac{1}{256 \times .05}$

$\approx$ .078 [1/μ]

Maximum frequency = 128 x 78 x 10^{-3} [1/μ] $\overset{\sim}{=}$ 9.984 [1/μ]

i.e. $\overset{\sim}{=}$ .1μ   (because of the symmetry).

This is <u>below</u> the resolution of the microscope (d $\overset{\sim}{=}$ .26μ)

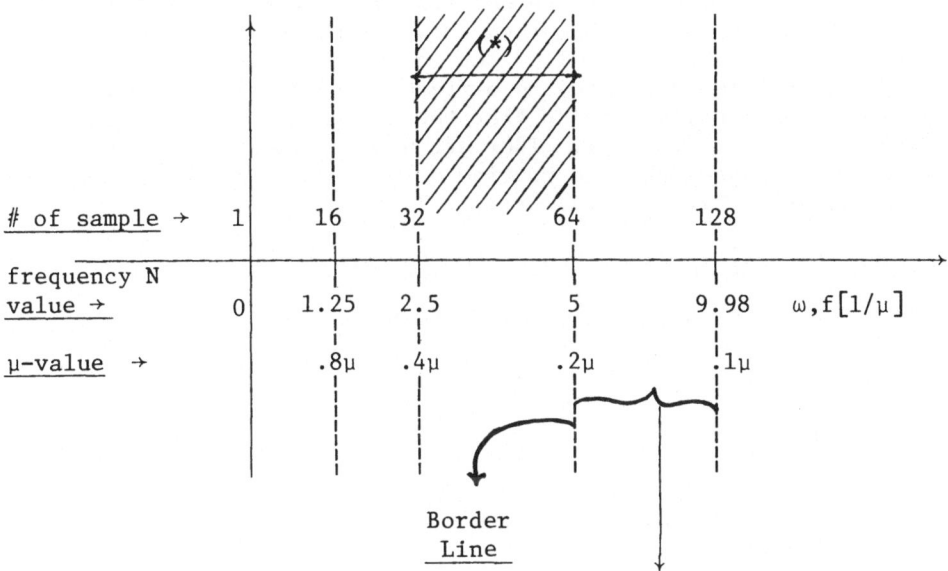

Border
Line

Region not detectable
because below the
resolution.

(*) For example this region could be the frequency window in which
are present the so called "higher ordered fibers" of the chromatin
(0.16 ÷ 0.3μ) detected by freeze-fraction (S. Bram) and wet
replica (S. Basu) techniques.

It can be shown (see Reference 3) and 4)) how the spectrum estimate obtained either through the indirect (through the auto-correlation function of the signal and then taking the DFT Of the result) or the direct (taking the DFT of the original signal and squaring it) method has some undesirable properties such as a very high variance, leakages, i.e. contributions to a given frequency estimate from distant parts of the spectrum. To reduce or minimize the most disturbing effect we intend also to weight the autocorrelation estimate or the raw data with a suitable function (the "window") chosen in a proper way.

COMPUTATIONAL CONSIDERATIONS AND THE FAST FOURIER TRANSFORM (FFT)

In order to implement the Fourier techniques, it is convenient to utilize a computational method known as the Fast Fourier Transform. These methods allow one to greatly decrease the computational time necessary to perform Fourier calculations. These methods usually require that the number of points be a power of 2 (note $256 = 2^8$). They are now standard on almost all computing systems. The routine we utilized was developed for the PDP11/40 by Professor F. Bertora (reference 4).

APPLICATION OF THE FFT TO HIGH SPEED CORRELATION

Given 2 finite sequences of the same length N

$$x_1(n) \text{ and } x_2(n), \ 0 \leq n \leq N-1$$

their linear correlation is defined as:

$$Z(m) = \sum_{n=0}^{N-1} x_1(n) \ x_2(m+n), \ 0 \leq m \leq 2N-1$$

Z is sum of lagged products.

Correlation, within the classical Fourier theory, is easily evaluated in the frequency domain. The same property is true in the DFT case with gain in speed compared with direct computation. The product of the suitably computed DFT's of x_1 and x_2, one of which must be conjugated, is the DFT of their linear correlation. It has to be pointed out that we cannot take simply the DFT's of $x_1(n)$ and $x_2(n)$, one of which conjugated, to obtain the DFT of their linear correlation. The result of this procedure would be, in fact, the circular and not the linear correlation of x_1 and x_2. (See Reference 2). DFT = Discrete Fourier Transform.

In order to obtain correct result by still operating in the frequency domain we <u>must</u> force a periodic correlation to give a linear correlation. This can be achieved by generating two new sequences $x_1'(n)$ and $x_2'(n)$ of length 2N that coincide with the original sequences for the first half points and are zero elsewhere:

$$x_1'(n) = x_1(n) \qquad x_2'(n) = x_2(n) \qquad 0 \le n \le N-1$$

$$= 0 \qquad\qquad = 0 \qquad\qquad N \le n \le 2N-1$$

The IDFT of $X_1'(k) \cdot X_2'(k)*$ (* means conjugate) is the periodic or circular correlation of $x_1'(n)$ and $x_2'(n)$, i.e. the correlation of the <u>periodicized sequences</u> x_{p1}' and x_{p2}' of period <u>2N</u>, i.e. the linear correlation of $x_1(n)$ and $x_2(n)$.

ALGORITHM FOR CROSSCORRELATION COMPUTATION

$$x_1(n) \quad x_2(n), \; 0 \le n \le N-1$$

$$Z(m) \triangleq \sum_{0n}^{N-1} x_1(n) \; x_2(m+n), \; 0 \le m \le 2N-1$$

1. <u>Perform</u>

$$x_1'(n) = x_1(n) \qquad x_2'(n) = x_2(n) \qquad 0 \le n \le N-1$$

$$\text{"} = 0 \qquad\qquad \text{"} = 0 \qquad\qquad N \le n \le 2N-1$$

3. <u>Take</u>

$$X_1'(k) = \sum_{0n}^{2N-1} x_1'(n) \; e^{-j(2\pi/2N)kn} =$$

$$= \sum_{0n}^{2N-1} x_1'(n) W_{2N}^{-kn} \qquad 0 \le k \le 2N-1$$

$$X_2'(k) = \sum_{0n}^{N-1} x_2'(n) W_{2N} \qquad 0 \le k \le 2N-1$$

3. <u>Take</u>

$$X_2'*(k) \quad \text{(complete conjugate of } X_2'(k))$$

4. $X_3(k) = X_1'(k) \cdot X_2'*(k)$

5. <u>Inverse Transform:</u>

$$\mathcal{F}^{-1}\{X_3(k)\} = Z(m) = \frac{1}{2N} \sum_{0k}^{2N-1} X_3(k) W_{2N}^{km} \qquad 0 \le m \le 2N-1$$

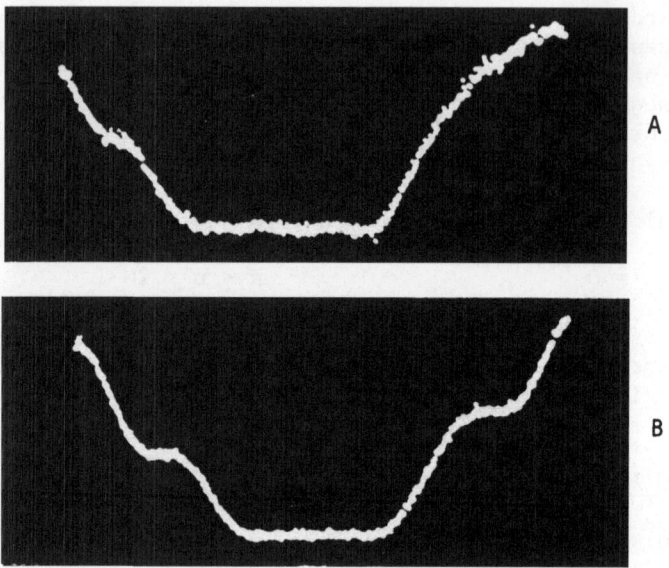

Figure I - Linear crosscorrelation between two orthogonal
(A) and parallel (B) central scan lines 10 pp apart, each
averaged 100 times for image III (mitotic cell). Similar results
are found if parallel lines are 30 or 50 pp apart.

SIGNIFICANCE OF CORRELATION AND SEARCH FOR PERIODICITIES OF CHROMATIN MORPHOLOGY

In using the above results, we have taken two approaches.
The first involves taking scans across either the center or
borders of the nuclei and comparing the Fourier Transform of
the resulting arrays of points (Fig. 8 of the text)

The second involves taking cross correlations of scans
either parallel to each other or perpendicular to each other
(Fig. I). The reason behind this approach is as follows:
given that any two dimensional pattern is obtained from the actual
three dimensional chromatin pattern by "squashing" the nucleus,
any periodicity found should be independent of the orientation of
the cell (and therefore the scanned lines).
Preliminary results indicate that either the linear (Fig. I)
or the circular (not shown) cross correlation yield similar
results either between two parallel (Fig. IB) or orthogonal
(Fig. IA) scan lines. Furthermore, if cells of different
phases have different configurations, we might expect to
observe different correlational patterns between these cells

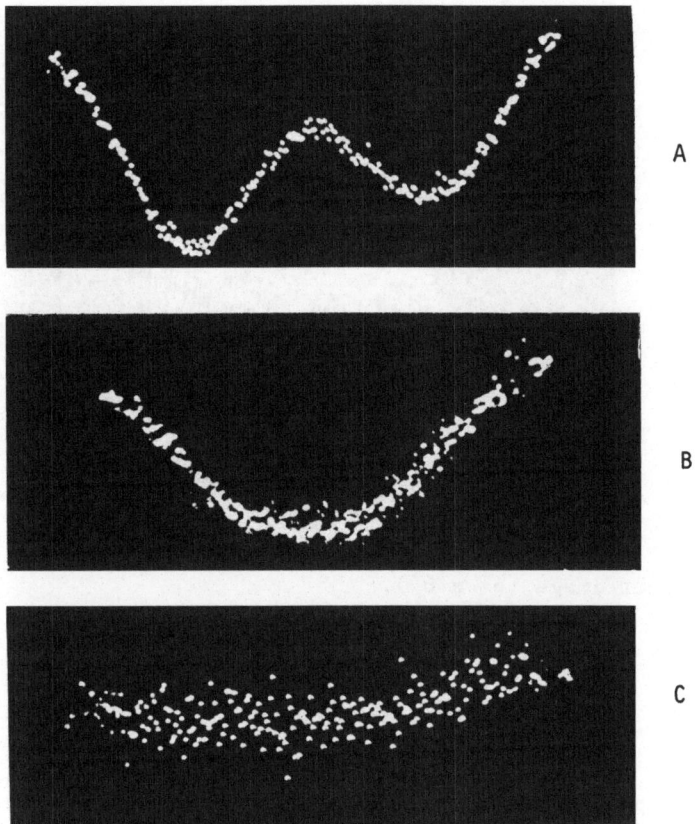

Figure II - Circular crosscorrelation between two parallel lines
10 pp apart (100 times averaged) central scan lines of mitotic (A),
middle G1 (B), and early S (C) He-La nuclear images.

as we do for early S, middle G1 and M phases (Fig. II). Indeed
if the correlational patterns were substantially different within
a given cell, such differences would have thrown doubt on the
cycle phase differences just discussed.

Of course, we should not be surprised that peripheral and central
scans of the same cell will yield different results, but they
should be uniform with respect to cells in the same phase, as
they are in the study so far conducted. For instance, middle-G1
cells do consistently show periodic spikes along the nuclear
border while mitotic cells do not (Fig. III).

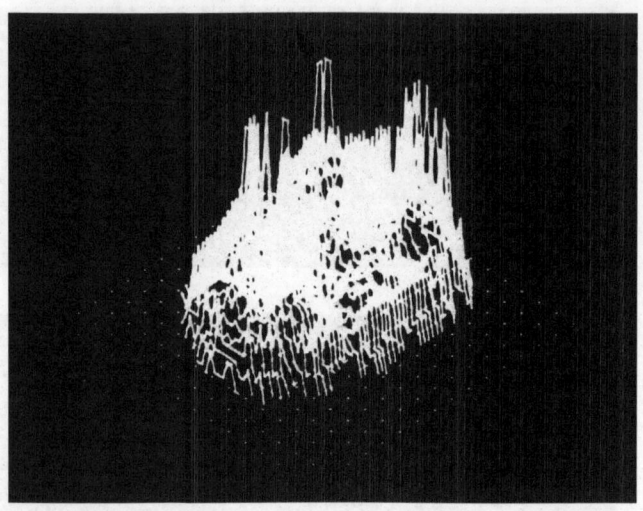

Figure III - Optical Density - overall display of mitotic (M)
HeLa nuclear image (64 x 64).

REFERENCES

1) A. Papoulis, The Fourier Integral and Its Application,
    McGraw Hill, 1962.

2) A. V. Oppenheim & R. W. Shafer, Digital Signal Processing,
    Prentice Hall, 1975.

3) G. D. Bergland, A Guided Tour of the Fast Fourier Transform,
    IEEE Spectrum, July 1969, Volume 6, #7 p. 41-52.

4) F. Bertora, C. Braccini, G. Gambardella and Musso,
    Study on the use of the Fast Fourier Transform in
    spectral analysis, ESOC Contract n. 486/73/T,
    Instituto di Electtrotecnica Universitá di Genova - Italy.

CHROMATIN STUDY *IN SITU*: II. STATIC AND FLOW MICROFLUORIMETRY

C. Nicolini, S. Parodi, S. Lessin, A. Belmont,
S. Abraham, S. Zeitz and M. Grattarola

Division of Biophysics
Temple University Graduate School
Philadelphia, Pa. 19128

I. INTRODUCTION

A recent review (1,2) indicates that significant correlation exist between the structure of chromatin and the extent of cell proliferation (aging, serum stimulation, virus transformation, cell-cycle phases, etc.). Chromatin, of course, refers to the diffuse interphase form of chromosome isolated from eukaryotic cells and the question remains open as to whether observed chromatin changes are reflected in the intact cell.

The techniques described in this report, by studying the dye-concentration dependence of mean fluorescence per cell, is intended to detect these differences as fast and as physiologically as possible, opening new approaches to the utilization of microfluorimetry in the characterization of chromatin, *in situ*.

In recent years, microfluorimetry has become extremely popular among cell biologists causing a chain reaction of scattered papers in a variety of related fields. Regardless of these efforts, a review of all pertinent literature of both *static and flow microfluorimetry* reveals little awareness of: a) the optimal staining conditions (which are frequently presented with slightly more dignity than a "cooking recipe"); b) the physico-chemical mechanisms and macromolecular organization, *in situ,* determining the spectral emission and fluorescence quantum yield; c) the different binding processes between the various dyes commonly utilized and cell components, as RNA, chromatin-DNA, and protein; d) the unique spectrofluorimetric properties of each dye when interacting with double or single stranded nucleic acid and proteins, as functions of ionic strength.

This review is aimed to address the above points in a coherent and comprehensive fashion, in order to prove that, *under proper staining condition, it is possible to differentially study chromatin conformation in situ* with dyes such as ethidium bromide (EB) and acridine orange (AO). Specifically, extending our original work with ethidium bromide (3), in this overview we have the goal of establishing *a bridge between the quantitative work done in solution with EB (or AO) and polynucleotides, and the cell system.*

II. BIOPHYSICAL-CYTOCHEMICAL CONFIGURATION

A. *FLOW SYSTEMS*

Cell fluorescence (green, red, and total) and low angle forward light scatter are measured on a flow microfluorimeter using an ion laser of various powers (between 35 mW and 4 watt) at various wavelengths, ranging between the far-ultraviolet up to the visible. The most commonly utilized in the study related to chromatin-DNA, is an argon-ion laser with optimal excitation wavelength of 488 nm. Usually a sample of fluorescently tagged cells may be injected into the center of a flowing stream. The stream after emerging from the nozzle tip, is about 50 microns in diameter. As shown in Fig. 1, as cells pass through the laser beam, different cells will give different but characteristic signals, both in fluorescence and in scattering. These differences manifest themselves in variations in the intensity of the light from each cell, which, in turn, are converted into electrical pulses of varying amplitudes by the photodetectors. The scatter signal is detected by observing the illumination in the forward direction. The scatter signal can be observed over a wide angle (up to about 10 degrees on either side of the laser beam) or over a narrow angle (down to 2 or 3 degrees on either side of the beam). Wide-angle scattering is comparatively insensitive to the effects of flow and instrument artifacts. Narrow-angle detection is dependent primarily upon cell or particle cross-sectional area, and not upon its refractive index relative to the medium in which it is immersed, nor its probable lack of optical density - a feature that is sometimes of importance in the observation of unstained biological materials.

The fluorescence characteristics of the cell are determined simultaneously with the scatter signal. Cells that have been tagged by means of a suitable fluorescent material can be detected by means of their emission, when excited by the high-intensity light from the laser. As shown in Fig. 1, in order to perform multiparameter analysis with our flow systems (either an Ortho cytofluorograf or a B-D cell sorter) on-line with our PDP11/40 computer, various electronic modifications to allow multiple fluorescent and/or scatter measurement on each cell were made (4). After

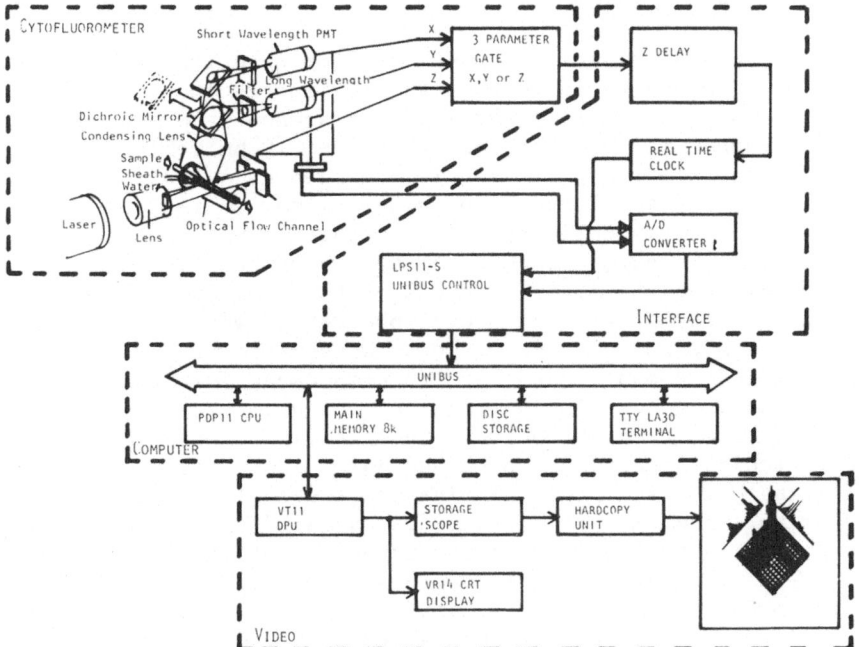

Fig. 1. Schematic diagram of our own *"in house"* automated multiparameter flow cytofluorimeter configuration (4), from data acquisition (cyto-fluorimeter) up to display (video).

collection, the data are sorted and stored in a two-dimensional 50x50 array on a Deck Disc for further analysis. To process the acquired data, various software routines have been written which allow for easy manipulation of the data (4). This routine includes the capability of plotting the two-dimensional cytogram of any specified window on the data as well as the simultaneous plotting of the 2-one dimensional histograms obtained by projecting the cytogram on each axis. The statistics on each window are available along with the capability of storing projection for the different windows, to allow the analysis to be performed by our computer, in real time. Unfixed cell suspensions either from *in vivo* or *in vitro* systems were stained with EB or AO at 1-4 x 10^{-5}M either directly at the final R (μM dye/μM DNA) = 1-5 (3,5) or, in order to prove the fallicy of certain arguments, after pretreatment with Triton X100 and chelating agents (6). Occasionally, in both stainings, cells were incubated with RNAse A (10 units/ml; Freehold, New Jersey) for 30 minutes at 35°C.

B. *STATIC SYSTEMS*

Epi-illumination is employed for quantitative fluorescence measurements as it reproduced the fluorescence excitation conditions automatically. The epifluorescence is measured with a Zeiss Ultraphot microscope, equipped with 510 nm reflector. Fluorescence-exciting radiation is provided by a 100 watt high pressure mercury arc lamp, with proper filter to isolate the blue line (450-485 nm). Color estimate of the nucleus and the cytoplasm is carried out by direct comparison of the fluorescence emission against the color diagrams of the Commission International de l'Eclairage. A fluorescence photograph was taken immediately after each experiment to serve as a permanent record. Our color assignment was constantly cross-checked, in terms of wavelength fluorescence emission, against an automated recording Zeiss Microscope Photometer 03. Quantitative corrected fluorescence emission spectra are obtained using interference contrast optics and blue light for illumination: a cell is selected and the variable photometric aperture is adjusted to encompass a fixed area either in the nucleus or the cytoplasm of the individual cell. The cells grown *in vitro* are stained either directly on the coverslips before and after fixation or as smears after detachment by trypsin or scraping. The cells grown *in vivo* are stained either directly after a mechanical-trypsin dissociation procedure (1) or after subsequent fixation. The mildest fixation procedure is with 2% glutaraldehyde in Krebs-Ringer Phosphate Buffer (pH 7) at 37°C for 1 hour; which gives completely permeable cells. Static fluorescence study was conducted in a 5 x 35 x 16 μm chamber as described elsewhere. (15)

III. ETHIDIUM BROMIDE STAINING

Ethidium bromide is a specific intercalating dye for nucleic acid, both RNA and DNA, which has been extensively studied both in *solution* and *in situ*.

A. *IN SOLUTION*

The ethidium bromide is important for at least two reasons: 1) the ethidium bromide molecule can be used as a probe of the DNA conformation in such structures as the closed circular DNA of phages and viruses and the chromatin of eukaryotic cells; and 2) the ethidium bromide interaction with DNA, probably an intercalation between two couples of paired bases, can be considered a typical model for a certain class of intercalating agents (7). The ethidium bromide molecule is an interesting probe also for the structure of RNA molecules. Important steps in the understanding of the ethidium bromide-DNA interaction are the discovery of a metachromatic shift in the absorption spectrum near 500 mμ (8),

the observed large increase in the fluorescence quantum efficiency of intercalated ethidium bromide (7), the observation that
the induced optical activity near 308 mμ and its direct effect on
the DNA spectrum is simply linearly proportional to the amount of
intercalated dye (9). Scatchard plot analysis (10) of DNA - EB
interaction at low ionic strength and 26°C yields an association
constant for the *primary sites* of 1.3×10^6 (moles/l), which
decreased by about a factor 10 at higher ionic strength (0.2M).
EB may intercalate with double stranded RNA, only with slightly
different association constants. *Secondary sites,* weakly bound
outside the nucleic acid, may also yield fluorescence, but with a
quantum yield equal or only slightly larger than free dye, while
primary sites do have a fluorescence enhancement of about factor
10, as can be easily verified by spectrofluorimetric studies (7).

B. *IN SITU*

        In order to discriminate between the binding process
(primary or secondary) of either DNA or RNA, we evaluate the
mean fluorescence per cell as a function of added dye. This could
allow us to determine the amount of bound and free dye at any
concentration of ethidium bromide. By analogy with the Scatchard
plot analysis (10) we can then *determine the association constant
and the number of primary binding sites in the intact cell.*
Fig. 2 shows the mean fluorescence per cell as a function of the
ratio R of added dye per unit of DNA from WI-38 human fibroblasts
confluent (G_o), mature duck erythrocytes (G_o), and HeLa S3 cells
in G1 and G2 phases. These curves show that, at saturation, the
mean fluorescence per cell reaches a plateau value which is directly related to the amount of DNA per cell; 7.0 pg for WI-38
human fibroblasts, 10.3 pg for WI-38 human fibroblasts, 10.3 pg
for G1 HeLa cells, and 20.6 pg for G2 HeLa cells. The exception
is constituted by mature duck erythrocytes (2.65 pg/cell) which
do not have any sizable amount of RNA: (a) the saturation occurs
at a lower R (0.5-0.6), and (b) the ratio of the mean fluorescence per WI-38/erythrocytes (or HeLa/erythrocytes) is, at
saturation (R>3.0), larger than what would be expected from a
pure DNA content ratio. In all other cell lines, the ratio of
observed mean fluorescence, at saturation, corresponds closely to
the expected value from pure DNA content per cell, perhaps because
in these examples the increase in RNA closely parallels the increase in DNA content per cell. Under our experimental conditions, the chromosomes maintain their conformation intact, such
that mainly *at low R* (<1.0), a *variation in conformation* could be
reflected in variation of fluorescence per cell. This becomes
apparent, when by analogy with Scatchard plot analysis, we evaluate the association constant between the dye and the intact cell,
and the relative number of binding sites available to the dye in
the intact cell (both primary and secondary with DNA and RNA).
Utilizing the observed intensity of mean fluorescence at very low
dye concentrations (R<0.20) where all ethidium bromide added can

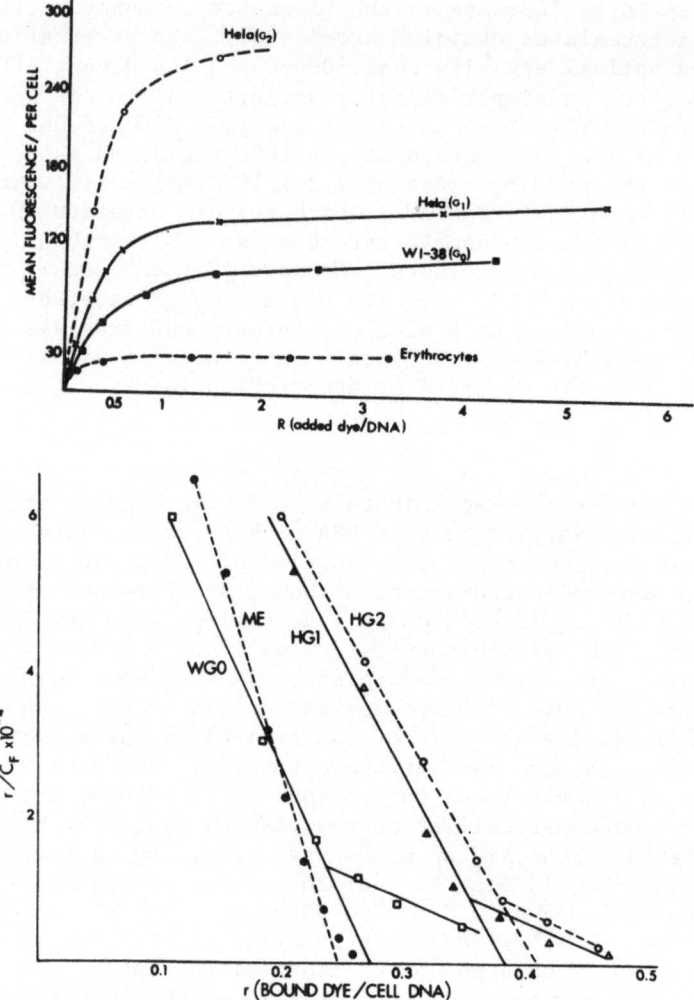

Fig. 2. (above) Mean fluorescence/cell as a function of R
(added dye/DNA) from duck erythrocytes (●---●) and WI-38 human
fibroblasts confluent (■——■). HeLa cells in G_1 (X——X) and G_2
phase (O---O). The assignment of cell-cycle phase for HeLa
cells was obtained from the frequency distribution of fluores-
cence/cell from a log phase growing in suspension. The so-called
"G_2" cell population also presumably contains M-phase cells.

        (below) r/C_F vs r (bound dye/DNA) for the binding
process between the intact living cell and the ethidium bromide
from mature duck erythrocytes   (ME,●---●), WI-38 confluent
(G_0,□ ---□), HeLa in G_1 phase (HG$_1$, Δ---Δ) and G_2 phase
(HG$_2$,O---O).

be considered bound, we obtained for each given cell a function
which gives the expected value (E) for the mean fluorescence per
cell versus the μmols of added dye/umols of DNA (in a given cell),
if all dyes added should bind to the nucleic acid. These functions are:

$$E_E = 100 \ R \ \text{for duck mature erythrocytes}$$
$$E_W = 240 \ R \ \text{for WI-38 human fibroblasts}$$
$$E_{HG_1} = 331 \ R \ \text{for HeLa cells in G1 phase}$$
$$E_{HG_2} = 652 \ R \ \text{for HeLa cells in G2 phase}$$

The ratio (r) of μmols of bound dye/μmols of DNA (in a given cell)
is obtained by

$$r = R \ (OB/E)$$

where OB and E are respectively the observed and expected mean
fluorescence per cell at a given R (added dye/DNA). The amount
(μmols) of free dye (C_F) per DNA, normalized to 100 μmols of DNA.
for a given cell population is then given by

$$C_F = \frac{(E-OB)}{E} \times 100 \ R$$

The lower panel of Fig. 2 shows the plot of r/C_F vs r(bound dye/
DNA), computes as previously indicated for the same cell lines.
From this, it appears (see Table I) that *mature duck erythrocytes
and confluent WI-38 human fibroblasts have approximately the same
number of primary binding sites available (even if the amount of
DNA per cell differs by a factor of 2.6, while the number of
primary binding sites for HeLa G1 or WI-38 G1 (3,5) and HeLa G2
cells are also very similar (regardless of the two-factor difference in DNA content), but are substantially increased with respect
to the WI-38 (G_O) and mature duck erythrocytes.* The relative
association constant K can be obtained (see Table I) by

$$\text{I)} \qquad \frac{r}{C_F} = K \ (n-r)$$

where (n) is the abscissa intercept and represents the number of
binding sites (in the intact cell) per DNA phosphate, (r) is the
ratio of bound dye per DNA phosphate, and C_F is the concentration
of free dye in umols. The *"primary sites"* association constant
for EB is of the same order of magnitude (3.2×10^5) computed (see
following pages) for AO primary sites in situ (2.6×10^5), and in
perfect agreement with EB association constants in solution, at
0.2 M NaCl. Parallel studies, conducted with RNA-ase treated
cells, seems to indicate that the *primary sites are related to
an alteration in chromatin DNA conformation,* as during G0-G1
transition of several *in vitro* (3,5,11) and *in vivo* (12) systems

TABLE I

Association constants ($\ell \cdot mol^{-1}$) and numbers of primary sites for chromatin-DNA IN SITU, for ethidium bromide primary (K^P) and secondary (K^S) binding sites. Data were obtained from a plot of r/C_f versus r obtained by staining unfixed cells with increasing concentration of E.B.

	n'	n"	K^P	K^S
HeLa G_1	0.38	0.47	3.2×10^5	6.8×10^4
Mature Duck Erythrocytes	0.25		5.7×10^5	
WI-38 GO	0.27	0.44	4×10^5	7×10^4
WI-38 G_1	0.37	0.43	3.0×10^5	6.3×10^4
HeLa G_2	0.40	0.48	3.3×10^5	6.5×10^4

TABLE II

Time (T) necessary for EB to enter 100% of CHO cells versus absolute cell concentration expressed in M DNA. CHO cells were stained with EB=10^{-3}M directly on the coverslip. Cell viability was monitored by 0.2% Trypan Blue exclusion, after EB entered all cells.

DNA	T (Minutes)	Percentage of Viable Cell
4.2×10^{-5}	2	90%
3.2×10^{-6}	3	90%
2.11×10^{-6}	5	50%
1.05×10^{-6}	3	90%
1.0×10^{-6}	4	80%
5.3×10^{-7}	6	80%
5.3×10^{-7}	5	80%

The orange-red fluorescence intensity is weak at very early time (1-2 minutes), constantly increasing until reaching a plateau level around 8-10 minutes corresponding to the emission obtainable after fixation. During the same lag period, Trypan Blue uptake does not change appreciably. The orange emission remains unchanged between 10 and 60 minutes when most cells loose viability (Trypan Blue, TB, enters). The same cell, which do uptake TB instantaneously, do also uptake EB instantaneously and intensely.

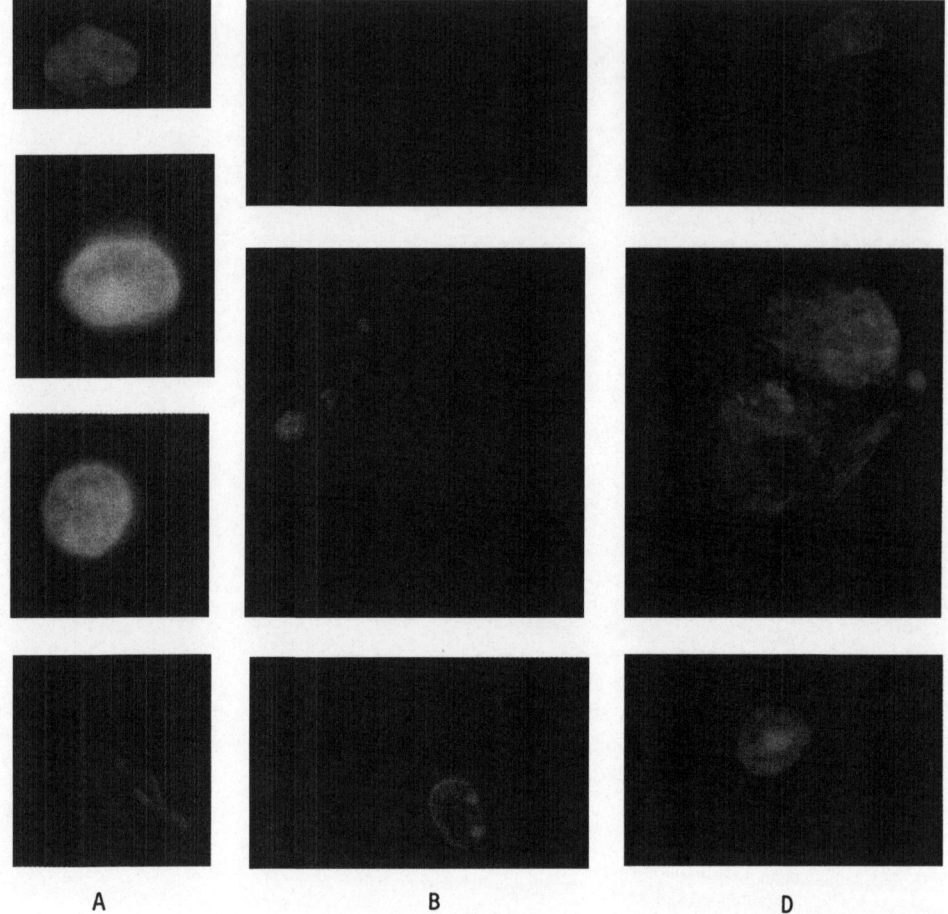

A   B   D

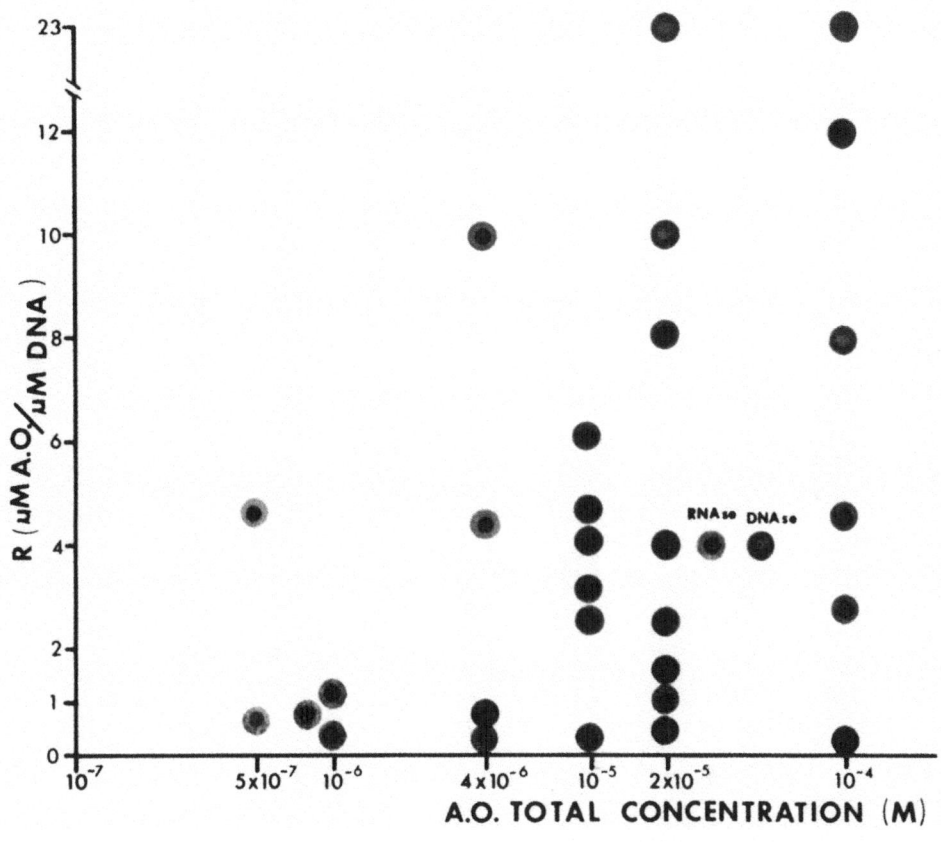

C

Fig. 5a. WE-38 cell, stained with Ethidium Bromide at R=4 and 10^{-5}M absolute dye concentration. As evident, cell maintains its original morphology and viability (as determined by Trypan-blue exclusion).

Fig. 5b. Static fluorescence emission of Melanoma B16 tumor cells, stained with Acridine Orange at 2.5 x 10^{-5}M and R=0.16 (below); R=4 (middle); R=10 (above).

Fig. 5c. Phase diagram of color emission for nuclei and cytoplasm from the same melanoma cells, stained with Acridine Orange at various AO total concentration and various molar ratios (μM AO/ μM DNA).

Fig. 5d. Effect of DNA-ase (center) and RNA-ase (above) digestions on the same melanoma cells, stained at R=4 and AO-2.5 x 10^{-5}M (below, for untreated cell). Cells were unfixed and immediately AO stained. DNA digestion was conducted for 1 hour at 37°C, with 3160 U/ml of DNA-ase, RNA digestion for 1 hour at 37°C using 7000 U/ml of RNA-ase.

(see Fig. 3), and compatible with findings on chromatin isolated from the same cell lines (2) (see also chapter by C. Nicolini on chromatin structure, pp. 613). Static fluorescence studies (see Fig. 4) indicate that the EB fluorescence is mostly confined in the nuclei (both at R = 0.16 and R = 4) while the cytoplasm is weak (at any R)which may be due to the low number of primary sites in cytoplasmic RNA and/or to different quantum yield arising when EB intercalates different base pairs or base sequence (13).

C. *CELL VIABILITY AND EB UPTAKE*

Frequently cell death has been correlated with EB uptake. We now address such correlations, even if cell death (eventually occurring in unfixed cells during the procedure) is not at issue here, since it does not alter DNA organization and subsequent fluorescence emission, within the short period of time between cell staining and analysis. A careful time sequence on CHO line *in vitro* cells, directly stained on their coverslip with EB at saturation, show that within 2-6 minutes (depending on cell concentration) *all viable cells exhibit clear orange fluorescent nuclei while preserving their morphology* (see Fig. 5a, Table II). Furthermore, when stained at R = 4 and absolute EB concentration of 4 x 10^{-5}M, not only do viable cells pick up the dye, but they (both CHO and M3 cells) maintain their capability to grow after quick dye removal (Table III).

IV. ACRIDINE ORANGE STAINING

Acridine orange (AO) is widely used for flow microfluorimetric (FMF) studies and for static fluorescence (SF) studies of the intact cells. However, until recently (15) we were lacking a rigorous approach to the AO staining, where, even if we are dealing with cells and *not* with nucleotides in solution, we have a comparably good control and understanding of the entire dye-cell interaction and subsequent fluorescence emission. The basic approaches to AO staining in SF are essentially of two types: *overstaining-differentiation* and *equilibrium staining* methods. A typical example of the first type is the widely used method of Rigler (16), where the cells are overstained with 10^{-4}M AO and then differentiated in a citric-acid-phosphate buffer. The overstaining-differentiation approach is essentially the classical histological approach. Other authors have used in SF equilibrium-staining methods (17) which are indeed the only one possible for FMF studies. In the latter cases, both for FMF and SF, we have found in the literature either *empirical* protocols (12,13,15) or quantitative but *misleading* approaches (6,8). For this reason, we have explored (15) a large range of R $\frac{\text{(total dye)}}{\text{(total DNA)}}$ ratios and final AO concentrations, and found that the results *in situ* even if more approximate, *are essentially in agreement with the studies in solution,* both in SF and FMF studies.

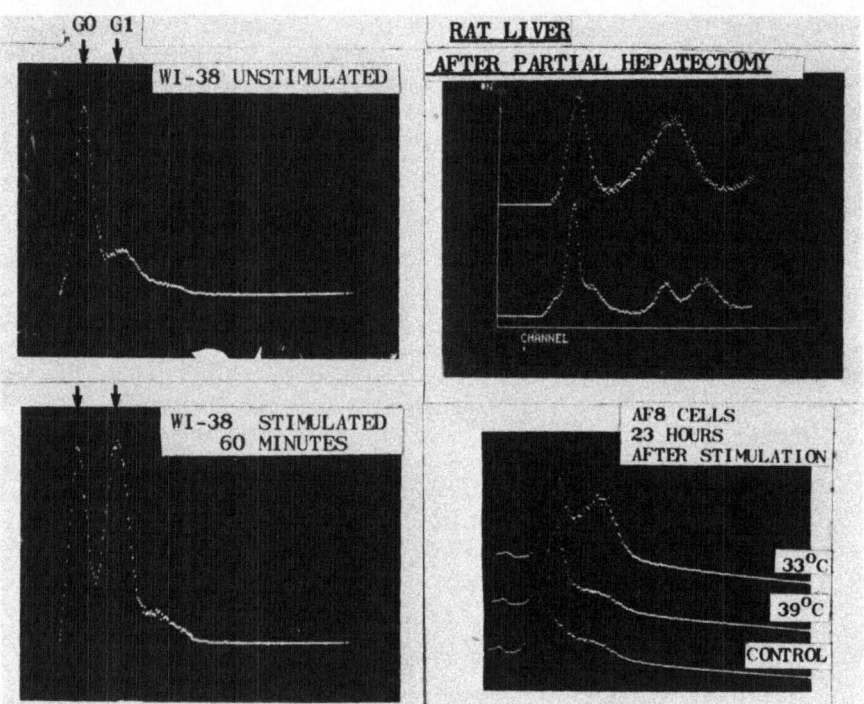

Fig. 3. Green (or total) *fluorescence emission of various cell lines, stained with EB at 10⁻⁵M and R=0.75 during G0-G1 transition in vitro* (WI-38 serum-stimulated; AF-8 temperature sensitive mutants at the permissive temperature, 33°C and non-permissive temperature, 39°C, at 23 hours after stimulation) and *in vivo* (rat liver after partial hepatectomy). In all three systems "G0 and G1" cell population have the same amount of DNA, and DNA synthesis, i.e., 3H-thymadine incorporation begins several hours later.

B 16 STATIC FLUORESENCE

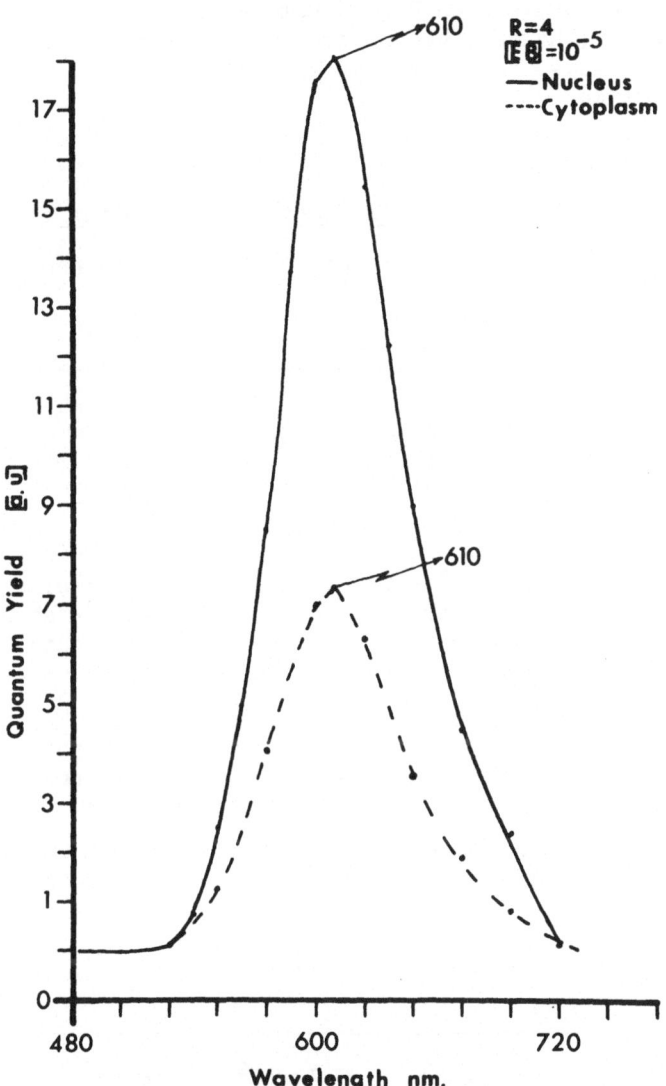

Fig. 4.   "Static" *Spectral* fluorescence emission of the same WI-38 cell shown in Fig. 4a, stained with EB EB=10^{-5}M and R=4, as determined by a Zeiss microfluorimeter.

TABLE III

CHO cells were incubated at EB = 4 x 10^{-5}M and R = 4, then 10 minutes later (with 100% of cells showing orange emission but 0% Trypan blue uptake) the EB was removed by 2 consecutive washes. The same cells (who received EB treatment) were then incubated at 37°C for 3 days. A control of untreated CHO cell was monitored at the same time.

	Number of Cell Plated	Number of Cell 3 Days After Staining
EB treated	45,000	322,000
Control	45,000	315,000

TABLE IV

Binding constants (ℓ.mol^{-1}) for AO-DNA and AO-RNA interaction in solution (at high ionic strength, i.e., 0.2M) and *in situ*. Primary sites refer to the green emission, while secondary sites refer to dimer and higher aggregate formation outside nucleic acids.

	AO-DNA		AO-RNA	
	Primary	Secondary	Primary	Secondary
SOLUTION (at high ionic strength	1.3×10^5	0.8×10^4	1.3×10^5	–
IN SITU	2.6×10^5	0.72×10^4	2.6×10^5	0.43×10^5

A. *AO-NUCLEIC ACID INTERACTION IN SOLUTION*

From the experimental work done with AO-DNA, the association constants K_1 for primary sites and K_2 (secondary sites) in solution can be computed using a model proposed by Armstrong, et. al. (18), with

$$K_1 = \frac{2 (\beta_1 - \beta_2) (1-2\beta_1)}{C_M (1-4\beta_1)^2} ; \quad K_2 = \beta_2/C_M(\beta_1-\beta_2)$$

where C_M, β_1 and β_2 are respectively the concentration of free dye, fraction of available primary and secondary sites occupied per unit DNA (or RNA). As summarized by Table IV, it appears that: 1) at very low primary sites, i.e., $\beta<0.1$, we have essentially monomers of AO intercalated between base pairs, the quantum yield of the 540 nm fluorescence being approximately 3 times the quantum yield of the free AO monomers (19) and at high ionic strength (0.2M) the *association constant of this monomer for native DNA is of the order of $1.3x10^5$* (18). 2) In the range of R = 0.1 - 0.3, another molecular species appears, tentatively a dimer. The formation of the dimer induces a red shift, with a progressive decrease in the quantum yield of the 540 nm fluorescence and probably the dimer is practically unfluorescent at 540 nm. At *high ionic strength* (0.2M) the *association constant* for these *second binding is of the order of $0.8 x 10^4$* (18). For R>0.3 higher orders of aggregation can be found (see *in situ* studies) responsible for the red fluorescence at 640-670 nm. The studies in solution at R>0.3 are, however, made difficult by the tendency of the complex to precipitate (18).

B. *STATIC MICROFLUORIMETRY*

In order to evaluate the AO stainability properties of nucleus and cytoplasm in presence or absence of cell pretreatments, we explore a large range of R (μMAO/μMDNA-P) from 0.05 to 100 with special attention to the range 0.1-10. This range of R was explored, by varying also the final AO concentrations between 10^{-7}M to 10^{-3}M. Fig. 5b shows a typical result obtained in melanoma B16 tumor cells at final AO concentration of $2.5x10^{-5}$M, by varying the molar ratios R: *the nucleus and the cytoplasm are both green at R = 0.16, respectively green and reddish orange at R = 4, and respectively yellow-orange and red at R = 10.* These data are confirmed by the quantitative spectral analysis of these cells' fluorescence by means of static microfluorophotometry in terms of actual emission wavelength (Fig. 6). Similar dependence upon molar ratios R(ranging between 0.1 and 23) are found in a wide range of total AO concentrations (10^{-4} - 10^{-7}M), as shown in Fig.5c where a color code (see Fig. 6 for corresponding emission wavelength) is utilized to visualize in a comprehensive and compact fashion the differential staining of nuclei and cytoplasm.

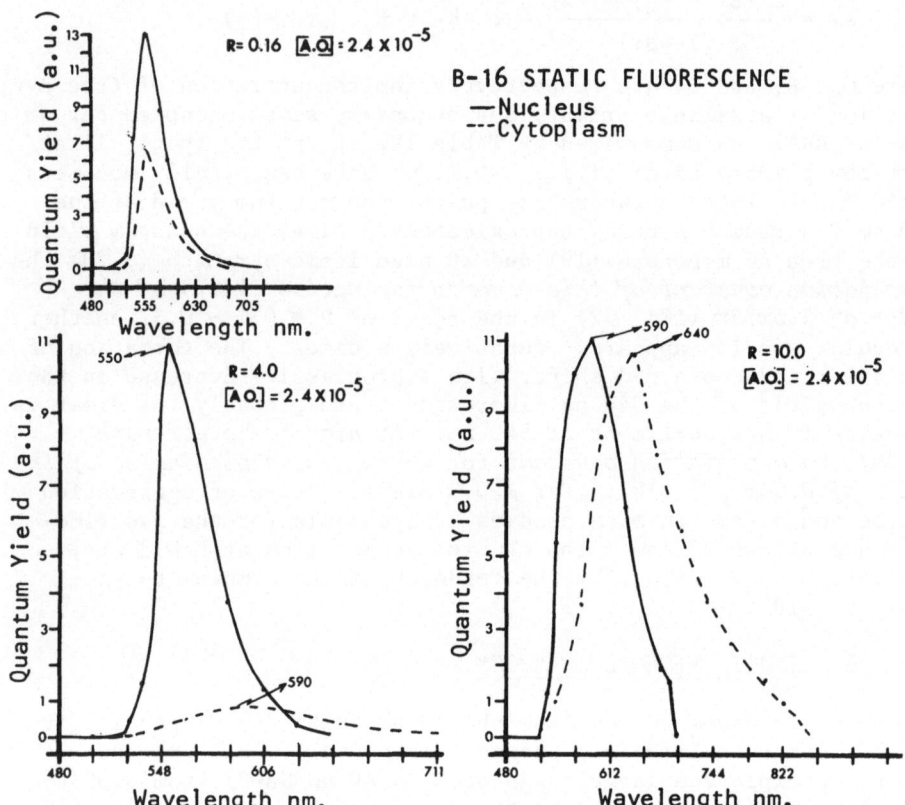

Fig. 6.   Static *spectral* fluorescence emission for
the same cell, as Fig. 5, at R=0.16, 4 and 10, as
determined by a Zeiss microfluorimeter (Acridine
Orange staining)

In these experiments we used amelanotic melanoma B16 cells dissociated by trypsinization and then *fixed* with 2% glutaraldehyde:
identical results have been obtained in *"unfixed"* cells, with the
exception that the cytoplasm stains slightly more intensely and
homogeneously after fixation. In order to determine what macromolecular species are predominant in the nucleus and the cytoplasm
and how they are related to differential spectral emission, we
have carried out RNAse and DNAse digestion in the same *"unfixed"*
B16 cells, at AO concentration of 2.5 x 10^{-5}M and R = 4:   under
these conditions, while the *nuclear green fluorescence is selectively affected* (80-90% reduction in intensity) *by DNAse,* the
cytoplasmic reddish-orange fluorescence is drastically *reduced by
RNAse digestion* (see Fig. 5d). These studies, confirmed in other
cell lines (as CHO and M3), *gave identical results whether with or
without cell pretreatment* by Triton X100 and chelating agents at
various pH (6). It is apparent that the *differential spectral
emission,* i.e., *nuclear DNA versus cytoplasmic RNA, strongly
depends from the dye/DNA ratio* contrary to previous statements
(6). Recent optical studies (18) of AO-DNA interaction *in solution* show, indeed, that denatured and native DNA have the same
association constant and *if* RNA was selectively denatured should
yield red cytoplasm even at R = 0.16. All these findings indicate
that the claim (12) for differential RNA and DNA denaturation by
chelating agents are quite erroneous, in agreement with the following FMF studies conducted in parallel (Fig. 9-12). Our results
indicate that at high ionic strength and AO total concentration in
the range 5 x 10^{-7} - 10^{-6}M, even at high R both nucleus and cytoplasm are green. This suggests that the association constants of
the dimer and higher aggregation sites are too low to generate
significant secondary binding (longer wavelength emission) even in
the cytoplasm. Actually, the nucleus becomes yellow (i.e., appearance of secondary sites) only at R = 10 and total AO equal
2.5 x 10^{-5}M, remaining green even at 1 x 10^{-5}M; this yields (see
Table V) an *association constant for chromatin red fluorescence in
the order of 0.72 x 10^4 in striking agreement* with previous determination of *DNA secondary sites in solutions* (18). In the cytoplasm, however, the red fluorescence begins to appear for R around
4, at AO concentration about 10 times lower (4 x 10^{-5}M), suggesting an affinity 10 times larger (0.5 x 10^5). Above concentrations of 4 x 10^{-5}M even the nucleus can become red, at high R.
Table V *summarizes* the relevant parameters and computed associa-
tions constants for the primary and secondary sites of nuclear DNA
and cytoplasmic RNA *in situ.* Our static fluorescence data (Table
V and Fig.5c) indicate that the association constants for chromatin-DNA green fluorescence is 2.6 x 10^5, while for the chromatin-
DNA red fluorescence (secondary sites) and for the dimer formation
in solution, are identical (0.8 x 10^4), suggesting that we may
assume the same association constant for all secondary sites
(dimer or higher aggregate). In summary, the fluorescence data in

TABLE V

Summary values of fractions of AO primary (β_1), secondary sites (β_2), for the molar ratio of total dye/DNA (R) and absolute dye concentration (AO), at which green or red fluorescence originally arise in nuclear DNA or cytoplasmic RNA (see also Fig. 3).

	NUCLEAR DNA		CYTOPLASMIC RNA	
	Green	Red	Green	Red
(AO (M)	10^{-7}	2.5×10^{-5}	10^{-7}	4×10^{-6}
R	0.1	10	0.11	11
β_1	0.01	0.21	0.01	.15
β_2	0.00	0.031	0.00	0.022
C_m (M)	$.8 \times 10^{-7}$	2.4×10^{-5}	$.8 \times 10^{-7}$	3.80×10^{-6}
K_1	2.6×10^5		2.6×10^5	
K_2		0.72×10^4		0.43110^5

Association constants ($\ell . \text{mol}^{-1}$) are computed for each binding process according to $K_1 = \dfrac{2 \ (\beta_1 - \beta_2) \ (1 - 2\beta_1)}{C_m \ (1 - 4 \ \beta_1)^2}$; $K_2 = \dfrac{\beta_2}{C_m \ (\beta_1 - \beta_2)}$

where free AO concentration C_m can be directly computed by subtracting the bound from the total dyes, i.e.,

$$C_m = AO \left[1 - \frac{(\beta_1 \ \beta_2)^{DNA}}{R^{DNA}} - \frac{(\beta_1 \ \beta_2)^{RNA}}{1.1 R^{DNA}} \right]$$

and for primary sites, $\beta_1 = .01$ corresponds to appearance of primary sites (green flourescence); $\beta_2 = .15 \beta_1$ corresponds to appearance of secondary sites (yellow) when $\beta_1 = 0.21$, i.e., saturation primary sites.

Note   We calculated K_2 as follows: We assumed   % red fluorescence $= \dfrac{3.333 \beta_2}{\beta_1 + 3.333 \beta_2} = .25$ which gives yellow color.   Then from K_1 we compute β_1 and therefore $\beta_2 = 0.15\beta_1$.   Using those values, we compute K_2.

solution and *in situ* are in striking agreement (Table IV). We
may then assume for the AO staining of intact cell that both
double stranded DNA (B form) and RNA (A form) possess primary
(green) and secondary (red, related to multimer AO formation)
binding sites, with different association constants (see Table IV):
*at low R, only primary sites with both DNA and RNA are occupied;
at intermediate R, only RNA exhibits secondary site binding; at
high R, DNA also saturates its primary sites and exhibits secondary
sites.*

To obtain an independent estimate of the association constant, at high ionic strength, related to green nuclear DNA fluorescence, we utilized data obtained by others (19) in mouse leuko-
cytes (see also ref. 15). Our computations tend to suggest that
at low β and high ionic strength, the association constant for AO
primary sites with DNA *in situ* is about $1.4 - 1.8 \times 10^5$M in agreement with our own data (2.6×10^5M): this number is furthermore
of the same magnitude reported in solution. It is curious to note
that the same author (17) obtained the same number, throughout a
series of compensating errors; for instance, West computed the
molarity of bound AO by assuming (17) a nuclear volume so small
(5 nm^3) that it would give an intracellular DNA concentration of
4.2 M (!?).

C. *PHYSICO-CHEMICAL MODEL FOR AO-BINDING IN SITU*

This qualitative description of AO binding can be
formulated in a more quantitative framework using a model proposed
by Armstrong, et. al. (18). Assuming two types of binding processes, *primary* consisting of dye intercalation and secondary consisting of formation of a dimer by the binding of an external dye
molecule with the intercalated molecule, and assuming that the
binding process does not change the electrostatic potential of the
nucleic acid polymer, Armstrong, et. al., arrive at the following
relationships:

$$K_1 = \frac{2 (\beta_1 - \beta_2) (1 - 2\beta_1)}{C_M (1 - 4\beta_1)^2}$$

$$K_2 = \frac{\beta_2}{C_M (\beta_1 - \beta_2)}$$

where β_1, β_2 and C_M are fractions of available primary and secondary sites per unit DNA, and free AO concentration. These two equations, in addition to the requirement $(AO)_T = (AO)_{bound} + C_M$,

allow for the solution of β_1 and β_2 as function of AO_T and R, given constants K_1 and K_2.

In the case of the intact cell, assuming that for every μM of DNA, we have 1.1 RNA, for the nucleus (n)

we have $\beta_1^n = \frac{1}{4} - \frac{1}{2}\sqrt{\frac{1}{4} - (\frac{2Y^n}{1-8Y^n-X})}$; $\beta_2^n = X^n\beta_1^n$ (I)

where $X^n = \frac{K_2^n C_M}{1+K_2^n C_M}$; $Y^n = \frac{K_1^n C_M}{2}$ (II)

$$AO = C_M + \frac{\beta_1^n+\beta_2^n}{R} \cdot AO + \frac{\beta_1^{cyt} + \beta_2^{cyt}}{1.1 \times R} \cdot AO \qquad (III)$$

where β_1^{cyt} and β_2^{cyt} refer to the primary and secondary binding in the cytoplasm, computed as (I-II). Using these equations we arrived at the following values for β (monomer) = β_1 - β_2 and β (dimer) = (2 x β_2), shown in Fig.7.

We can now relate these numbers to the percentage of red fluorescence, under the premise that the monomer emits in the green and the dimer in the red, with their relative quantum yield (actually quantum yield X absorption). Actually in the real situation, the species responsible for red fluorescence is most likely a large aggregate while in the dimer, fluorescence is probably quenched. The fact that the equilibrium conditions for dimer formation appear to be identical to those for appearance of red fluorescence is probably a reflection of the critical step in aggregate formation, being dimer formation. If we incorporate this change in our model by assuming β_D = β_{AG} and including this change in our expression for C_M as a function of AO_T, we obtain the results shown in Fig. 8, for aggregate sizes of 20 (which yields the optimal fit to the experimental data). Of course, what we have presented above a "yes or no" phenomenon (dimers → aggregate, green fl. → red fl.) seems to be really a *continuous* process (as indicated by solution studies—our own data), whereby there is a progression in aggregate size accompanied by progressions in quantum yield and fluorescence emission spectra. Thus taking these factors into account, our present treatment, although primitive, does yield the basic description of the binding phenomena and supports our present understanding of the equilibrium binding process applied to intact cells. In the model, we have assumed

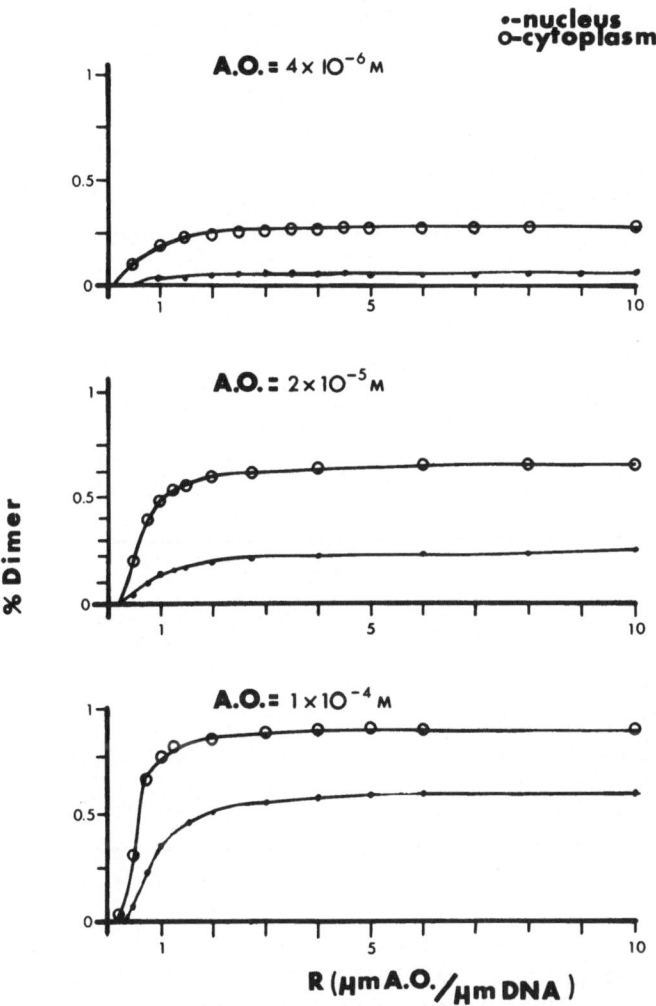

Fig. 7. Acridine Orange dimer formation
according with the model described in the text,
as function of molar ratios R, at 3 different
AO final concentration.

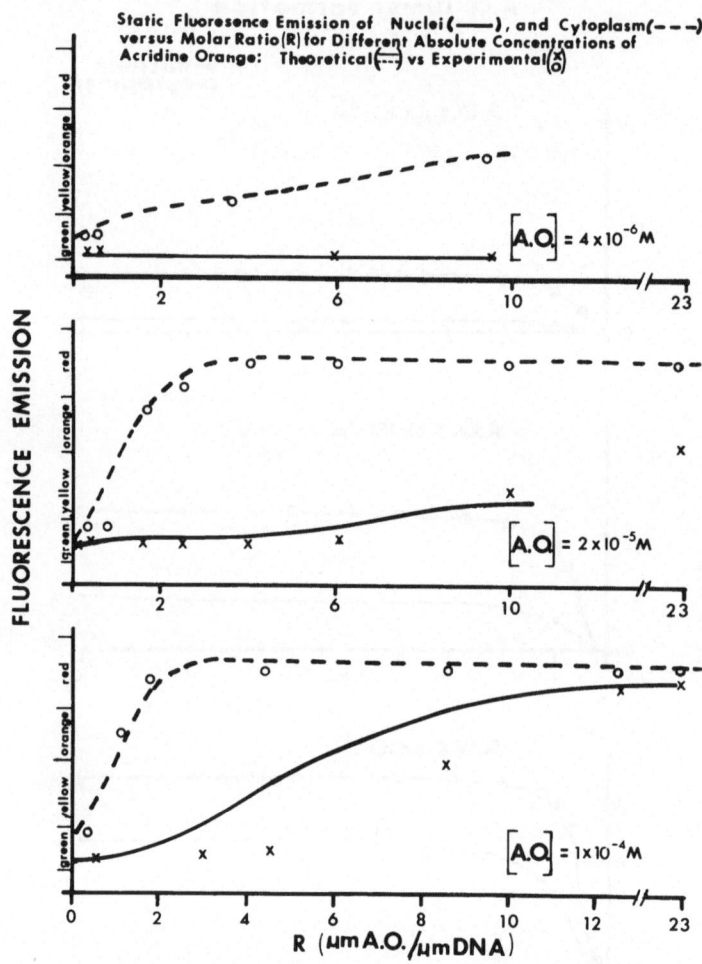

Fig. 8 . Experimental color fluorescence emission
of AO-stained melanoma cells, for nuclei (X) and
cytoplasm (0) as function of molar ratio, at 3 different final AO concentration. The dotted (cytoplasm)
and continuous (nucleus) lines represent the theoretical prediction of the physico-chemical model
outlined in the text.

that *"primary sites"* (bound AO monomer) yield an enhancement (19) of quantum yield of 3 with respect to free AO monomer, with the *"secondary sites"* (bound multimers) yielding 10 (see Table V).

D. *FLOW MICROFLUORIMETRY*

Frequently in cell biology, the cell populations to be characterized are quite heterogeneous, which when combined with the macromolecular species' heterogeneity (as previously shown) require *automated multiparameter* data acquisition for a meaning-ful characterization. The heterogeneity of our melanoma cell suspensions can be made apparent, using a compressed logarithmic scale for the scatter, by the computer two-parameter histogram of scatter vs. fluorescence shown in Fig. 9 . Cells were directly stained with acridine orange at R = 4 with or without RNAse treat-ment. This compressed scatter scale allows simultaneous identifi-cation of debris (low scatter and low fluorescence); necrotic melanoma cells (low fluorescence and intermediate scatter); lym-phocytes (low scatter and intermediate fluorescence), and viable melanoma cells (large scatter and various levels of fluorescence intensity). This assignment to various cell types has been achieved by analyzing the morphology of the Wright-stained "sorted" cell sub-populations (using our B-D cell sorter) (15). As shown also for cells pretreated with Triton-X100 and chelating agents (15), after direct staining with acridine orange, the green fluorescence distribution of cells of any size is unaffected (or slightly affected) by the RNAse treatment. On the contrary, RNAse treatments do cause a four-fold decrease on the mean red fluores-cence of melanoma cells (largest scatter). We should remember that the gain on our photomultiplier was greater for the red than for the green signal since we only intend to determine relative effect of RNAse on the two spectral emissions. Moreover, *let us clarify a semantic problem:* due to the photocathodes-filters utilized in the cytofluorograf the actual spectral response of the FMF configurations are quite broad, both in the so-called *"green"* (between 515 and 585 nm) and *"red"* (between 520 and 750 nm): i.e., we may have a *Gaussian orange emission* (as for ethidium bromide) centered around 600 nm, *being detected both as a "green" and "red" in FMF language* (Fig. 10) (1,3,5,6). Considering the spectral emission of single and double stranded nucleic acid, at R = 0.1 - 1.0 molar ratios most signals are detected in the "green" FMF region (15), but even in the presence of mostly DNA primary sites, a signal properly magnified may be detectable by the "red" FMF configuration (Fig. 10). At AO concentration $1-3 \times 10^{-5}$M and R = 1-10, while the *green emission clearly reflects only vaia-tions in chromatin-DNA primary dinding sites* with AO *in situ* the *red emission partially* relates to the *amount of RNA present* in the intact cell (either as single or double strand in the A form) *but* reflects *also variations in secondary weak-binding sites with DNA (at higher R)* and in the formation of AO multimers bound to acid

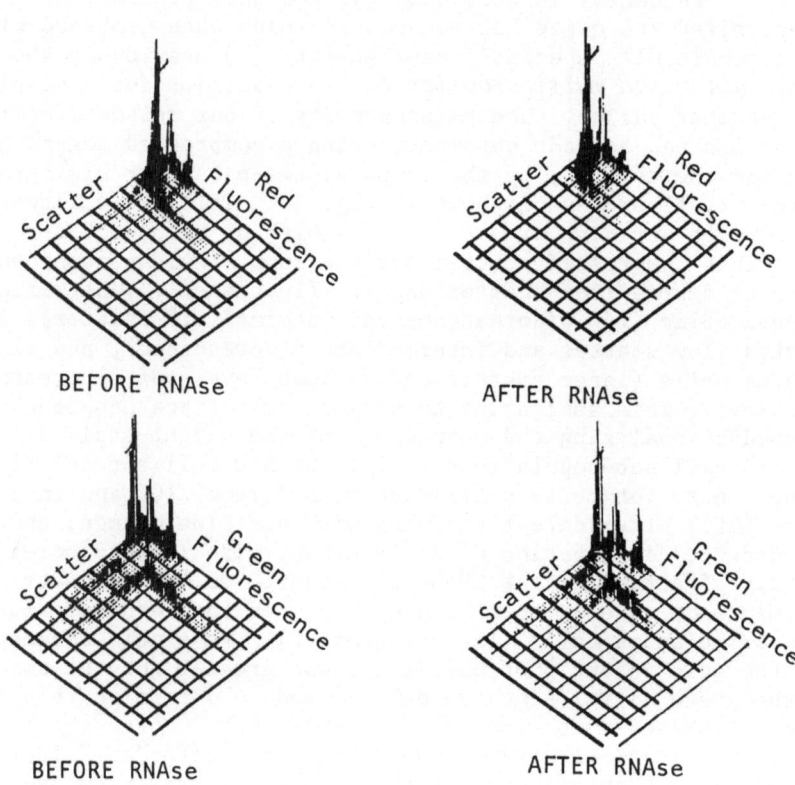

BEFORE RNAse   AFTER RNAse

BEFORE RNAse   AFTER RNAse

Fig. 9.   Computer-drawn two parameters hystograms
of scatter versus red and scatter versus green fluorescence, for melanoma B-16 cells, stained with AO at R=4
and final AO concentration 2.5×10^{-5}M. To display the
effect of selective removal of RNA, the same cytograms
are shown before and after RNase digestion at 37°C for
1 hour.

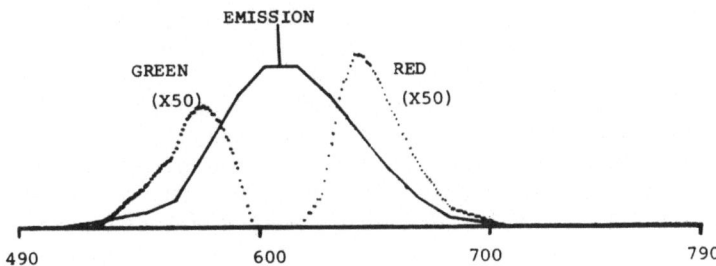

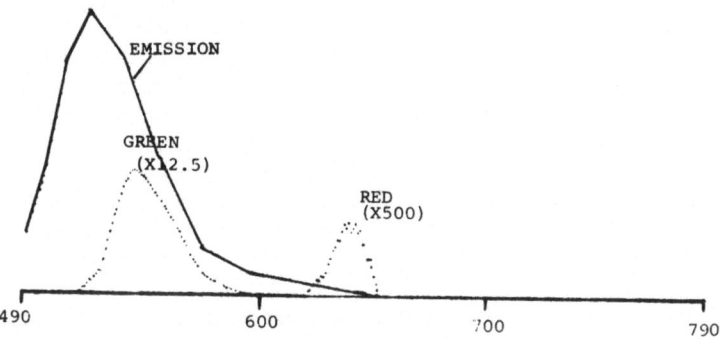

Fig.10.   Spectral emission (———) of DNA-bound
ethidium bromide (above) and Acridine Orange (below)
both determined by a Spectrofluorimeter at molar
ratio R=1 and dye final concentrations 10^{-5}M. The
dotted line represent the quantum efficiency for the
emitted light, of the "red" and "green" standard
cytofluorograph configuration, as determined by the
filters and photocathodes utilized.

macromolecules as polysaccharides, and is consequently strongly
dependent upon the dye concentration. If we now trigger, using
our simultaneous three parameter acquisition, only on large
scatter, we may obtain the red versus green distribution only for
melanoma B16 cells (15). If we now subdivide the *"pretreated"*
melanoma cell population into subpopulations, in terms of their
red fluorescence (Fig.11), we found that the cells with G1 fluorescence or larger seems to have green fluorescence distribution
typical of a log phase population with the presence of a small
peak   (about 6% of the total) with *green fluorescence lower than
G1 and a red fluorescence equal or larger than G2 cells.* Sorting
this cell population by means of our fluorescence-activated cell
sorter appears to confirm this assignment as *mitotic cells* in
agreement with the recent findings on PHA-stimulated lymphocytes (6). The *most striking example, which confirm both the
existence of a differential emission for RNA versus DNA at the
proper AO molar ratio and our previous EB data on the G0-G1
transition of WI-38 fibroblast,* is shown in Fig. 12. WI-38
human diploid fibroblasts have been grown up to 23 days into
"deep" confluency with weekly changes of the medium. Two days
after plating, WI-38 cells show a log-phase distribution for the
green fluorescence with a population of cells having the same
scatter (respect to G1) but quite lower chromatin primary sites
(green) and lower amount of RNA (red): This subpopulation, likely
relating   to *non-cycling cells* (that, after plating, even if
adherent to the plastic, did not start growing), drastically
reduces at 5 days   (5%) and then progressively increases at 12
days ($\sim$30%) and 23 days (up to 95%). This reduced lack of proliferation is apparent also by the *red vs   green* cytograms, where
the red fluorescence (RNA), under the same staining and instrument   conditions, progressively decreases with time after plating, going below the minimum threshold at 23 days. It is comfort-
ing that at 2 days G2 cells with twice as much DNA (green) show
also twice as much RNA (red) in respect to G1. At the same time, the
green fluorescence distribution show the constant presence of a
G1 peak (around channel 20) progressively decreasing in amplitude
at the expense of a "quite lower fluorescence G0 peak (constantly around #5). These findings are compatable with previous
observation, using EB, that the *transition* between a proliferating
G1 and non-proliferating *G0* cells is *not a continuum,* but rather a
quantum jump (5). An intermediate chromatin AO-uptake, appears at
12 hours, where WI-38 cells already reached confluency: (This
data, to be yet confirmed by more observations.) This could refer
to a chromatin difference between a readily reversible (G0) and a
deeper GO (or Q) non-cycling cells. Occasionally most WI-38 are
in the "so-called" deep GO, already at 12 days, under similar
nutritional conditions. We have to stress, indeed, that a large
variability does exist in WI-38 cells stimulated to proliferate
(regardless of previous "superficial" optimistic statements), *that
only proper automated multiparamenter analysis of cells stained*

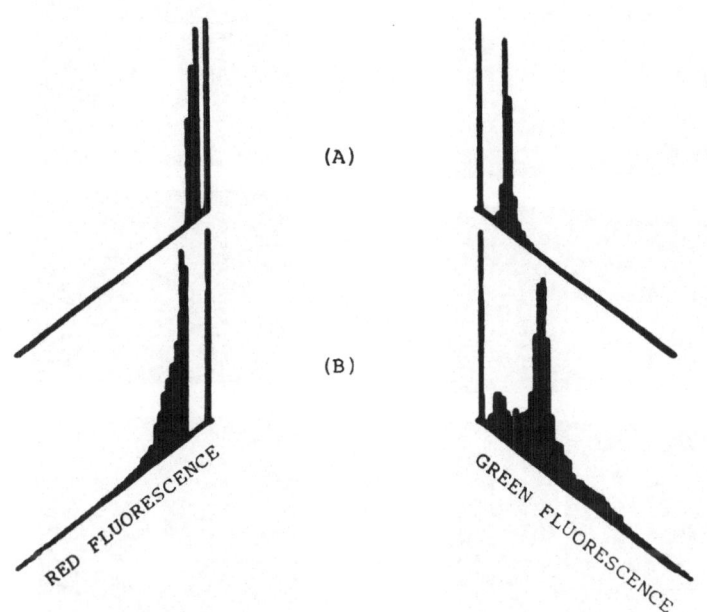

Fig. 11. *Green* fluorescence distribution of melanoma cells with small (A) and large (B) red fluorescence. Cells are stained with AO at R=4.

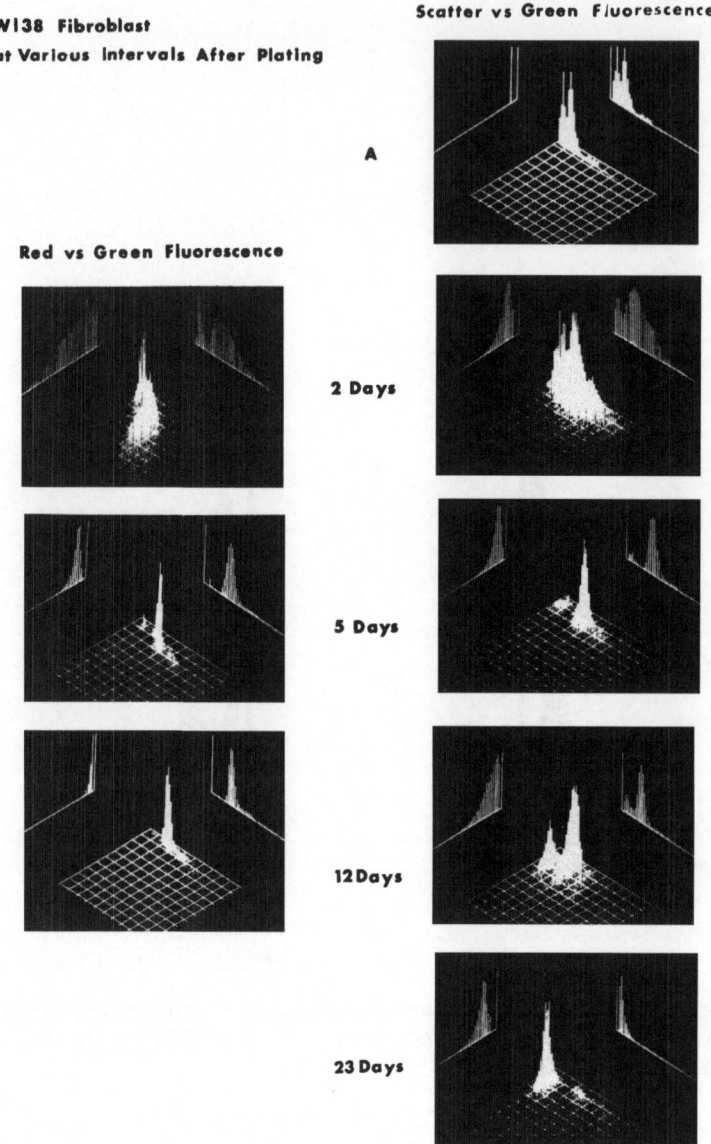

Fig. 12. Two parameter hystograms of WI-38 at 2, 5, 12 and 23 days
after plating (Phase II, grown in parallel). Cells were stained
with AO = 3×10^{-5}M (final) and R = 4-5. All cytograms were obtained
with green PMT = 4.6 and red PMT = 4.8 Scatter=Medium in a cytofluorograf on line with PDP11/40. Cytogram A was obtained when
channel 3 on the scatter axis was isolated from the 2 days after
plating cytogram. The medium was changed weekly - 95-98% of the
WI-38 cells are viable, as shown by Trypan blue exclusion.

remembering the mass action-law, may quantitatively monitor.

Sorting, by means of a fluorescence activated cell sorter, of the GO and G1 fluorescence peak either *in vitro* or *in vivo* (Fig. 13), shows that the two cell subpopulations are indistinguishable by light microscopy, compatible with the fact that the larger dye uptake relates to molecular alterations in the tertiary-quaternary chromatin structure (and RNA synthesis). In this respect is significant a recent comment of the Editors of an international Journal to the effect that *"we cannot understand the authors' claim that they can distinguish proliferating and non-proliferating cells by machine in the face of nothing distinguishable by eye."* This reflects a common empirical attitude among life scientists, where an automated image analyzer or flow microfluorometer (*"machine"*) is approached with a suspicion typical of a fifteenth century inhabitant of a South Pacific island approaching an automobile.

We should remember that experiments based on subjective hands and/or eyes left most problems of life science *yet unresolved* (actually the literature is "floated" with controversial empirical observation); while hard physical scientists were able to split the atom (without ever seeing a proton or a neutron with their eyes, being less than 10^{-14} cm), produce electricity and the most sophisticated electronic devices, as computer without ever seeing an electron), predict the presence and exact location of a planet by a series of equations (without ever dreaming of one).

V. CONCLUSION

In conclusion, AO allows us to selectively discriminate between chromatin-DNA and cytoplasmic RNA but *only* at proper R and absolute AO concentration and *without* need of any cell pretreatment. This is based on the *differential spectral emission of AO primary (green) and secondary (red) binding sites which are on the contrary only quantitatively (quantum yield) but not qualitatively (same orange emission) different for EB.* However, we have to stress that both for AO and EB the *same mechanism* is mostly *responsible for the larger dye uptake in G1 with respect to GO* (or Q) cells: namely, the *alteration in chromatin structure* i.e., its two-order superhelical configuration as originally and frequently proved by our laboratory (see chapter by C. Nicolini on Chromatin, pp. 613). The same conformational changes during the cell cycle originally monitored in isolated chromatin from M, G1 and S phases (2) are responsible for the reported *decreased primary binding sites for metaphase cells.* These studies, both by static and flow microfluorometry gave *identical* results, either before (*"unfixed"*) or *after fixation* with glutaraldehyde (8), either *with* or *without pretreatment* by Triton-X100 and

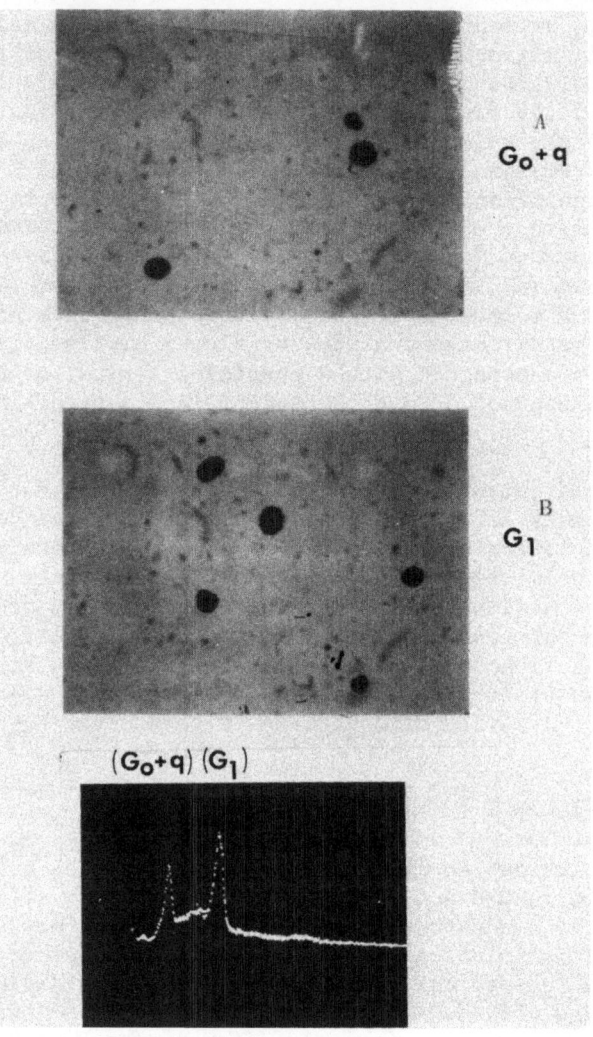

Fig. 13 . Light microphotograph of Wright stained
B-16 melanoma cells. $G_0 + Q$ cells (A) and G_1 cells
(B) were sorted from $G_0 + Q$ and G_1 peaks of green
fluorescence vs. scatter histogram (bottom) obtained
from B-D FACSII cell sorter.

chelating agents, at various pH. Investigators should pay more attention to the *"mass-action law"* instead of randomly exploring all possible combinations of chemical environment and pH. *Unfortunately, the mass-action law and other basic physico-chemical laws are usually ignored by most investigators working in either classical or automated cytology, who use intact cell staining with slightly more dignity than a "cooking" recipe.* For what concerns the static fluorescence, the method of overstaining and differentiation (5,7) is widely used. However, this method appears quite *limited* when quantitative measurements are desired, considering that the *rediffusion* process of the excess AO is *difficult* to control. Within a slide, variation in the local concentration of cells can be reflected in variation of the diffusion gradient and consequently of the destaining process; indeed, this has been frequently reported, even without a clear understanding of the reasons. Also, different degrees of convection in the solutions are very likely from experiment to experiment. In our opinion this procedure of *overstaining and differentiation* with reversible dyes, quite popular in classical histology, has such a degree of *empiricism* and *variability* that it should be *substituted* with *equilibrium* staining. As above reported, a difference between AO and EB can be expected in their kinetic interaction with native highly superhelical chromatin-DNA in viable cells, since the directional entrance into DNA appears to be from the narrow groove with EB intercalating and from the wide groove with AO (14). EB may bind first to the kink in DNA and then subsequently *"slip into"* the interior of the double-helix; AO may first bind to the sugar-phosphate chain at the kink and then intercalate as DNA straightens and base-pairs separate. If we may speculate, these differential binding processes already detectable in dinucleotide monophosphate crystalline complexes (14), can be further enhanced with consequent kinetic alterations by a highly *"kinked"* supercoiled configuration at the levels of *tertiary, quaternary* and *quinternary* structure in native chromatin-DNA in the intact cell. A final reflection: *biological phenomena are of such complexity that empirical recipes and qualitative analysis only contribute to further confusion.*

ACKNOWLEDGEMENT

This work was supported by Grants CA18258 and CA20034 from the National Institutes of Health.

REFERENCES

1. Nicolini, C., Biophys. Biochem. Acta, 458, 243-282 (1976)
2. Baserga R., and Nicolini, C., Biophys. Biochem. Acta, 458
    109-134 (1976)
3. Nicolini, C., Desaive, C., Kendall, F. and Fried, J., Canc.
    Treatment Rep., 60, 1819-1826 (1976)
4. Wu, S., Toton, S., Zietz, S., Kendall, F. and Nicolini, C.,
    Pulse Cytophotom., III, 57-76 (1977)
5. Nicolini, C., Kendall, F., DeSaive, C., Clarkson, B.,
    Fried, J., Exp. Cell Res., 106, 111 (1977)
6. Darzynkiewicz, T., Traganos, F., Sharpless, T. and
    Melamed, M., Canc. Res., 37, 4635 (1977)
7. LePecq, J. and Paoletti, C., J. Mol. Biol., 87-105 (1967)
8. Waring, M., J. Mol. Biol., 27, 87-109 (1965)
9. Parodi, S., Kendall, F. and Nicolini, C., Nucleic Acid Res.,
    2, 477-486 (1975)
10. Scatchard, G., Ann. N.Y. Acad. Sci., 51, 660-671 (1949)
11. Linden, W., et. al., Pulse Cytophotom., 277-289 (1977)
12. Nicolini, C., Linden, W., Zietz, S. and Wu, S. Nature, 270,
    607-609 (1977)
13. Krugh, T. and Beinhardt, C., J. Mol. Biol., 97, 133-162,
    (1975)
14. Sobell, H., et. al., J. Mol. Biol., 119, 333-365 (1977)
15. Nicolini, C., Parodi, S., Lessin, S., Zietz, S. and Belmont A.
    submitted for publication
    Parodi, S., Lessin, S., Fang, M., Zietz, S. and Nicolini, C.,
    Biophys. J., 24, 97a (1978)
16. Rigler, R., Acta Physiol. Scand., 67, suppl., 267, 1-291
    (1977)
17. West, S., in Physical Techniques in Biological Research,
    Pollister, A., ed., Vol. III part, C., 1 Academic Press,
    N.Y.-London (1969)
18. Armstrong, R., Kurmsev, R. and Strauss, U., J. Am. Chem. Soc.,
    92, 3174-3181 (1970)
19. Kubota, Y. and Steiner, R.F., Biophys. Chem., 6, 279-289
    (1977)

CHROMATIN STUDY IN SITU: III. DIFFERENTIAL EFFECTS OF

FEULGEN HYDROLYSIS

W. A. Linden, S. M. Fang, S. Zietz and C. Nicolini

Department of Physiology and Biophysics
Division of Biophysics, Temple University
Health Sciences Center, Philadelphia, Pa.   USA

ABSTRACT

Smears of synchronized HeLa S3 cells at 1, 3, 5, 8 and 12
hours after mitosis were Feulgen-stained with 1N HCl at 60° C for
15, 60 and 120 minutes. For each cell population the integrated
optical density, nuclear perimeter and area, measured by using the
image analyzer Quantimet 720-D on line with a PDP11/40 computer,
show a striking differential dependence upon hydrolysis time. IOD
frequency distributions of log-phase HeLa cells are statistically
significant at the three hydrolysis times; specifically, both the
coefficient of variation of "G1" peak and the fraction of cells
with "G1 DNA content" increase with hydrolysis time. Laser flow
microfluorimetric studies were also conducted on the same log-
phase HeLa cells, Feulgen-stained with acriflavine, respectively
at 1N HCl at 60°C for 5, 10, 20, 60, and 120 minutes; 4N HCl at 25°C
for 15, 60 and 120 minutes and 5N HCl at 37°C for 15, 60 and 120
minutes. The decompositon of the DNA histograms yields dramatic
increases of "G1" fractions, accompanied by increases of the coefficients of variation of the "G1" peaks, whenever the modal
fluorescence per cell decreases as a result of variation in HCl
normality, temperature or time of hydrolysis. This differential
hydrolysis dependence of the Feulgen reaction is compatible with
the difference in chromatin supercoiling for cells in different
phases and subphases of the cycle. Optimal hydrolysis conditions
are outlined, stressing the limitations of utilization of Feulgen
reactions for a quantative assay of DNA content and for FMF cell
cycle analysis.

INTRODUCTION

The Feulgen reaction is generally considered as a reasonably specific and quantitative method for measuring DNA content (1,2). However, there are indications that the amount of Feulgen stain is not only depending on the actual amount of DNA present in the nucleus but that it is reflecting the state of the deoxynucleoprotein complex and the degree of chromatin condensation (1). Thus, different somatic cells require different time of hydrolysis (3). Still more important are differences in the hydrolysis time dependence reported by Böhm and Sandritter for normal cells and mouse ascites tumor cells (4). Comparative studies of mesothelial cells, lymphocytes and tumor cells performed by these authors showed that, always in respect to the diploid value of the lymphocytes, the staining of the tumor cell population varied during different times of hydrolysis from triploid to hypotetraploid values (4). The same holds true for granulocytes and lymphocytes, which may yield DNA values 5-15% under the expected diploid value (5, 6, 7).

With the advent of flow microfluorimetry the fluorescent Feulgen method has been widely used for the estimation of the fraction of cells in each phase of cell cycle (8,9). This is accomplished by computer evaluation of the cytophotometric DNA distributions (10,12,11). Previous physico-chemical studies conducted either in isolated (16) and "in situ" (15) chromatin from synchronized HeLa cells indicate drastic changes in chromatin conformation during the entire cell cycle. A differential staining effect, for each cell cycle phase, of variation in temperature and acid concentration during the Feulgen hydrolysis reaction, due to the difference in chromatin structure, would have significant implications for these studies. Thus, we tried to evaluate the influence of hydrolysis time, temperature and acid concentration on Feulgen stained HeLa cells in different phases of the cycle using automated image analysis and flow microfluorimetry.

MATERIALS AND METHODS

Cell Culture

Logarithmically growing HeLa S-3 cells were maintained in suspension culture in Joklik - modified Eagles Minimum Essential Spinner Medium supplemented with 3.5% each of the calf serum and fetal calf serum (13).

Automated Image Analysis

The cells were synchronized by selective mitotic detachment, as described previously (13). Approximately 90% of the detached cells were observed by phase-contrast microscopy to be in mitosis immediately following harvesting. The detached cells were maintained in culture at 37°C for 18 hours. During this period, smears were prepared in triplicate from the same culture at 1, 3, 5, 8, 12, 15 and 18 hours after synchronization. The triplicate smears for each post-detachment time were dried for one hour and then fixed for 30 minutes in a mixture containing 85% methyl alcohol, 10% formalin and 5% glacial acetic acid. Hydrolysis treatment was performed in 1N HCl at 60°C for 15, 60 and 120 minutes respectively. The smears were stained with the Schiff reagent for 60 minutes following the procedure of DeCosse and Aiello (14). After staining, the samples were mounted in Canadian Balsam. The smears were prepared with particular attention since prolonged hydrolysis time could lead to solubilization of DNA, quite sensitive to fixation and air drying. Furthermore, in order to avoid any systematic differences due to reagents, all fixatives, hydrolysis, and staining of triplicate specimens were physically conducted in parallel in the same reagents, and the same reagent stocks were used for triplicate specimens sampled at other times. Subsequent analysis proved that the geometric and densitometric data, obtained from the three smears for each sample, were highly reproducible.

Illumination was provided by a 100w tungsten halogen lamp and a highly regulated direct current power supply using a 546nm filter (40nm half-backwidth, Fish-Schurman, New Rochelle, N.Y.). Shading error was minimized by use of a shade corrector. A flat field was produced by use of a 100x planar achromatic oil objective of 1.25 n.a. and open iris with internal magnification of 10x produced by a Reichlert high-quality magnification changer. The coefficients of variation for IOD (integrated optical density) and area measured for a single nucleus positioned at 6 locations around the border of the field and in the center were 1.0 per cent for I.O.D. and 2.5 per cent for area. Each slide was used as its own blank to define 0.0 O.D. and densitometer calibration was then checked by means of a neutral density filter. Both the gain of the video amplifier and the densitometer zero level were continuously referred to peak white during measurement by means of a Servo system which can typically maintain measured values within 1% over a twenty-four hour period. Besides IOD several geometrical parameters were measured (15): nuclear area and perimeter will be presented in this paper. At least 100 cells were evaluated per slide. The raw data were loaded on a PDP11/40 computer for further analysis. A detailed description of our image analysis procedures as well as the data processing had been given elsewhere (15, 16, 21).

Flow Microfluorimetry (FMF)

HeLa S-3 cells logarithmically growing in suspension culture
were prepared for measurement in the Cytofluorograph (Bio/Physics
Systems Inc., Mahopac, N.Y.) using the fluorescent Feulgen reaction adapted for flow microfluorimetry by Truillo et. al. (8).
Basically, the cells were removed from culture medium by centri-
fuging at 250xg for 4 minutes. The pellet was washed in cold
saline, the cells were resuspended in saline containing EDTA
(0.5 mM) and trypsin (0.1 mg/1.0 ml saline) and fixed for 18 hours
at 0°C. Similar results were found using either pure alcoholic
fixative (80% methanol) or 20% formaldehyde. Several hydrolysis
treatments were applied, specifically 1N HCl at 60°C for 5, 10,
20, 60 and 120 minutes, 4N HCl at 25°C for 15, 60, 120 minutes and
5N HCl at 37°C for 15, 60, and 120 minutes respectively. The
cells were stained for 20 minutes at room temperature with acriflavine HCl (500 mg $K_2S_2O_5$, 10 ml 0.5 N HCl 90ml 0.02% aqueous
acriflavine HCl Allied Chemical). Non specifically bound dye was
removed afterwards by washing three times with acid-alcohol (1ml
12N HCl in 100 ml 70% ethanol). The performance of the FMF was
checked before and after each series of measurements by means of
acridine orange-stained calf thymocytes. This procedure was used
to insure that no detectable instrumental drift had occurred
during the run and to insure that the instrument itself produced
a constant coefficient of variation of approximately 5.5% for all
of the experiments herein reported. Each hydrolysis treatment was
conducted in parallel on 3 different samples: measurements on
each stained sample were then repeated 3 times at various time
intervals, yielding a peak position and coefficient of variation
quite reproducible (with an overall variation of 0.5%). Changes
in fluorescence intensity due to internal staining variability,
were therefore minimal.

Computer Determination of Cell Cycle Parameters

In order to decompose the histogram to determine the fraction
of G_1, S and G_2+M cells, we took our basic method from the work
of J. Fried (12). Basically the method involves fitting the
experimental histogram by a sum of Gaussians, one for G_1 cells,
one for G_2+M cells, and a predetermined number for cells in S
phase. The problem becomes one of estimating the means, standard
deviations, and amplitudes of the Gaussians. As previously
described (12), we assume that the ratio of the mean G_2+M (DNA)
channel position to the mean G_1 position is 2, and that the coefficient of variation (C.V.) of all the Gaussians is constant.
After determining the C.V. by taking ½ of the full width of 0.67
of the height of the G_1 peak, the heights of the Gaussians are
determined by utilizing a global optimization routine, which

allows the entire method to be programmed on our PDP11/40 computer.
More details are presented in other communications (18). Since the
basic mathematical model is ill-conditioned (18) we can show that
equally good fits, both by visual observation and chi-square analysis, can be obtained by changing the initial values of the
parameters (G_1 peak position and functions). Furthermore, mathematical analysis of the problem shows that as the C.V. of the
Gaussians increases, the variance of the estimated parameters
drastically increases (18). Thus, several runs have been conducted both by varying the fit of the model to the same set of
data or by the same fit to different samples from the same Feulgen-
stained population, and report the mean and standard deviation
of the parameters obtained by the method (See Tables I-II).

RESULTS

Image Analysis

The dependence of integrated optical density on hydrolysis
time using 1N HCl at 60°C is demonstrated in Fig. 1. The IOD
values are means of approximately 100 cells. The error bars
designate 95% confidence intervals. The 4 curves correspond to
HeLa cells at 1 hour, 5 hours, 8 hours and 12 hours after selective mitotic detachment. At these times the main fractions of the
cell population were respectively in early G_1, late G_1, early S
and middle-late S phase respectively (15,17). The curves show a
monotonic decrease of mean IOD with increasing hydrolysis time
but reveal a statistically different significant dependence from
hydrolysis time for different times after mitosis, suggesting that
cells in different phases (or sub-phases) of the cell cycle are
differentially affected by Feulgen hydrolysis, perhaps related to
a specific difference in chromatin morphometry (15,16).

Figures 2 and 3 demonstrate that both the geometric parameters, nuclear area and perimeter also show a unique differential
hydrolysis dependence for each time interval elapsed after mitosis.
These two geometric parameters are differently affected by increasing hydrolysis time: i.e., at the same time after mitosis,
the area of cells in early S-phase (8 hours) decreases slightly
with increasing hydrolysis time (Fig. 2) while perimeter shows a
marked increase (Fig. 3).

These differential effects of hydrolysis time on cells in
different phases of the cell cycle have an impact on quantitative
evaluations of integrated optical density (DNA content) as it
appears evident in Fig. 1 and 4.

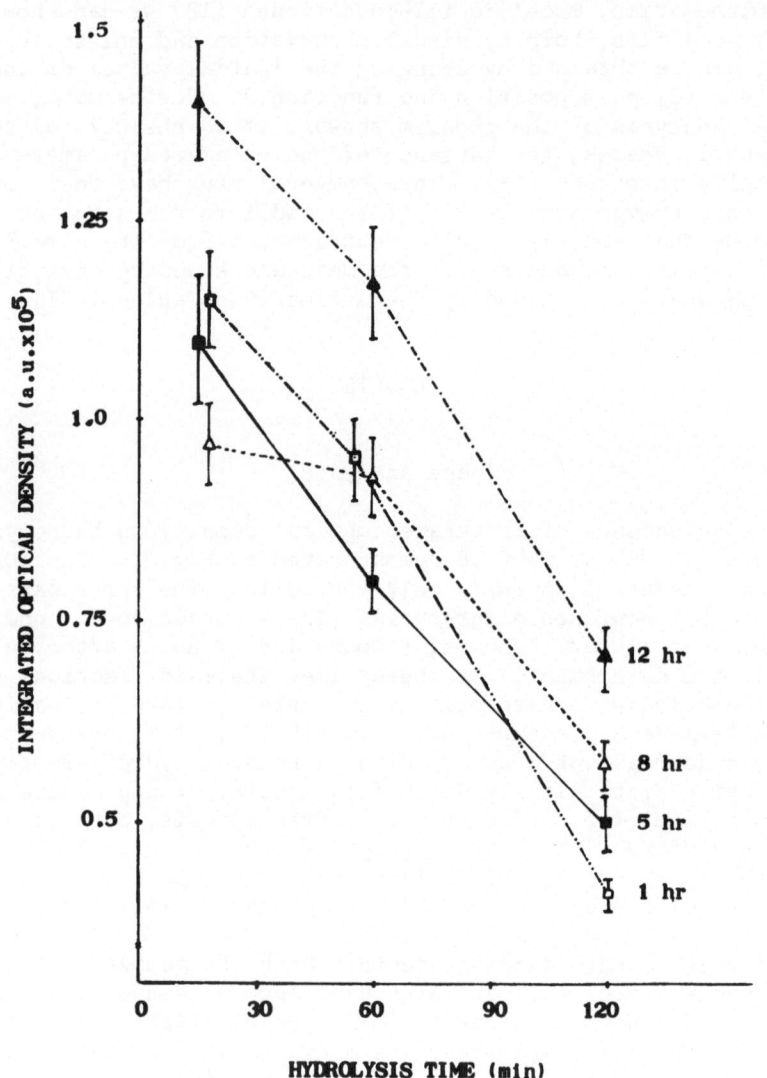

Fig. 1. Integrated optical density (means ± 95% confidence
intervals) of HeLa cells versus hydrolysis time at different
times after mitotic detachment: 1 hour (◻---◻), 5 hours (◼---◼),
8 hours (△---△), 12 hours (▲---▲). Hydrolysis was performed in
1N Hcl at 60°C.

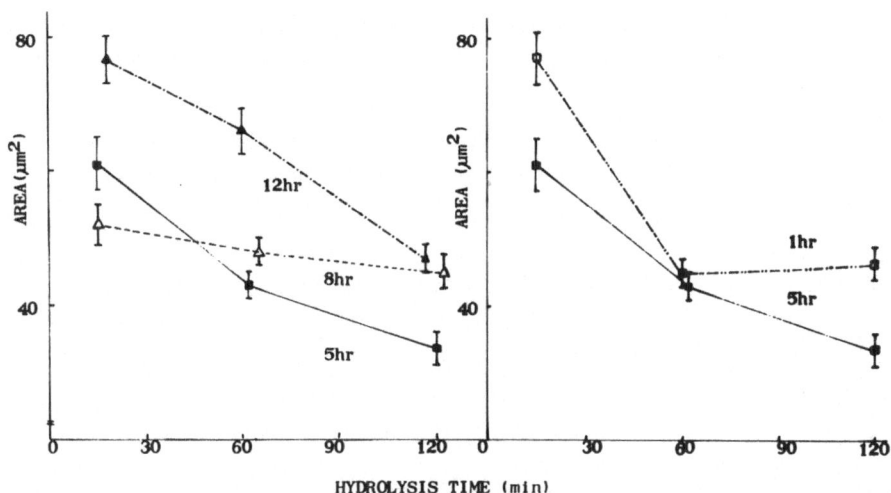

<u>Fig. 2</u> Nuclear area of HeLa cells (means ±95% confidence intervals)
versus hydrolysis time in 1N HCl at 60°C at different time intervals
after mitosis. (□–••–□) 1 hour, (■——■) 5 hours, (Δ---Δ) 8 hours,
and (▲–•–•–▲) 12 hours after mitosis.

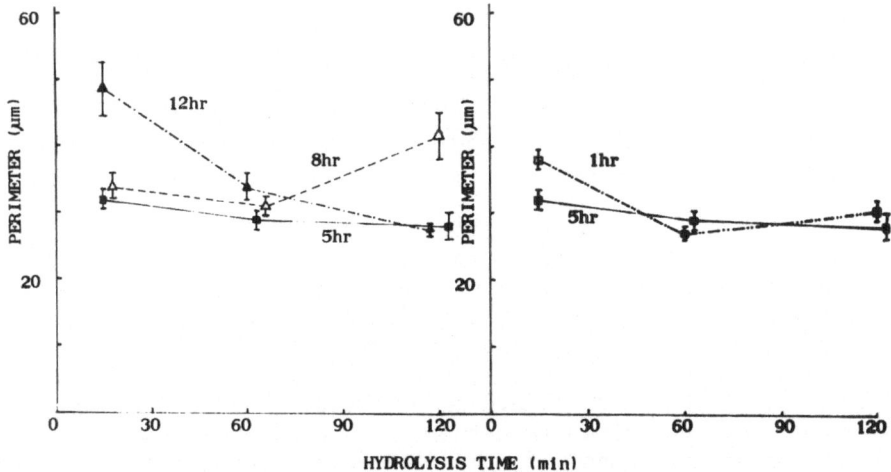

<u>Fig. 3</u> Nuclear perimeter (means ± 95% confidence intervals) of HeLa
cells versus hydrolysis time at different times after mitotic
detachment, as Figure 2.

Figure 4 presents IOD (DNA) distributions of HeLa cells 12 hours after mitosis using 15, 60 and 120 minutes of Feulgen hydrolysis. The number of cells is plotted versus the IOD (expressed in arbitrary units). A Kolmogorov-Smirnov statistical test carried out on the three frequency distribution proved that significant differences exist between these IOD histograms obtained at 3 different hydrolysis times. The first peak of each histogram corresponds to G_1 cells (DNA content X), the second peak to (G_2+M) cells (DNA content 2X). The cells recorded between the two peaks are S cells with variable DNA content between X and 2X. Figure 4A gives the IOD histogram in the same scale showing the decrease of IOD with increasing hydrolysis time, while in Figure 4B the IOD scale is normalized in order to display the G_1 peaks at the same vertical position (unit 15). As is well known (17), at 12 hours after mitosis the degree of synchrony has decreased such that a major fraction of the cells is in S phase, but there are considerable fractions of cells also in the G_1- and (G_2+M)-phases. Figure 4B reveals that with increasing hydrolysis time the IOD distribution is changed drastically; the fraction of cells with higher IOD (DNA content 2X) decreasing with increasingly hydrolysis time (see Table I). The implications of this finding for a quantitative estimation of DNA content, using Feulgen staining on cells with the same ploidy level, but different degrees of chromatin supercoiling, are evident. In this communication, we further explore the implication of this finding on the quantitative cell cycle analysis from cytophotometry of Feulgen-stained cells.

Laser Flow Microfluorimetry

The performance of the FMF was checked before and after each series of measurements using acridine orange-stained calf thymocytes in order to maintain a constant coefficient of variation (C.V.) of approximately 5.5% throughout all experiments herein reported.

Decomposition of the DNA histograms (See Materials and Methods) yields the fraction of cells in each phase of cell cycle: regardless of both the low χ^2 and the perfect overlap of theoretical and experimental frequency distribution of fluorescence per cell (18), these model's estimates are mostly given to infer trends in the data. Table II shows that C.V. of the G_1 peak for log-phase HeLa cells Feulgen-stained in 4N HCl at 25° C for 5 minutes is about 11.3%, compatible with the 10-11% previously reported under the same conditions with a similar cytofluorograph (9). Better C.V. can be obtained using our B-D Cell Sorter but the FMF distributions show a similar dependence from Feulgen hydrolysis. Figure 5 demonstrates that in agreement with our image analysis data and previous findings (19), the modal fluorescence

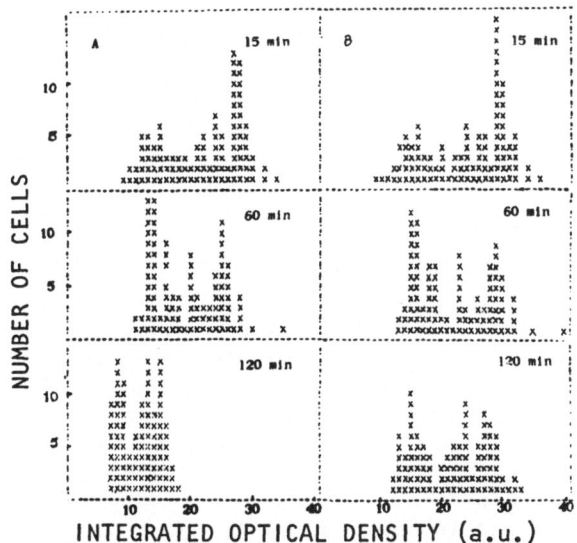

Fig. 4 IOD (DNA) frequency distributions of HeLa cells 12 hours after mitosis: A) IOD scale in absolute value (a.u.), B) different IOD scales in order to have the G1 peaks lined vertically.

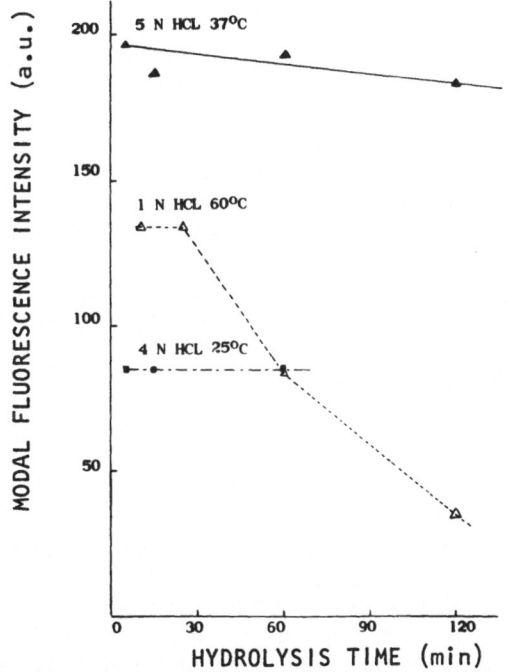

Fig. 5 Modal fluorescence intensity of log-phase HeLa cells versus hydrolysis time for 1N HCl at 60°C (Δ----Δ), 4N HCl at 25°C (■---■), and 5N CHl at 37°C (▲---▲). The ordinate gives the channel number (a.u.) of the G1 peak in the FMF fluorescence (DNA) histograms).

intensity of the acriflavine Feulgen-stained HeLa cells is also
strongly dependent on HCl normality and hydrolysis time. The
values were taken from FMF DNA histograms of the HeLa cells in
log-phase. The ordinate gives the channel number on the multi-
channel analyzer of the FMF for the G_1 peak of the fluorescence
intensity (DNA) histogram, the abscissa gives the hydrolysis time.
For 1N HCl, 60°C we find a rapid decrease of fluorescence intensity
corresponding to the rapid decrease in IOD shown in Figure 1.
After hydrolysis of 4N HCl at 25°C and 5N HCl at 37°C, the fluorescence intensity is respectively constant or slowly decreasing
during the time interval studied, the hydrolysis in 5N HCl at
37°C yields approximately double fluorescence intensity, compatible with previous absorbance measurements (19).

Figure 6 shows few representative histograms of log-phase
HeLa cells after acriflavine Feulgen-staining using different
HCl normality and hydrolysis time; respectively, in order of de-
creasing fluorescence intensity: A) 5N HCl at 37°C for 5 minutes,
B) 1N HCl at 60°C for 10 minutes, C) 4N HCl at 25°C for 60 minutes,
D) 1N HCl at 60°C for 120 minutes.

Visual analysis indicates that a decrease in the relative
intensity of the G_1 peak, with increasing HCl normality and/or
hydrolysis time is accompanied by a relative decrease of the
fraction of cells in the second (G_2+M) peak with corresponding
increasing of the fraction of cells in the G_1 peak using the same
log-phase HeLa cell population. Simultaneously, the coefficient
of variation of the G_1 peak has increased considerably, with
decreasing modal fluorescence. All these empirical observations
are confirmed by a rigorous computer analysis of the same Feulgen-
stained HeLa cell population, hydrolyzed at different temperatures,
HCl normalities and times. The mean values for each fraction of
cell cycle phases are obtained by a least square fit of the DNA
histograms to the mathematical model previously described (see
Table II); they are intended mostly to infer trends in the data,
since occasionally the C.V. is larger than 15% (not due to a
reduced performance of the cytofluorograph, which was constantly
monitored against the standard calf thymocytes). The same trends
are indeed substantiated by automated image analysis, which does
not suffer the same pitfalls.

All these studies have been carried out in the wide range of
temperature, time and acid concentration, commonly utilized;
particular attention was given to the short time scale (5 - 10
minutes), changing temperature and HCl normality.

Figure 7 summarizes the findings obtained, either by laser
flow microfluorimetry (upper panel) or automated image analysis
(lower panel), on the effect of varying Feulgen hydrolysis
(either time, temperature or HCl normality) on the same log-phase

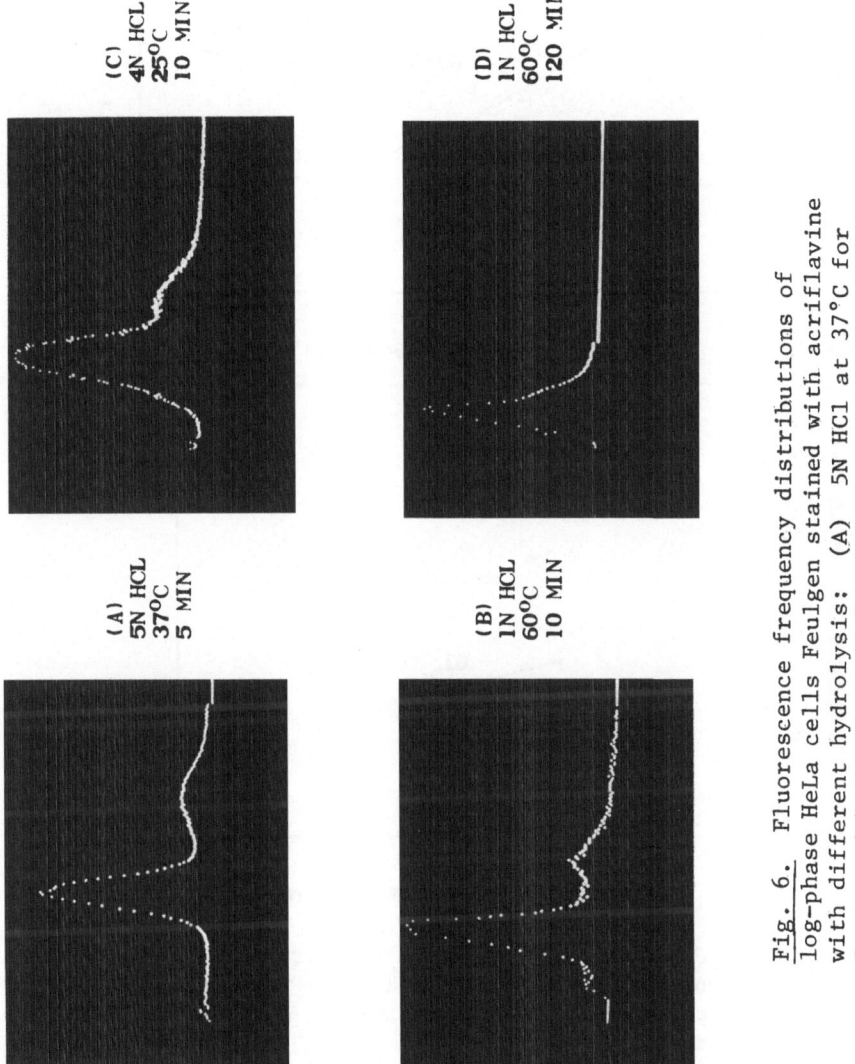

Fig. 6. Fluorescence frequency distributions of log-phase HeLa cells Feulgen stained with acriflavine with different hydrolysis: (A) 5N HCl at 37°C for 5 min., (B) 1N HCl at 60°C for 10 min., (C) 4N HCl at 25°C for 60 min., (D) 1N HCl at 60°C for 120 min.

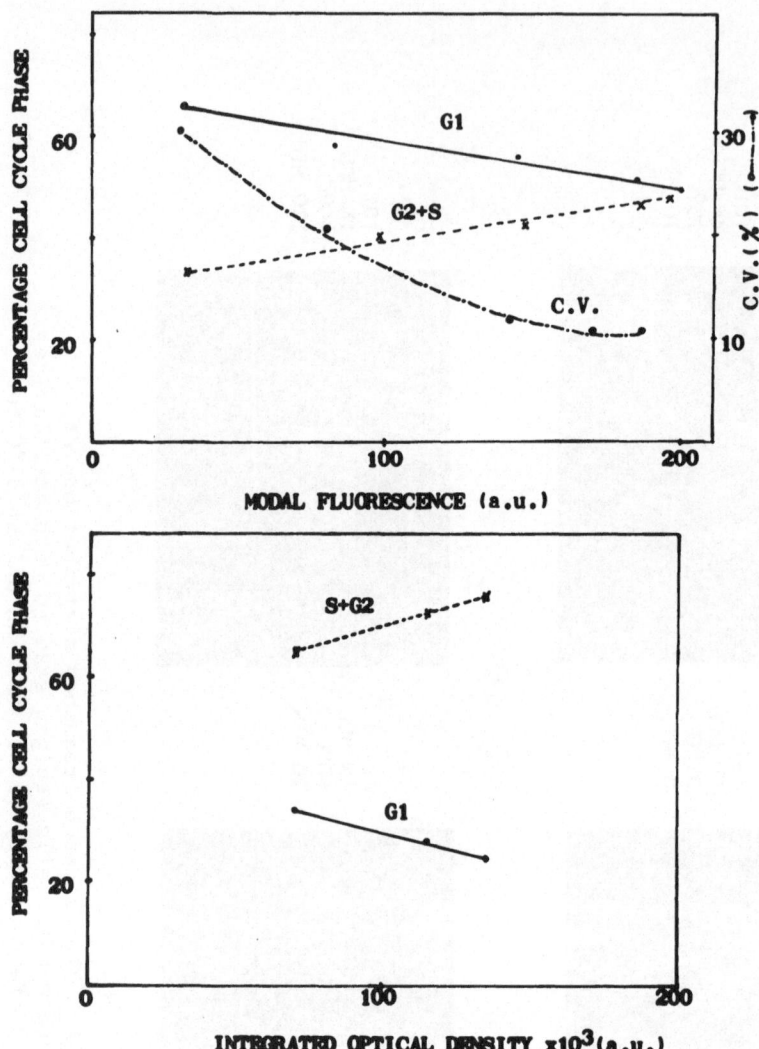

Fig. 7 Upper: Percentage of cells in Gl (._____.), G2+S (x----x),
and coefficient of variation of Gl peak (o-.-) versus modal fluorescence of the Gl peak, obtained by the same log-phase HeLa cell population hydrolized at different extents (Fig. 5 and Table II   using
laser flow microfluorimetry).

Lower: Percentage of cells in Gl (._____.) and G2+S (x----x) versus
integrated optical density, determined on the same HeLa cell population 12 hours after mitosis, using automated image analysis.
The fraction of cells in mitosis in both cases is less than 1%.

HeLa cell population, containing a constant mixture of cells in different phases of the cycle.

A decrease of the modal fluorescence per cell measured by laser flow microfluorimetry (Fig. 7, upper panel), is accompanied by an increase in the "computed" fraction of G_1 cells at the expense of a significant decrease in the computed "G_2+S" fraction. Identical phenomena is observed by the independent method of automated image analysis (Fig. 7, lower panel), which measures integrated optical density, i.e. DNA absorbance.

Another common feature of these two independent measurements is the dramatic increase of the coefficient of variation of the G_1 population: (Fig. 7 upper panel) for laser flow microfluorimetry, and Table III for image analysis.

DISCUSSION

The idea that Feulgen-staining may be influenced by chromatin has been around for a long time as it can be seen from recent excellent reviews on the subject (1,2). This combined study of Feulgen-stained densitometric texture analysis (15,16,21) and laser flow microfluorimetry (8,9), specifically characterizes the differential staining effect of Feulgen hydrolysis on the various compartments of the cell cycle. Several profound implications can be drawn from these findings. First, the Feulgen stain, which also reflects the state of chromatin (1), cannot be considered a quantative method for measuring DNA content when cells with different metabolic activity and extent of proliferation are compared. G_2 cells, which have twice the amount of DNA with respect to G_1 cells, show a decreased ratio in their relative absorbance, with increasing hydrolysis (Fig. 4), or fluorescence (Fig. 6), as indicated by the more pronounced decrease in "G_2" peak and consequent apparently increased fraction of G_1 cells in a frequency distribution of a log-phase population (Table I-II): this is compatible with the differences in chromatin morphometry between G_2 and G_1 cells, which have been previously reported (15,21). Similarily, cells with the same G_1 DNA content show a differential hydrolysis dependence of densitometric (Fig. 1) and geometric (Fig. 2 and Fig. 3) parameters of Feulgen-stained synchronized HeLa cells at 1 hour (early S) after mitosis: this differential effect is responsible for the increase in the coefficient of variation (Table II-III) of the G_1 population and is compatible with a significant variation of chromatin morphometry and supercoiling (23) at the late G_1- early S transition and within the G_1 phase, which can now be subdivided into three subphases (15,21). This indicates that the Feulgen stain is related not only to the amount of DNA, as previously suggested, but also,

in a specific manner to DNA conformation within the intact nucleus.
Indeed, this property can be properly utilized to objectively
characterize, at the optimal hydrolysis time and acid concentration,
cells with the same DNA content, but different chromatin function
and structure, i.e., proliferating (G_1) and non-proliferating
(G_0) cells (16). Our findings suggest reasons for the strong
hydrolysis dependence reported for tumor cells with respect to
lymphocytes (4) and the lower yield of DNA absorbance of granulocytes and lymphocytes (up to 15% under the expected diploid value
(5-7)).

Finally, the differential hydrolysis dependence of Feulgen
reaction, due to differences in chromatin supercoiling within the
cell cycle phases and subphases (15-17) has implications for flow
microfluorimetric (FMF) analysis of cell kinetics. As can be
visualized by Figures 5-7, not only the relative fluorescence intensity of the G_1 peak (Fig. 5), but also the frequency distribution of fluorescence intensities (Fig. 6) and the computed cell
cycle parameters (Table II) drastically change by varying either
the HCl normality, the temperature or the time of hydrolysis on
the same HeLa cell population.

This implies that cell kinetic data collected in various
laboratories on similar biological systems, by FMF analysis on
acriflavine Feulgen-stained cells, can be compatible only if obtained under the same hydrolysis conditions: the optimal condition
is achieved with 5N HCl at 37°C; where A) the fluorescence intensity per cell is larger and slightly dependent on hydrolysis
time; B) the coefficient of variation of the G_1 peak is minimum
(extremely important for an accurate computation of cell cycle
parameters). The 5 N HCl at 37°C also gives the closest correspond-
ence between autoradiographic (not shown) and FMF estimates of
the cell cycle parameters.

A general trend can be found in both laser flow microfluorimetric (Fig. 5-6, Table II) and densitometric image analysis
(Fig. 1-4, Table I and III), regardless of the normality, temperature and time: the decrease in relative intensity in either
fluorescence (Fig. 7, upper panel) or absorbance (Fig. 7, lower
panel) is systematically accompanied by a significant decrease of
the computed G_2+S fraction, with consequent increase of computed
G_1 fraction from the same HeLa cell population. In both cases
the spread of the G_1 population significantly increases, compatible
with the existence of at least 3 subcompartments in G_1 phase with
different chromatin structure and therefore with different hydrolysis dependence (15,21). Furthermore, the most commonly utilized
Feulgen-staining, i.e., using 4N HCl at 25°C (in FMF analysis) and
1N HCl at 60°C (in cytopathology), seem to be quite limiting because of the lower yield of fluorescence (or absorbance), the

higher coefficient of variation and (for 1N HCl) the critical dependence on hydrolysis time. Since identical data were obtained
in alcoholic fixatives, either with or without formalin, this
conclusion is compatible with recent studies (20), using alcoholic
fixatives containing no formalin, on the exposure and removal of
aldehyde groups during Feulgen acid hydrolysis, which seems optimal
at temperatures slightly above room temperature in combination with
high acid concentration (20). Finally, maybe as a consequence of
the significant variations in chromatin supercoiling (23) and
nuclear morphometry (15,21) between and within cell cycle phases
the inherent limitation of densitometric and FMF determination of
cell cycle parameters appears evident in Feulgen-stained cells.
In that this limitation, accompanied by the pitfalls of the various
mathematical models (10-12, 18) (including ours which is used only
to infer trends in the data), makes imperative the need for alternative approaches to cell kinetic studies, particularly by means of
physical techniques which provide more direct physical observables,
by mathematical models which more faithfully reflect the characteristics of the various subphases of cell cycles, and by the mathematical analysis of time sequence of FMF distribution in perturbed
and unperturbed cell populations (22, 24, 25).

TABLE I

Percentage of cells by automated image analysis in a given
phase (G_1, S, G_2+M) of the cell cycle from the same HeLa cell population 12 hours after selective mitotic detachment (17), for
each hydrolysis time in 1N HCl at 60°C (Fig. 4).

The mean value and standard deviations are obtained as described in Materials and Methods.

HYDROLYSIS TIME	G_1 (%)	S(%)	G_2+M(%)
15 minutes	25 ± 3	40 ± 3	35 ± 6
60 minutes	28 ± 1	58 ± 3	14 ± 2
120 minutes	35 ± 2	62 ± 3	3 ± 1

TABLE II

Percentage of Cells in G_1, S, G_2+M phases by a least square fit to a mathematical model (see methods) of various Feulgen stained HeLa - S3 cells in log phase (laser microfluorimetry).

	G_1	S	G_2 + M	C.V.
5N HC1-37°C 5 minutes	49.9 (± 4.1)	36.2 (± 4.0)	13.9 (± 1.3)	10.3
5N HC1-37°C 60 minutes	50.9 (± 0.8)	37.1 (± 2.1)	11.9 (± 1.8)	12.0
5N HC1-37°C 120 minutes	61.7 (± 2.9)	24.0 (± 2.0)	14.3 (± 1.5)	11.9
1N HC1-60°C 10 minutes	57.7 (± 2.4)	24.5 (± 1.5)	17.8 (± 0.8)	12.3
1N HC1-60°C 120 minutes	67 (± 3.5)	26.5 (± 1.5)	6.5 (± 1.5)	31.5
4N HC1-25°C 5 minutes	52.5 (± 2.0)	30.8 (± 7.5)	16.7 (± 1.7)	11.3
4N HC1-25°C 60 minutes	56.0 (± 3.0)	36.9 (± 4.9)	7.1 (± 1.1)	21.8

TABLE III

Time hydrolysis dependence of mean value and standard deviation (in percentage of the mean) of integrated optical density (a.u.) from a population obtained by combining HeLa cells with similar G_1 DNA content but taken 1 hour (early G_1), 3 hours (middle G_1), 5 hours (late G_1) and 8 hours (early S) after mitosis. The cell population was hydrolized in 1N HCl at 60°C for 15, 60, and 120 minutes. The data were obtained by automated image analysis (see fig. 1 and reference 15).

Hydrolysis Time	Mean (a.u.)	S.D. (%)
15 minutes	110×10^3	13.5
60 minutes	88×10^3	12.5
120 minutes	52×10^3	22.8

REFERENCES

1. Ringertz, N.R. (1969) in Handbook of Molecular Cytology, ed.
   A. Lima-De-Faria (North Holland Publishing Co. Amsterdam),
   p. 658.
2. Lillie, R.D., Fullmer, H.M. (1976) Histopathologic Technic and
   Practical Histochemistry (McGraw Hill Book Co., N.Y.) P. 17.
3. Sprenger, E. (1974) in Impulscytophotometric, ed. M. Andreeff
   (Springer-Verlag, Berlin), p. 5.
4. Bohm, N. and Sandritter, W. (1966) J. Cell Biol. 28, 1-7.
5. Bohm, N. and Sandritter, W. (1975) in Current Topics in
   Pathology, 60, ed. E. Grundmann and W.H. Kirsten (Springer-Verlag,
   Berlin) p. 156.
6. Garcia, A.M. (1964) Acta Histochem. 17, 249-257.
7. Mayall, B.H. (1969) Histochem. Cytochem. 17, 249-257
8. Trujillo, T.T. and Van Dilla, M.A. (1972) Acta Cytologic   16,
   26-30.
9. Nicolini, C., Kendall, F., Desaive, C., Baserga, R., Clarkson, B.,
   and Fried, J., Cancer Treatment Rep. 60, 1818-27 (1977).
10. Dean, P.N. and Jett, J.H. (1974) J. Cell Biol. 60, 523-527.
11. Baisch, H., Gohde, W., and Linden, W.A. (1975) Rad. Environ.
    Biophysics, 12, 31-39.
12. Fried, J., (1976) "Computers and Biomedical Research", 9, 263-76.
13. Stein, G. and Borun, T. (1972) J. Cell Biol. 52, 292-307.
14. DeCosse, J.J. and Aiello, N. (1966) J. Histochem. Cytochem. 14,
    601-604.
15. Kendall, F., Swenson, R., Borun, T., Rowinski, J., and
    Nicolini, C. (1977) Science, 196, 1096-1108
16. Nicolini, C., Giaretti, W., Desaive, C., and Kendall, F.
    (1977) Experimental Cell Research 106, 119-125.
17. Nicolini, C., Kozu, A., Borun, T., and Baserga, R. "J. Biochem."
    (1975) 250, 3381-3385.
18. Wu, C.T., Toton, S., Kendall, F., Zietz, S., Linden, W., Eisen, M.
    and Nicolini, C. (1977) Pulse Cytophotometry III, 51-62.
19. Fand, S.B., "Introduction to Quantitative Cytochemistry",
    (1970), ed. Wied, G.L. and Bohr. Vol. 2, Academic Press, N.Y.
    p. 209.
20. Kjellstrand, P.T.T. (1977), The J. Histochem. Cytochem. 25,
    p. 129-134.
21. Nicolini, C., Kendall, F., and Giaretti, W., 1977 Biophysical
    Journal, 19, 163-176.
22. Scherr, L. and Zietz, S., Radiation Research 67(3), 585, 1976.
23. Nicolini, C. and Kendall, F., Differential light scattering
    in native chromatin. Corrections and Inferences, combining
    melting and dye binding studies. A two-order superhelical
    model. Physiol. Chem. & Phy. 9(3) 265-83
24. Zietz, S. and C. Nicolini Cell Tissue Kinet. (submitted, 1978)
25. Zietz, S. and C. Nicolini in Biomathematics and Cell Kinetics
    (1978) ed. A.J. Valleron, Elsevier.

SCANNING AND FLOW PHOTOMETRY OF CHROMOSOMES

Mortimer L. Mendelsohn
Biomedical Sciences Division
Lawrence Livermore Laboratory
University of California
Livermore, California 94550, U.S.A.

The chromosome deserves a prominent place in a book on chromatin structure and function because in its condensed metaphase form the chromosome achieves the highest order of chromatin structure. Of course chromosomes are interesting for many other reasons. The formation and behavior of chromosomes in mitosis and meiosis continues to be a fascinating problem in biology, chromosomes provide a clinically useful window into human genetics, and chromosomal vulnerability to breakage and subsequent functional disruption by clastogens defines a type of genetic injury which requires careful study and control

This chapter describes two approaches toward biophysical photometric study of chromosomes: one based on scanning microscopy and one based on flow cytometry. They stem from two major achievements of Tobjorn Caspersson, the development of high resolution microscopic cytophotometry[1] and the discovery of chromosome banding with the associated insights into DNA fluorochromes.[2]

BACKGROUND

An average human metaphase chromosome contains several hundred million base pairs of DNA, enough for approximately a million nucleosomes divided equally between two identical chromatids. These paired 4 cm lengths of DNA compact into a 5 by 1 by ½ micron, dense, refractile, highly stainable, resilient

341

object. Lower orders of structure within the chromatids, such as
bands and major coils, are sometimes visible in the light
microscope.

    Many properties of the chromosome can be studied optically.
Basic staining is perhaps the oldest of these and takes advantage
of electrostatic binding to DNA phosphate. UV absorbance is a
property of the bases and when carefully applied can be an
excellent measure of DNA content. Deoxyribose is selectively and
quantitatively stained by the Feulgen reaction. In this
half-century-old method, the DNA is depurinated by strong or hot
hydrochloric acid and the uncovered sugar aldehydes are stained
by one of many chromophoric Schiff reagants. The Feulgen
reaction was the standard method for cytophotometric DNA
measurement until its recent displacement by DNA fluorochromes.
Some DNA fluorochromes, such as ethidium bromide and acridine
orange, intercalate into DNA; others, such as chromomycin A3 and
Hoechst 33258, bind to the grooves. The fluorochromes are
comparatively easy to use, their few constraints allow them to be
combined readily with other cytochemical methods, and as will be
shown below, they have useful and interesting differences in
their affinities to various DNA's. Fluorescent antibodies to (a)
single stranded DNA, (b) BrdU in DNA and (c) thymine dimers and
other photoproducts of DNA can be used to stain these DNA
components specifically. Repetitive DNA sequences can be marked
by RNA or DNA complementation using stains or autoradiography.
At an even higher level of organization, chromosomal bands are
elicited by quinacrine staining and by several absorbent or
fluorescent stains combined with enzymatic or chemical
degradation of the chromatin. Finally, the chromosome itself can
be measured for length and area and can be classified to varying
degree by its size and the location of its centromere. Size
properties unfortunately are highly relative because chromosomes
continue to compact (i.e., become shorter and denser) as cells
are held in metaphase.

    Over the past 15 years, first at the University of
Pennsylvania and now at the Lawrence Livermore Laboratory, the
approach my colleagues and I have taken to chromosome measurement
is to seek relatively invariant chromosomal properties,
particularly properties based on DNA content, and to use these to
identify chromosomes and to study their stability under normal
and abnormal conditions. We began with image-analytic,
absorbance measurement of chromosomes on slides.

CHROMOSOMES ON SLIDES

    Analysis of DNA content of individual chromosomes of a well
flattened, properly stained, metaphase cell involves application

of standard photometric principles to objects which have been morphologically isolated by image analysis.[3]

The Beer-Lambert Law relates intensities (I_0 = incident intensity, I = transmitted intensity) of monochromatic light to the concentration (c), pathlength (l) and absorptivity (k) of a uniformly distributed chromophore.

$$\frac{I}{I_o} = e^{-kcl}$$

The equation is often used in its logarithmic form

$$- \log \frac{I}{I_o} = OD = kcl$$

where OD is the optical density or absorbance. For cytophotometry it is convenient to restate the Law in terms of the mass of chromophore (m) in the measuring field and b the area of the field.

$$OD = k \frac{m}{b}$$

$$m = OD \; \frac{b}{k}$$

Obviously objects such as chromosomes do not have uniformly distributed chromophore. The presence of resolvable differences of grayness within a measuring window is equivalent to averaging intensities; because of the exponential relationship in the Beer Lambert Law, such averaging results in a negative error in measured mass of chromophore. This distributional error increases with heterogeneity and optical density and can well be 30% for deeply stained chromosomes. Scanning microscopy essentially eliminates distributional error by making each measurement of optical density in a spot or window which is at the limit of resolution of the optics and hence shows no resolvable internal differences in grayness. Many such measurements cover the entire object in a regular raster and the sum of the optical densities gives a true measure of stain content (M) of the object

$$M = \frac{r \cdot b}{k} \, \Sigma \; OD$$

where r corrects for overlaps or gaps in the scanning raster.[4]

Image analysis is required in this process to limit the summation of optical densities to the region representing the single chromosome or chromosome part. The image in this case is

a high resolution digital image of a metaphase cell. In our work
it consists of a 200 by 200 raster of 40,000 points; each point
is separated by 0.25 μ and contains up to 8 bits (256 levels) of
optical density information. We obtain the image from a
microscope slide in six seconds using a flying-spot cathode-ray
scanner operating near the limit of optical resolution in the
blue region of the visible spectrum. The high quality of such a
digital image is shown in Fig. 1.

A well prepared metaphase cell is a compromise between
keeping all the chromosomes sufficiently close to establish that
they are from the same cell and sufficiently separated to
minimize overlapping and touching chromosomes. Because
chromosomes are inherently unbounded objects and are seen through
diffraction limited optics, they present broad grayness profiles
that are roughly Gaussian.[5] The tails of these profiles extend

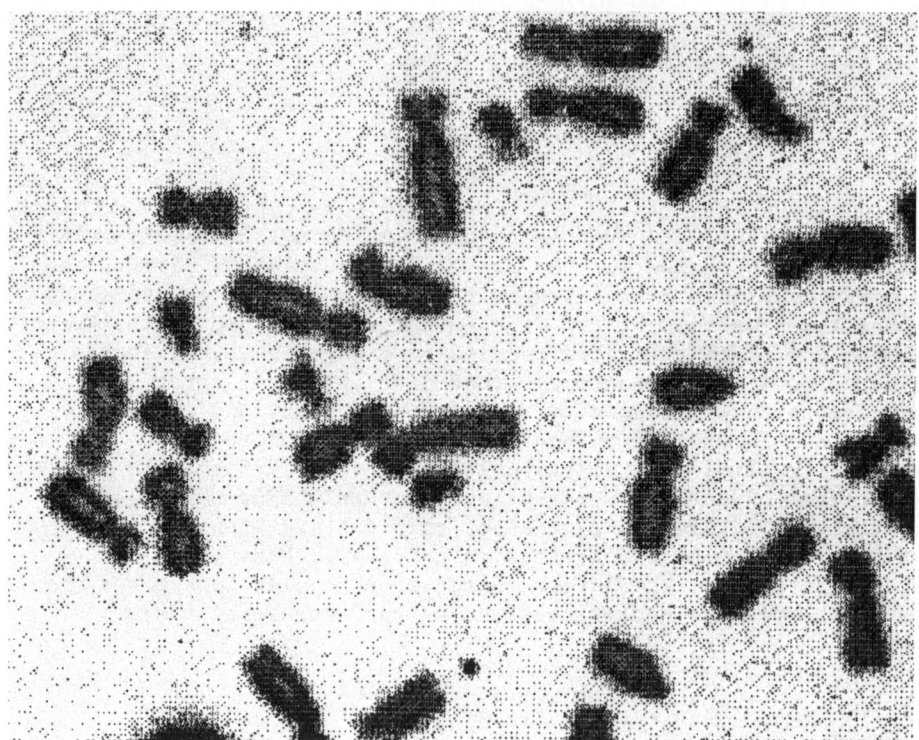

Fig. 1.   A digital image reconstituted from a partial scan of a
          human metaphase cell. The chromosomes are stained with
          gallocyanin-chrome alum and their grayness represents
          DNA content. This image is near the limit of optical
          resolution of the light microscope.

a micron beyond what appears to the eye to be the chromosome core
and actually contain a third of the chromosome's optical
density. We program a general-purpose digital computer to core
the chromosomal images by finding the grayness level that
corresponds to the inflection point of each chromosome's grayness
profile. At this point, human editing is used interactively with
the computer to correct occasional touching or fragmented cores.
A one-micron region around the core is then added to each
chromosome, and points claimed by more than one core are assigned
to the nearest core. Finally a sampling of unclaimed points in
the vicinity of each chromosome is used to estimate the
background. The computer is now ready to sum the optical
densities, subtract out the background and store the definitive
estimate of relative mass of chromophore.

The centromere of a chromosome can be identified either as
the region of narrowest width or by tracking the chromatids back
to their intersection. We prefer yet a third method based on the
relative paucity of DNA in the centromeric region. The computer
finds the axis of the chromosome, divides the chromosome into
strips perpendicular to the axis, sums the optical density in
each strip, searches for the dip in optical density in the
centromeric region, fits a least-squares quadratic to the dip,
and uses the minimum of the function to define the centromere to
a resolution of 0.1 micron.[6]

Typically, we analyze ten metaphase cells (460 chromosomes)
from an individual. The cells are obtained from blood cultures
and prepared by conventional methods. They are first stained
with quinacrine, photographed and analyzed by eye using banding
patterns to classify each chromosome. They are then destained,
treated with ribonuclease and restained with gallocyanin-chrome
alum, a basic stain with good stoichiometry and reasonable
specificity for DNA phosphate. The scanning and computer
analysis follow, and finally the data are stored and manipulated
in a large computer file. The throughput is slow: about one
individual can be studied per week.[3]

A typical result is shown in Fig. 2. By combining DNA
content and centromeric index (the ratio of DNA contents of the
long arm and the total chromosome), the computer can characterize
the chromosomes well enough to classify them into 20 groups.
This performance is better than the eye can do on non-banded
chromosomes, but is not competitive with banding which generally
allows classification of each chromosome. This is a secondary
issue in our studies because the chromosomes are already
classified and the main point is to cumulate and interpret the
quantitative data on DNA content.

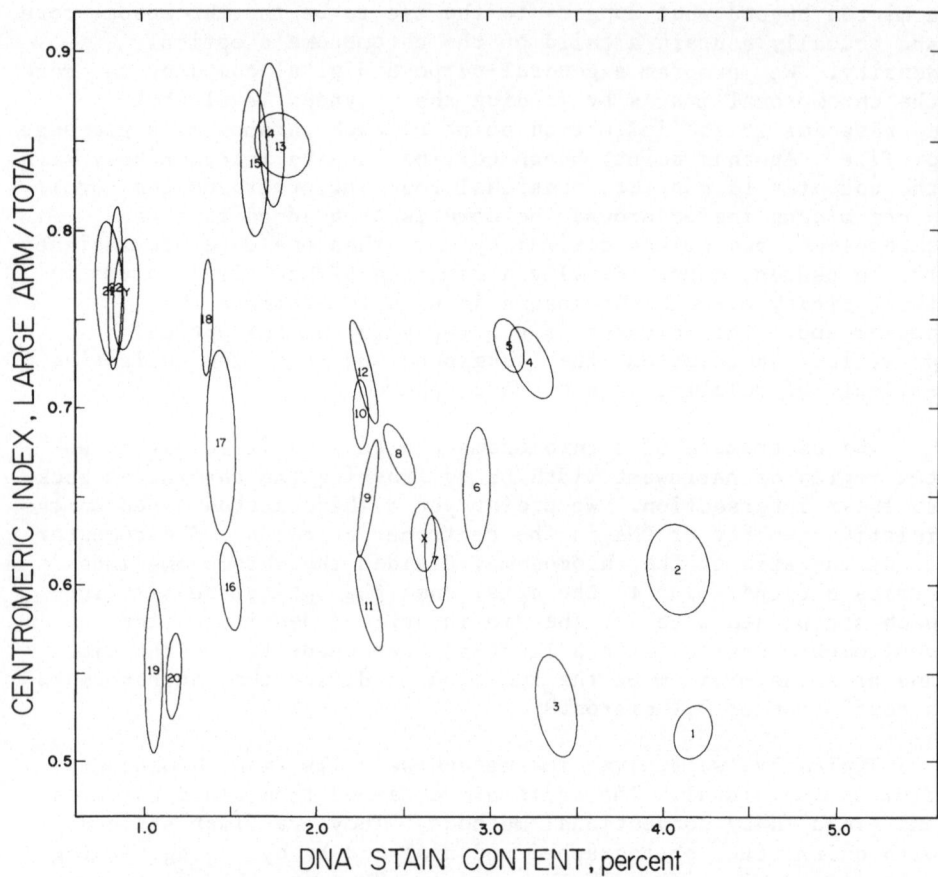

Fig. 2   The DNA content and centromeric index of the chromosomes
         from 10 metaphase cells of a normal man.   Each
         preidentified chromosome type is shown as the 50%
         confidence ellipse centered on the mean for the type.
         This general pattern is highly stable among normal
         humans.   On the abscissa 100% is the sum of DNA contents
         of the 44 autosomes.

How good are these measurements?   In replicate scans, made
after repositioning the cell, multiple measurements of the same
human chromosome give an average standard deviation of
replication of 2.5% for DNA content and 2.3% for centromeric
index.   Standard deviations between homologs or among chromosomes
of the same type are 4.4% for DNA content and 3.5% for
centromeric index.   Standard errors for means are correspondingly
smaller.   Thus we estimate that in a sample of 10 cells we can
detect mean deviations in DNA content of average chromosomes of

three-ten thousandths of a genome or 3 million base pairs. By molecular-genetic standards we are still talking about a gross property, but by cytological standards this is extraordinary resolution. To achieve this error rate we must normalize the measurements within each cell to the cumulative stain content of all 46 chromosomes. This is because cellular DNA stain content varies by 6.9%. Such large variation is typical of cytochemical measurements on slides; it is poorly understood but perhaps reflects the subtle effects on stoichiometry of drying, attachment to the glass, surrounding cytoplasm and fixation of cells and chromosomes.

The table shows the means and the standard deviation of the means within individuals and among individuals from our standard human data set.[7] The means differ only slightly from corresponding data on chromosome length, but the error bars for photometry are much smaller than for length measurements. Perhaps the most intriguing and perplexing result of the photometry is the appearance of several outliers in the chromosomes from each individual. About half of the outliers are associated with a morphologically identifiable change, such as in banding or in centromeric heterochromatin. For example, in one normal individual a large, brightly fluorescent satellite is

STANDARD VALUES FOR DNA-BASED MEASURES OF THE HUMAN KARYOTYPE

	Normalized optical density			Centromere index		
	Mean	SD within	SD among	Mean	SD within	SD among
1	4.295	0.141	0.198	0.518	0.012	0.016
2	4.190	0.129	0.163	0.612	0.016	0.022
3	3.482	0.116	0.128	0.541	0.018	0.022
4	3.336	0.131	0.121	0.727	0.019	0.017
5	3.183	0.130	0.089	0.730	0.019	0.018
6	2.984	0.101	0.161	0.647	0.018	0.021
7	2.769	0.105	0.114	0.625	0.019	0.016
8	2.515	0.120	0.132	0.687	0.025	0.021
9	2.371	0.108	0.173	0.651	0.024	0.023
10	2.355	0.112	0.065	0.696	0.023	0.034
11	2.335	0.079	0.080	0.600	0.023	0.025
12	2.319	0.096	0.080	0.731	0.021	0.033
13	1.896	0.031	0.156	0.857	0.024	0.048
14	1.781	0.090	0.139	0.852	0.025	0.039
15	1.728	0.077	0.010	0.843	0.024	0.052
16	1.608	0.070	0.125	0.599	0.028	0.041
17	1.471	0.064	0.080	0.695	0.031	0.049
18	1.395	0.061	0.061	0.765	0.030	0.033
19	1.082	0.053	0.087	0.554	0.032	0.041
20	1.160	0.062	0.091	0.581	0.042	0.046
21	0.830	0.056	0.074	0.772	0.036	0.039
22	0.888	0.053	0.116	0.771	0.038	0.046
X	2.659	0.114	0.154	0.624	0.033	0.041
Y	0.918	0.056	0.086	0.766	0.055	0.018

present in every cell on one chromosome 21 and associates with a
20% increase in DNA content of that chromosome. But for the
other half of the deviants, we see no change in the chromosome
and have only the DNA measurement to go on. These variants occur
at or below the 0.01 level of significance at a frequency of 1%
and hence could be statistical in nature. However, preliminary
studies in families suggest that over 90% of the outliers in
children can be traced back to one or the other of the parents,
indicating the high heritability and the possible reality of most
of the deviant chromosomes. We presume the deviations are due to
repetitive and other non-informational DNA. Whatever their
cause, they provide both an opportunity and a difficult challenge
to our approach. The opportunity is to understand more about the
phenomenon of DNA constancy and its exceptions, and the challenge
is to learn how to set normal limits and apply chromosomal DNA
measurements in the face of such stable, apparently normal
outliers.

CHROMOSOMES IN SUSPENSION

Beginning four years ago[8], we have been exploring a new and
dramatically different approach to cytogenetics based on flow
cytometry and sorting of fluorescently stained chromosomes.

Fluorescence is a more complex and information-rich process
than absorbance. Its measurement is affected by the phenomena of
fading, quenching and energy transfer, all of which have no
counterpart in absorbance; but because fluorescence provides a
positive signal against a near-zero background it can be much
easier to measure than absorbance. In fluorescence, the
absorbance of exciting light follows the Beer-Lambert Law and is
potentially a source of distributional error; fortunately, good
measurements with many fluorochromes can be made at such low
optical densities of excitation that distributional error is
essentially non-existent. In flow cytometry one can assume that
excitation is proportional to stain content and fluorescence is
proportional to excitation.

A simplified schematic of a flow cytometer is shown in Fig.
3. Fluorescently stained objects in suspension flow rapidly
through an exciting beam of light and the resulting burst of
fluorescence is collected by a lens, detected by a
photomultiplier and measured by a pulse-height analyzer. Signals
are proportional to stain content and typical rates of
measurement are 1000 objects per second.

In flow sorting (Fig. 4), objects are measured the same way,
but those with predefined fluorescence (or other) values can be
separated from the rest of the population. The preferred way to
accomplish this is to have the flow stream break up into droplets

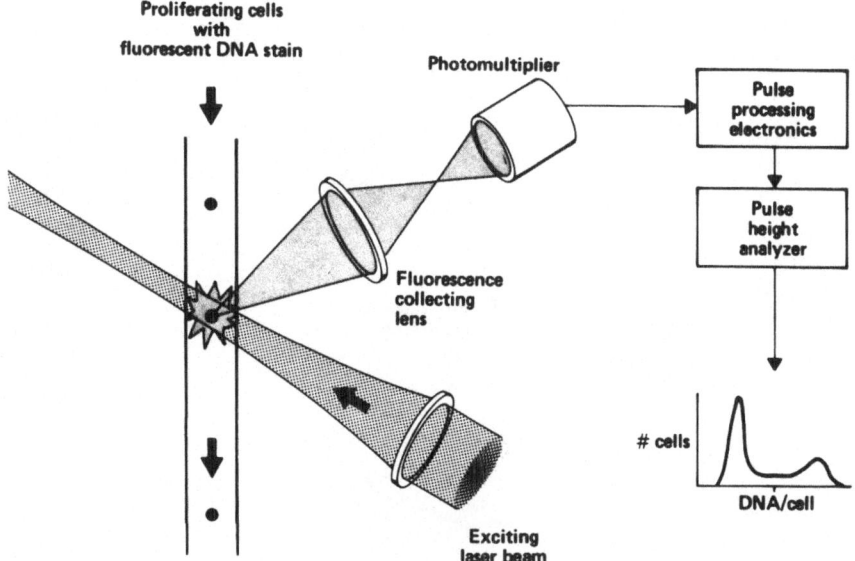

Fig. 3.   The basic flow cytometer. For flow cytogenetics replace
proliferating cells with disrupted metaphase cells and
DNA/cell with DNA/chromosome.   From Van Dilla and
Mendelsohn[9].

soon after passing through the measuring beam. The few droplets
likely to contain the desired object are electrically charged,
causing them to be displaced as they subsequently fall between a
pair of highly and stably charged deflection plates. Using
positive and negative charging of droplets, two classes of
objects can be sorted at rates of 1000 measured objects per
second.

Flow cytometry and sorting of chromosomes requires abundant
mitotic cells, but is otherwise remarkably straightforward.
Cultured cell lines are harvested after blocking the cells in
metaphase with colcemid. The cells are treated with hypotonic
buffer and are briefly sheared to disrupt the cells and suspend
the chromosomes. A DNA fluorochrome is added and the suspension
is entered into the flow system without purification or
manipulation.[10]

Flow cytometric results with one line of Chinese hamster
cells are shown in Fig. 5.[11] Each peak represents a single
chromosome or homologous pair of chromosomes. The location of

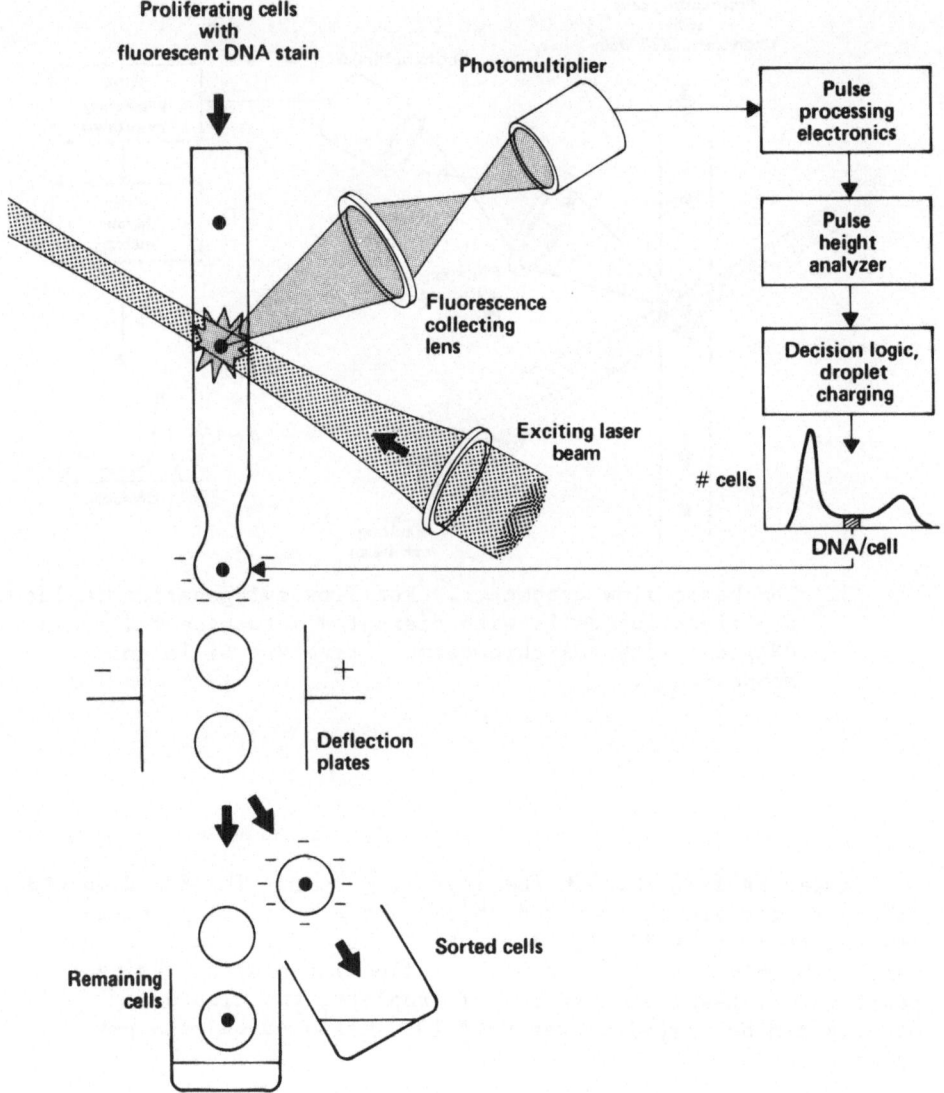

Fig.4.   The basic flow sorter.   From Van Dilla and Mendelsohn[9].

the peak indicates mean stain content, the area of the peak
indicates relative abundance of the constituents, and the shape
of the peak indicates variability of the measurement.
Corroborative experiments have confirmed the identities shown in
the figure; such experiments include sorting the chromosomes from
individual peaks, followed by visual identification, as well as
measuring corresponding metaphases by scanning photometry to

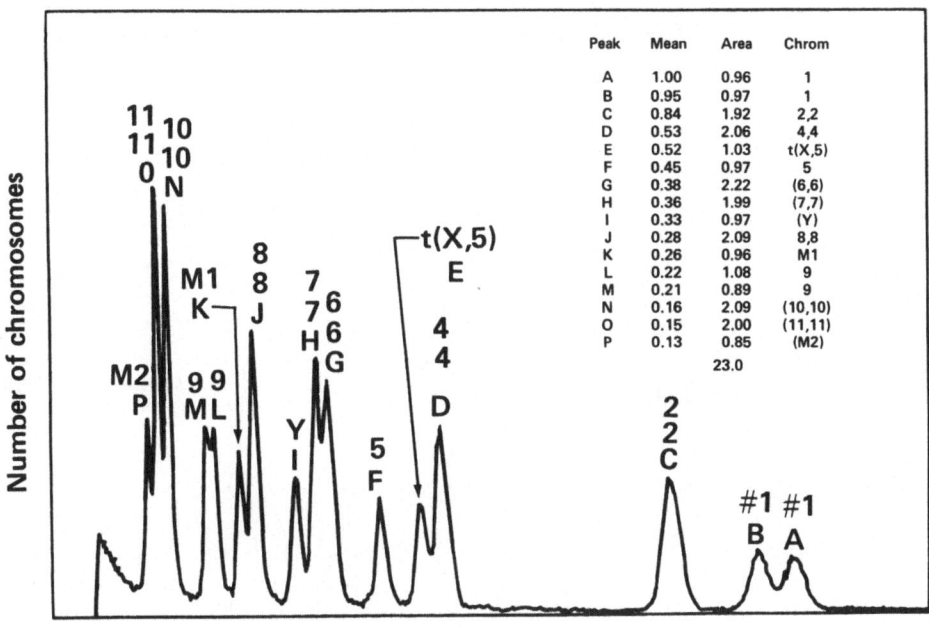

Peak	Mean	Area	Chrom
A	1.00	0.96	1
B	0.95	0.97	1
C	0.84	1.92	2,2
D	0.53	2.06	4,4
E	0.52	1.03	t(X,5)
F	0.45	0.97	5
G	0.38	2.22	(6,6)
H	0.36	1.99	(7,7)
I	0.33	0.97	(Y)
J	0.28	2.09	8,8
K	0.26	0.96	M1
L	0.22	1.08	9
M	0.21	0.89	9
N	0.16	2.09	(10,10)
O	0.15	2.00	(11,11)
P	0.13	0.85	(M2)
		23.0	

Chromosomal fluorescence

Fig.5.   A flow karyotype of Chinese hamster M3-1 line
chromosomes stained with Hoechst 33258. The scale of
the ordinate is such that the peak point J represents
thousands of chromosomes.   From Gray, et al.[11].

relate stain content with chromosome type. The areas of the
peaks correspond to expectation, indicating that all chromosomes
have an equal likelihood of entering suspension and being sampled
by the measurement. The coefficients of variation of the peaks
are around 2% and are decidedly smaller than the variability
observed with scanning measurements.

Comparison of several DNA fluorochromes in Chinese hamster
and other chromosomes has shown interesting differences.[12]
Ethidium bromide, a non-banding intercalator gives flow results
which closely mirror the gallocyanin-chrome alum measurements of
metaphases on slides. Hoechst 33258, a groove-binder with
preference for AT regions, produces quinacrine-like bands and in
flow measurements gives different peak means and smaller
coefficients of variation than ethidium bromide. Chromomycin A3,
another groove-binder but with GC preference, produces reverse
banding and has yet another pattern of peak means in flow
measurements.

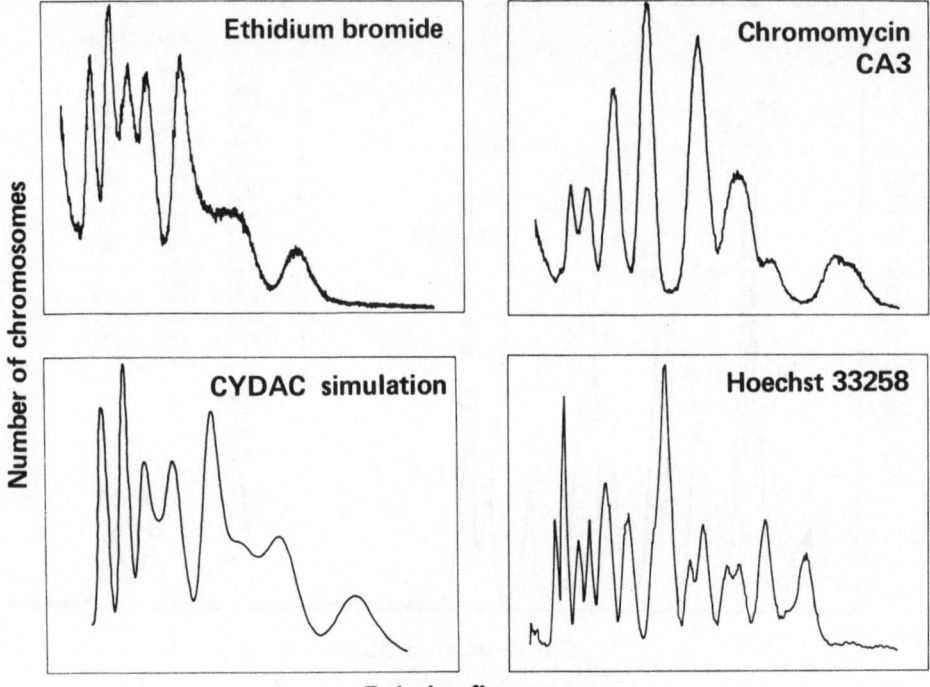

Fig. 6. Distributions for human chromosomes stained by four
         different methods. Metaphase chromosomes were prepared
         from human foreskin strain No. 706. In the CYDAC
         simulation, the chromosomes are stained with
         gallocyanin-chrome alum and measured on slides by
         scanning photometry. In the other three panels, DNA
         fluorochromes and flow cytometry are used. The two
         panels on the left are almost identical suggesting that
         ethidium bromide and gallocyanin-chrome alum reflect
         similar properties of DNA, whereas the panels on the
         right differ dramatically indicating that chromomycin A3
         and Hoechst 33258 each reflect distinctive properties of
         chromosomal DNA.

     These differences are particularly dramatic for human
chromosomes as shown in Fig. 6.[11] Differences among several
human cell lines are also seen with any one fluorochrome and are
akin to the differences we generally see among normal people with
scanning measurements.

     For diagnostic and mechanistic studies it is highly
advantageous to measure two fluorochromes in the same

chromosome. Studies show that stain pairs such as Hoechst 33258 and chromomycin A3 compete little for DNA sites and behave with good stoichiometry when present simultaneously.[11] However, they cannot be measured by taking advantage of their differing emission spectra because energy transfer from Hoechst 33258 to chromomycin A3 causes almost all emission to follow the chromomycin A3 spectrum. Fortunately these stains can be separated by their excitation spectra using a dual laser flow cytometer. In this device the chromosomes flow first through a laser beam at 351 and 364 nm and give a fluorescent signal representing Hoechst 33258; 20 microseconds later they flow through the second beam at 454 nm and the fluorescence from chromomycin A3 is recorded.[12]

A dual laser result with human chromosomes is shown in Fig. 7. The mountain-range effect is a dramatic indication of the resolution achieved by the complementary stoichiometry of Hoechst 33258 and chromomycin A3. It is our current working hypothesis that the relative contents of these two stains reflects the AT-GC ratio of specific chromosomes.

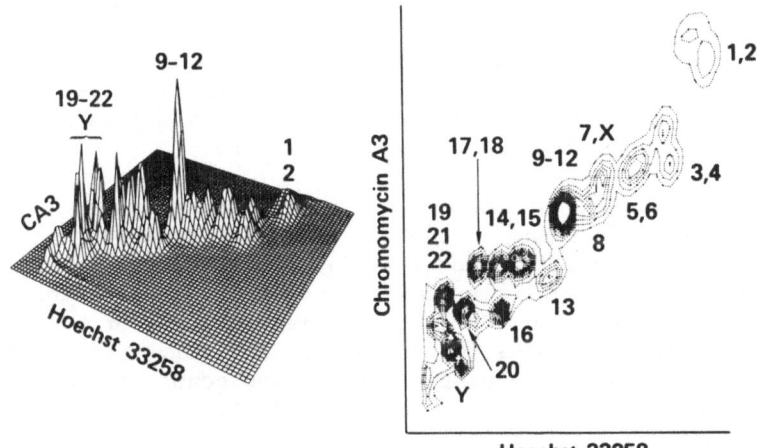

Fig. 7. Dual laser and dual fluorochrome flow cytometry of human chromosomes. Below are shown the 24 types of human chromosomes. On the upper left is a two-dimensional distribution of Hoechst 33258 and chromomycin A3 fluorescence of human chromosomes. The tendency for these stains to show some negative correlation is seen best in the upper right where a topographic representation of the dual fluorescence is shown.

The ability to sort chromosomes can be of enormous importance. Purified chromosomes are suitable for biochemical analysis, template activity, antigenicity and biological transduction. They make possible critical studies on linkage, chromosome structure, mutagenicity, gene control, evolution and genetic engineering in a variety of species including man. For well separated peaks, purities of 95% are available, and with present rates of flow analysis, an 8-hour sort of Chinese hamster chromosomes can produce 2.5×10^6 purified chromosomes. For the average-sized chromosome, this amounts to a microgram of DNA, a quantity that requires the utmost care in handling and is suitable for only the most sensitive techniques of analysis. Ongoing studies on sorted chromosomes include comparison of histones on different chromosomes, localization of viruses or genes to a particular chromosome and attempts to use purified chromosomes as antigens.

THE FUTURE

Future studies in quantitative cytogenetics will be oriented toward improving and applying the scanning and flow analysis of chromosomes. In scanning, a major thrust will be the development of fluorescence scanning to take advantage of banding and of the interesting affinities of fluorochromes for DNA. The nature of chromosomal deviants and their heritability needs further study, and attempts will be made to apply quantitative analysis of metaphase cells to relevant clinical problems.

In flow, we will continue the exploration of DNA fluorochromes, alone and in combination. The related cytochemistry of nuclear proteins will also be pursued. Methods are sorely needed to permit flow cytogenetics in other than cultured cell lines; we particularly need methods for cultured blood and tissue biopsies in the human. An active program is underway to use slit scanning to locate and count chromosomal centromeres. Finally, there are many potential and exciting applications of flow sorting of chromosomes to important problems in biology and medicine.

Work performed under the auspices of the
U.S. Department of Energy by the Lawrence
Livermore Laboratory under contract number
W-7405-ENG-48.

This report was prepared as an account of
work sponsored by the United States
Government. Neither the United States nor the
United States Department of Energy, nor any of
their employees, nor any of their contractors,

subcontractors, or their employees, makes any
warranty, express or implied, or assumes any
legal liability or responsibility for the
accuracy, completeness or usefulness of any
information, apparatus, product or process
discolosed, or represents that its use would
not infringe privately-owned rights.

REFERENCES

1. Caspersson, T. Cell Growth and Cell Function, Norton, New
    York, 1950.

2. Caspersson, T., Zech, L., Johansson, C., and Modest, E.J.
    Identification of Human Chromosomes by DNA-Binding
    Fluorescent Agents. Chromosoma 30, 215-227, 1970.

3. Mendelsohn, M.L., and Mayall, B.H. Chromosome Identification
    by Image Analysis and Quantitative Cytochemistry. (In) Human
    Chromosome Methodology (J.J. Yunis, ed.), Academic Press,
    311-346, 1974.

4. Mayall, B.H., and Mendelsohn, M.L. Errors in Absorption
    Cytophotometry: Some Theoretical and Practical
    Considerations. (In) Introduction to Quantitative
    Cytochemistry, Volume 2, (G.L. Wied and G.F. Bahr, eds.),
    Academic Press, 171-197, 1970.

5. Mendelsohn, M.L., Mayall, B.H., and Perry, B.H. Generalized
    Grayness Profiles as Applied to Edge Detection and the
    Organization of Chromosome Images. (In) Advances in Medical
    Physics, Second International Conference on Medical Physics,
    Inc., Boston, 327-341, 1971.

6. Mendelsohn, M.L., Bennett, D.E., Bogart, E., and Mayall,
    B.H. Computer-Oriented Analysis of Human Chromosomes. IV
    Deoxyribonucleic Acid-Based Centromeric Index. J. Histochem.
    Cytochem. 22, 554-560, 1974.

7. Mayall, B.H., Carrano, A.V., Moore II, D.H., Ashworth, L.K.,
    Bennett, D.E., Bogart, E., Littlepage, J.L., Minkler, J.L.,
    Piluso, D.L., and Mendelsohn, M.L. Cytophotometric Analysis
    of Human Chromosomes. (In) Automation of Cytogenetics (M.L.
    Mendelsohn, ed.) Asilomar Workshop, 30 Nov. - 2 Dec., 1975,
    ERDA Conf-751158, NTIS, Pacific Grove, Ca., 135-144 1976.

8. Gray, J.W., Carrano, A.V., Steinmetz, L.L., Van Dilla, M.A.,
    Moore II, D.H., Mayall, B.H., and Mendelsohn, M.L.
    Chromosome Measurement and Sorting by Flow Systems. Proc.
    Nat. Acad. Sci. U.S.A. 72, 1231-1234, 1975.

9. Van Dilla, M.A., and Mendelsohn, M.L. Resume (In) <u>Flow Cytometry and Sorting</u> (M.R. Melamed, P. Mullaney and M.L. Mendelsohn, eds), Wiley & Sons, expected 1978.

10. Carrano, A.V., Van Dilla, M.A., and Gray, J.W. Flow Cytogenetics: A New Approach to Chromosome Analysis. (In) <u>Flow Cytometry and Sorting</u> (M.R. Melamed, P. Mullaney and M.L. Mendelsohn, eds), Wiley & Sons, expected 1978.

11. Gray, J.W., Langlois, R.G., Carrano, A.V:, and Van Dilla, M.A. High Resolution Chromosome Analysis: One and Two Parameter Flow Cytometry. (in preparation).

12. Jensen, R.H., Langlois, R.G., and Mayall, B.H. Strategies for Choosing a Deoxyribonucleic Acid Stain for Flow Cytometry of Metaphase Chromosomes. <u>J. Histochem. Cytochem.</u> <u>25</u>, 954-964, 1977.

13. Dean, P.N., and Pinkel, D. Dual Laser Flow Cytometry. <u>J. Histochem. Cytochem.</u> in press, 1978.

DISCUSSION (PART II)

DR. YAGIL: Why is it necessary to add E. coli polymerase. Could not endogenous polymerase be responsible for transcription? Also, can one obtain specific hybridizable ^{32}P labelled globulin mRNA in intact foetal liver cells?

DR. GILMOUR: With endogenous polymerase we obtain amounts of messenger too small to be used for our hybridization experiments. There is evidence that E. coli polymerase transcribes sequences that are not transcribed in intact nuclei from endogenous polymerase, but it remains a problem to make hybridization experiments starting with this system.

DR. SARMA: Anti-globin RNA can arise by copying the contaminating endogenous globin mRNA or anti-globin gene. What evidence exists that globin mRNA serves as the template and not anti-globin gene? Can Hg-UTP by reacting with major groove (where histones are believed to occur) enable copying anti-globin gene which is not normally copied?

DR. GILMOUR: The copy of the message is not a complete copy. It looks like what attracts the polymerase is the polyA 3' end, but it does not go very far in copying the structural RNA.

DR. WINICOV: In view of your model which invokes an imbalance of two or more chromosomes in expression of a malignant state, would you comment on the cell fusion experiments from the laboratory of Dr. Carlo Croce, which seem to propose the involvement of a single chromosome and also show that in each case it does not have to be the same chromosome?

DR. SACHS: I do not think that such results invalidate the results of our experiments. In fusion experiments between malignant and nonmalignant cells you come out with malignant or nonmalignant progeny apparently depending upon which chromosomes are selected for.

357

DR. *PARODI:* Has somebody tried to hybridize cells which had incorporated large amounts of BUdR into their DNA with an untreated, different (as in differentiation) cell and thereafter tried to destroy the BUdR-DNA with U.V.? If the nucelo-proteins alone maintain differentiative potentialities we could perhaps detect this fact.

DR. *SACHS: That is a good experiment; at the best of my knowledge I do not know anybody who has done it.*

DR. *NICOLINI:* Do you not think that in the complex mammalian chromatin "in situ" we should think not only in terms of static genes but also in terms of the dynamic chemical-electromagnetic environment?

DR. *SACHS: You are asking for differentiating what is due to genes and what is not due to genes. It is important - however, it is difficult to do that.*

DR. *DIXON:* An apparently clear finding of the physical techniques is that in 2M salt when the DNA is dissociated from the histone core there is very little change in CD. This seems difficult to understand if the N-terminal tails of the histones have any sort of ordered structure in the nucleosome.

Would you like to comment on which sort of structure you might visualize the N-terminal tail as having in the intact nucleosome and if the structure reverts to a random coil in 2M salt, why is there no reflection in the CD spectra?

DR. *FASMAN: The conformation of the N-terminal tail either in complex with DNA in the nucleosome or after dissociation is not known.*

DR. *BRAM:* Dr. Fasman, your binding studies with acetylated histones are very interesting. On the other hand, you were careful to point out that changes in CD could arise either from a higher order "tertiary fold" or from changes in secondary structure. However, you presented the classical canonical A, B, C, scheme. Unfortunately DNA structure has been shown to be far more complicated than A, B, C. The A conformation is the only form of DNA which is more or less unique. For B my X-ray studies in solution and in chromatin show that they are families of structures where the helical parameters vary by about 15%. That means the angle between base pairs can change by up to 50% per b.p. In fact, Ivanov et al (Mol. Biol. $\underline{58}$, 277 1971) have shown that most changes in CD can be explained by a variation in angles between base pairs. Consequently it seems to me that all changes in CD

you report could well be interpreted by a modification of DNA
secondary structure.

DR. FASMAN: *I agree on what you said.*

DR. YAGIL: You demonstrated that two different condensed forms of
DNA both verified by X-Ray as being in the B conformation have
very different C.D. spectra. This indeed takes the basis for
C.D. to be indicative of conformation in the condensed state.
How can be extrapolate from the condensed to the dissolved
state, when C.D. is not unique in condensed phase?

DR. FASMAN: *I think there are two types of condensation: one
that involves changes in the C.D. spectra probably affects the
secondary structure and others, operating on a larger scale
(solenoid,superhelix), that perhaps do not affect significantly
the secondary structure.*

DR. SARMA: Chemical acetylation does not alter the C.D. spectra
of chromatin, whereas such acetylation alters the transcription
ability and DNase I digestibility. Do you want to comment
on this?

DR. FASMAN: *It is a discrepancy. Obviously, the two techniques
are measuring different aspects concerning the stability
of the structures, but I have no explanation of this observation.*

DR. HARRINGTON: Are any other H1 modifications observable in
the histone from C.D. spectra?

DR. FASMAN: *Yes. Acetylation shows similar effects to that
observed with phosphorylation.*

DR. BETTECKEN: Are the holes in the nucleosomes as visible in
E.M. micrographs or are they artifacts?

DR. OLINS: *Stain binds to negative charges, therefore you see
negligible structure. There might be a depression, but not
well visible.*

DR. BETTECKEN: Did you try E.M. microsamples of histone-enriched
nucleosomes?

DR. OLINS: *No.*

DR. MARLADI: Do you mean that the dense spot inside the nucleosome you see in your picture is a real depression as indicated
by scanning - tranmission E.M. (quoted) and that this depression
could be interpreted as the presence of a low density protein
core?

DR. OLINS: *Yes, it may be interpreted as a low density region due to the localization of histone just in the center of the nucleosome particle.*

DR. VAN HOLDE: Is there evidence that ribosomal genes that show chromatin structure in the E.M. also digest to give the 200 base pair repeat?

DR. OLINS: *No.*

DR. TRIFONOV: Are the rope like helical structures that you showed in some sperm chromatin made of some material different than histones?

DR. OLINS: *I honestly don't know what the protein is in the slides I showed. You are right in that the protein need not be histone; our example of this is the sea urchin sperm. However, I presented this data primarily to show how another method of electron microscopy can handle higher order structures.*

DR. BONNER: What is structure of the chromatin strand in that portion of Bombix chromosome which is being transcribed into fibrin message?(the EM picture presented was by Steve McKnight)

DR. OLINS: *The polymerases are closely packed and it is impossible to identify nucleosomes in the transcription units.*

DR. DIXON: You mentioned that in transcribing regions of Oncopeltus chromatin nucleosomes could not be seen, but in Drosophlia they could and you suggested that even where they could be seen where there were still 200 base pair repeats to micrococoal nuclease as a probe as well as sensitivity to DNA-ase I (in the core region presumably). How were these experiments done? Is it possible to assess whether the nucleosome structure is absent or very unstable? Is it possible to stabilize the nucleosomes that are transcrible or committed to transcription by chemical crosslinking before EM?

DR. OLINS: *The chromatin is already fixed in formaldehyde but the Oscar Miller technique does tend to extend chromatin and it could be particularly liable to extend especially labilized nucleosomes in transcribing regions. It would be nice however to find a technique for stabilizing transcribing regions of chromatin.*

DR. TRIFONOV: The change of sedimentaiton coefficient of nucleosome in acidic media is it not just a result of denaturation and collapse of DNA?

DR. D. OLINS: *The comparative study on DNA that was done was with large size DNA and not 140 base size DNA. However it is known that acidic denaturation of DNA takes place at considerably lower pH and although we have to maintain reservation in extrapolating to 140 base size, I think that probably DNA is not denaturated.*

DR. SHAW: You indicated that the nucleosome cores that you used for your physical studies were enriched in 140 base pair DNA by precipitatation with KCl to remove 160 base pair particles. In our laboratory we are not able to remove all 160 base pair particles using similar salt precipitation methods. There still remain considerable amounts of 160 base pair particles that lack H1 and H5. Exactly how pure were the cores that you used? If they contained 160 base pair contaminants could that account for the changes in sedimentation coefficient you observed by varying the ionic strength since there were extra tails of DNA?

DR. D. OLINS: *Yes there were considerable amounts of 160 base pair particles in the preparations used, as seen by gel electrophoresis. We now use the method of H1 removed first (as adapted by Tatcheel and Van Holde) to obtain clean 140 base pair cores. We have not used these in the studies above. It is quite possible that the extra 20 base pairs in the cores could influence the sedimentation changes with varying ionic strength.*

DR. PRUNELL: Was H1 present in your nucleosomes?

DR. HARRINGTON: *No. We worked with core particles.*

DR. BRAM: Your calculations suggest that banding of DNA is energetically favorable. Some time ago J. Brahms and I built CPK models with a flexible center support and found that the force of gravity was sufficient to band DNA into a conformation with an equivalent radius of curvature equal to about 30Å.

DR. HARRINGTON: *I perfectly agree if the charges are neutralized in the histones.*

DR. BETTECKEN: What evidence does exist for binding of H2A and H2B to undenatured portions of chromatin during thermal denaturation?

DR. WALKER: *H2A and H2B do not bind to denatured DNA. During the first half of the thermal transition, H2A and H2B dissociate and move onto the undenatured half of chromatin. Denaturation of chromatin is therefore "all-or-none". The binding of H2A*

and H2B to the native chromatin increases the Tm of the second
step. On cooling from Tm the transition is not reversible but
the full hypochromism is regained. The denatured fraction may be
separated from the native fraction and is found to contain only
H3 and H4.

DR. GIARETTI: The use of the parallel axis theorem relationship
to calculate the distances of subunits in a complex has been
criticized (Jacrot; Rep. Progr. Phys. 39 (1976) 911). Could
you please comment on this?

DR. PARDON: *I am not aware of this criticism.*

DR. DIXON: Your model showed bands of protein crossing over the
DNA gyres in eight places - do you consider these to be the
N-terminal fingers?

DR. PARDON: *Yes, but not definitely - this is only one possible*
way that the N-terminal fingers could interact with the DNA.
They could also be between the gyres of DNA in a somewhat
indefinite position.

DR. WILHELM: How did you obtain the H_{2a}, H_{2b}, H_3, H_4 tetramer?
Have you been able to study the octamer? and the $(H3H4)_2$
tetramer?

DR. PARDON: *By extracting chromatin with 20M NaCl, pH9.0*
essentially using the methods originally described by both
Weintraub et al and Thomas and Kornberg. We have examined
an octamer formed from tetramers by chemical crosslinking.
The radius of gyration (Rg) of this octamer is very similar
to both the (Rg of the isolated histone) tetramer and the
core particle under conditions where the protein dominates.
We have not studied the $(H_3H_4)_2$ tetramer.

DR. TRIFONOV: What kind of further detail could give maximum of
35Å in scattering curve of histone core (at 65% D_2O)?

DR. PARDON: *Possibly some distribution of α-helices in the*
histone core.

DR. HARRINGTON: If you have a polydisperse system in your
neutron scattering work on core histone complexes, what
kind of average Rg does the neutron scattering method yield?
A z-average?

DR. PARDON: *Probably a weighted average, but the Rg value changes*
very little between octamer and tetramer and also is independent
of concentration.

DR. *FASMAN:* Have you any evidence that crosslinking doesn't change the conformation of the histones (in the octamer)?

DR. *PARDON: No evidence - S values are approximately 3.7 to 4.2.*

DR. *HARRINGTON:* Are you aware of any applications of "magic angle" spinning techniques for nucleic acids or nucleoproteins?

DR. *TS'O: These are not yet sufficiently well developed to be applicable here.*

DR. *BETTECKEN:* What solvent conditions did you use for your nucleosome examinations?

DR. *TS'O: Either D_2O or H_2O.*

DR. *TS'O:* What is the ultimate limit of resolution of the cytochemical analyses by light microscope?

DR. *CASPERSON: 0.3 micron.*

DR. *LINDEN:* What is the resolution in the TV-system for fluorimetry?

DR. *CASPERSON: Up to now we have been working with photographs from chromosomes using the quinacrine fluorescent binding technique. These photographs are viewed by the TV system. We hope to be successful in developing a TV system for fluorometry with good resolution.*

DR. *LINDEN:* You mentioned the use of quantitative interferometry. Would you comment on the value of this technique compared to the use of cytochemical stains with absorption or fluorimetric techniques?

DR. *CASPERSON: This method still has its value because of its higher accuracy and reproducibility. We just did some studies on the protein content of paramecium with a reproducibility of 2 - 3% which is higher than what you get in normal cytophotometry.*

DR. *DIXON:* I was intrigued by your finding that you could separate the two homologues of chromosome 1 and 9 in your line of CHO cells. Would it be correct to assume that if you looked at other lines you might see for example, two identical 1 chromosomes of one class or two of the other class assuming that the differences are of paternal or maternal origin.

DR. MENDELSOHN: *Yes, one would expect such a situation but it has not been systematically looked for. The CHO line (3-1) is about 7 years old and may be from the same source as the CHO line of Stubblefield in which he has shown a 6% difference in the chromosome 2 homologues by 3H hymidine incorporation.*

DR. PARODI: Could the fact that you can distinguish between chromosomes of the same couple be related to the Lyon phenomenon?

DR. MENDELSOHN: *It is a possibility that we have to keep in mind.*

DR. KOVAL: Do you think a potential exists for eventually detecting damaged areas of DNA or chromatin with your machine as you have been able to resolve 10^6 base pairs already?

DR. MENDELSOHN: *Perhaps. I don't know. We would have to get the damages to fluoresce.*

DR. SAWICKI: What is the contribution of cytoplasmic DNA for chromosome measurement by pulse-cytophotometry.

DR. MENDELSOHN: *It may cause a very slight elevation of the background on which the chromosome peaks are sitting. The effect would not produce peaks in the chromosome region.*

DR. BAISCH: There have been problems in flow systems with particles deviating from spherical shape. Chromosomes are considerably different from spheres. Do you have problems from orientation of chromosomes while being measured?

DR. MENDELSOHN: *This refers to experiments that have been done with several cell types. We have done it with sperm and the Hertzenberg group has done it with red cells. When orthogonal flow systems are looking at flat cells the signal depends on how the cell is oriented. The operation one gets with a flat sperm cell is the most extreme degree of non-randomness that we have available. It can be as much as twofold. It is an interesting artifact and involves both the excitation and environmental parts of the phenomenon. With chromosomes we have information from high speed photographs that they line up in the flow direction. So we are talking about whether the two chromatids are in line as they look at the laser beam or whether they are at right angles. And this is probably going* to be a very small effect. It is down to the resolution of the beam and we don't see the effect in the histogram, so my guess is that it is not present.

DR. KOVAL: Concerning Janet Rowley's data, is it possible that a chromatin structural deficiency exists such that the chromosome is susceptible to breakage at a certain point?

DR. MENDELSOHN: Perhaps but it could also be that this certain phenotype results in the leukemia rather than vice-versa.

DR. BETTECKEN: Does it make sense to continue working with cells characterized and sorted in your flow cytometer, since the handling inside the instrument is very rough (staining, shear forces, exposure to high intensity light and high voltage) and might destroy the cells?

DR. MENDELSOHN: Experiments with stain Hoechst 33342 show that cells sorted by the flow-cytometer grow as well as untreated cells.

DR. NICOLINI: We have done the same in Melanoma B16, with low concentration of acridine orange. It works fine.

DR. DIXON: When you see that single red cells "light up" with a specific antibody to HbS, do you really know that the red cells contain HbS or merely that it contains a mutant HbS which cross-reacts with an anti-HbS antibody? In other words the mutation need not giving rise to identical HbS which alterates β chain Glu → Val but merely gives rise to a structure cross-reacting with anti-HbS antibody.

DR. MENDELSOHN: Yes, there could be more distal interactions which effectively produce a mutant Hb which could cross-react with anti-S.

DR. LINDEN: You mentioned your method of screening for carcinogens. I suppose you use fluorescence activated chromosome sorting. Would you please comment on that in some more detail.

DR. MENDELSOHN: We are looking for translocations. But we hope to apply the same technique for use in sickle cell anemias and hope to apply it to humans in the future.

DR. SARMA: Do you have any data in identifying preneoplastic cells in a normal population of cells.

DR. MENDELSOHN: We are working on these problems. For instance, we are working on Farber's model of inducing nodules and looking for cells by gamma-glutamyl transpeptidase activity.

DR. SARMA: Today many speakers were using the term "chromatin in situ". In fact in some of their studies the chromatin was hydrolyzed, possibly depurinated and been molested, if I may be permitted to say so. I think this term "chromatin in situ" is inappropriate to these studies.

DR. NICOLINI: This is true only in our image analysis studies. But this is merely the first step. For what concerns the semantic, I think that the term "Chromatin in situ" can be considered sufficiently accurate since Feulgen's reaction reflects DNA space and location, in situ.

DR. DIXON: Since the Heisenberg uncertainty principle it has always been a problem for scientists that the method of observation may perturb the system during the observation. Worcel observed that ethidium bromide could dissociate histones from chromating by intercalating and unwinding the DNA. Do you feel that your method of binding ethidium bromide and acridine orange might not perturb the chromating being observed in situ?

DR. NICOLINI: We don't think that the unwinding is a serious problem at the moment although we are aware that intercalation of ethidium bromide does induce unwinding. We must start somewhere and later we will explore this problem.

DR. SARMA: You correlated some changes observed by image analysis to increased transcriptional ability of the cells. This maybe true. My question is: What type of changes would you observe when the cells have been treated with carcinogens of a type that actually inhibits transcriptional ability? Maybe you will find similar changes. The reasons for asking this question is that carcinogen treated rat liver chromatin exhibits similar CD spectra to those of replicating rat liver chromatin-DNA - even though the carcinogen-treated cell has less transcription when contrasted to the replicating cell.

*DR. NICOLINI: In fact we have correlated differences in average optical density with both increased transcriptional activity (as in WI*38 cells stimulated to proliferate) and decreased transcriptional activity (as in stationary 2RA cells when compared to quiescent WI38 cells). Experiments of the nature that you suggested have yet to be completed but these are some of the directions in which we are working.*

DR. MENDELSOHN: In the two-dimensional density reconstruction of nuclei, there were spikes that looked to be one picture point wide. At a sample spacing of $0.1\mu m$, the modulation transfer function of the microscope would not permit optical responses of such high frequency. Can you comment?

DR. NICOLINI: *Core size constraints imposed by our minicomputer restricts us to a maximum display array size of 64 x 64. Consequently the displays were obtained with much lower magnifications than those which we employ for analytical work (when a single picture point represents an interval less than 0.1 μm). This low magnification data is subject to serious distributional error. We use it only for display purposes.*

DR. MENDELSOHN: What was the period length in microns of the sinusoidal correlations seen in G1 and S cells?   What was the relation to 60 cycles per second?

DR. NICOLINI: *The periodicity of the mid-G1 nuclear image corresponded to 13 micra and in the S-phase image to above 4.2 micra.   A single column scan represents about 4Msec. Inasmuch as the system clock was independent of the power line and the resulting frame rate was not integrally related by multiple or submultiple of the power line, any signal produced by this source would tend to drift through the image on repeated scans.   Each array was in fact scanned 100 times and the results averaged, so such effects would be negligible. It would be far more likely that such artifacts would result from mechanisms controlled by the system clock.   The fact that the periodicities observed were consistently dependent upon cell cycle phase rather than equipment set-up on a particular day indicates that they are probably not artifactual.*

DR. BETTECKEN: Did you consider the elctric field caused by the power lines?   Did you shield?

DR. CHIABRERA: *The wave form shown in the slides is the measurement of the rotational component of the electric field, which produces the current in the solution.   Then the contribution of the power line is negligible.   In any case, by comparing the behavior of control cells with the irradiated ones changing the amplitude and the repetition rate of the signal, any 60 Hz effect can be excluded.   We checked the effects of a shield, and we concluded that it was not necessary nor useful.   But in fact, eddy currents may be induced in the shield by the signal itself, thus affecting the actual electric field waveform in the solution.*

DR. TS'O: Who makes the final decisions about the positive or negative diagnoses?

DR. PLOEM: *The pathologist.*